The Ecohydrology of South American Rivers and Wetlands

edited by

Michael E. McClain

Department of Environmental Studies, Florida International University, Miami, Florida 33199, USA

This publication was financially sponsored by

the UNESCO Regional Bureau for Science in Europe
established in Venice, Italy

IAHS Special Publication no. 6

(December 2002)

IAHS Special Publication no. 6

ISBN 1-901502-02-3

Published by the International Association of Hydrological Sciences 2002
IAHS Press, Centre for Ecology and Hydrology, Wallingford, Oxfordshire OX10 8BB, UK

British Library Cataloguing-in-Publication Data. A catalogue record for this book is available from the British Library.

Editors for the Series of IAHS Special Publications:

H. Salz
The International Institute for Infrastructural, Hydraulic and Environmental Engineering (IHE), Delft, The Netherlands

Zbigniew W. Kundzewicz
Editor *Hydrological Sciences Journal*, Research Centre of Agricultural and Forest Environment, Polish Academy of Sciences, Poznan, Poland

IAHS is indebted to the employers of the Editor for the invaluable support and services that enabled him to work effectively and efficiently.

The formatted pages were produced at IAHS Press, Wallingford, UK, by Penny Kisby.

IAHS Special Publications are available only from:
IAHS Press, Centre for Ecology and Hydrology, Wallingford, Oxfordshire OX10 8BB, UK
tel.: +44 1491 692442; fax: +44 1491 692448; e-mail: jilly@iahs.demon.co.uk

Printed in The Netherlands by Krips BV, Meppel.

Preface

The term ecohydrology was popularized in the early 1990s and quickly adopted by fields of study ranging from plant physiology to aqueous geochemistry. The enthusiastic and widespread use of this term, however, has diluted its meaning and many are left wondering "what is ecohydrology really?" In this book we subscribe to the concept of ecohydrology embodied in the Fifth Phase (1996–2001) of UNESCO's International Hydrological Programme (IHP). The Ecohydrology Project stimulated a broad spectrum of creative research integrating the physical processes of hydrology with the biological processes of ecology. Together, these processes serve to regulate environmental conditions within aquatic systems, maintaining energy levels, water quantity and water quality within ranges suitable to the native flora and fauna. While the research objectives of the Ecohydrology Project centred on understanding fundamental processes, there was also an explicit objective of evaluating how these natural processes could be incorporated into water management programmes. In more developed nations these processes might serve to enhance the highly engineered systems already in place, while in less developed nations these ecohydrological processes might serve as the primary means for managing water quantity and quality through natural attenuation and ecological engineering.

One continent with great potential to capitalize on the natural attenuation effects of ecohydrological processes is South America, the wettest and most ecologically diverse continent on Earth. South America's annual average precipitation of 1600 mm is more than twice that of any other continent, producing an unrivalled waterscape of giant rivers and wetlands. The continent is also rich in indigenous and colonial cultures. Today more than 300 million people inhabit the continent, distributed among 13 countries originally established as Spanish, Portuguese, French, Dutch, and British colonies. The health and prosperity of South America are closely linked to the abundance and quality of its water resources. According to the World Resources Institute, hydropower accounts for nearly 80% of South America's electricity supply and natural rainfall is the sole source of water to 92% of the continent's agricultural lands. South Americans also turn to their large rivers and wetlands for drinking water, food, transportation, recreation, and waste disposal.

South Americans have greatly benefited from the resources derived from their rivers and wetlands, but these aquatic systems have not generally benefited from their association with humans. Dam building, dredging, and canalization are expanding across the continent, eliminating the natural flow regimes of many river reaches and draining wetlands. Much of the riparian vegetation lying along river margins in more developed portions of the continent has been removed, and even in rural areas humans tend to focus their activities at river and wetland margins where soils are often more fertile. The disturbance of riparian zones has changed the rates of material exchanges from land to aquatic systems, generally increasing erosion and contaminant fluxes. Uncontrolled waste discharges have also degraded water quality in many urban rivers and wetlands, and this poor water quality has had strong negative feedbacks on the health of many urban and downstream rural poor.

The content of this book stems mainly from the International Symposium on Hydrology and Geochemistry of Large-scale River Basins (with special emphasis on the Amazon and other tropical basins), held 15–19 November 1999, in Manaus Brazil. However, each chapter goes beyond the scope of the material presented in the symposium and seeks to provide a broader overview of ecohydrological processes operating in South America's most important aquatic systems. Chapters describe river reaches ranging from pristine to heavily impacted and detail processes operating in channels, wetlands, and in riparian environments. The continent's three largest rivers, the Amazon, Orinoco, and Paraná, receive the greatest attention, but the condition of important rivers in the states of São Paulo (the Piracicaba River) and Rio de Janeiro (the Paraíba do Sul River) is also evaluated. Together these chapters convey a message of urgent need for action bolstered by the realization that many spectacular ecosystems remain to be preserved and the natural attenuation processes operating in these ecosystems stand to aid South America in achieving its goal for sustainable use of its resources. The authors and I hope that through these insights into our current understanding of the interactions of hydrology–water quality, and biological communities, this book might contribute to better management and conservation of these unique ecosystems and resources.

We owe a great deal of thanks to the many individuals who reviewed drafts of the chapters and the agencies and foundations that funded the individual research efforts; they are acknowledged in the chapters. Maciej Zalewski chaired the Scientific Steering Committee of the IHP Ecohydrology Project and provided much appreciated encouragement at the beginning of this project. This book was made possible through the generous financial support of UNESCO's Regional Bureau for Science in Europe, and the individual efforts of Philippe Pypaert. The symposium on which several of the chapters were based was organized by the HiBAm (Hydrology and Geochemistry of the Amazon Basin) Project, under the leadership of Jean-Loup Guyot. Invaluable assistance was provided by Penny Kisby and Madelyn Mateo in preparing the manuscript.

Michael E. McClain
February 2002, Miami, USA

Contents

Contributors

ANTHONY AUFDENKAMPE, Stroud Water Research Center, 970 Spencer Road, Avondale, Pennsylvania 19311, USA aufdenkampe@stroudcenter.org

MARIA V. BALLESTER, Centro de Energia Nuclear na Agricultura (CENA), Universidade de São Paulo, Avenida Centenário 303, 13416-000 Piracicaba, SP, Brazil

SORAYA BARRERA, Museo Nacional de Historia Natural, CP 8706, La Paz, Bolivia sbarrera@ceibo.entelnet.bo

BÉNÉDICTE BARRÈRE, Institut de Recherche pour le Développement, Misiòn IRD en Ecuador, AP 17 12 857, Quito, Ecuador; *formerly at:* Institut de Recherche pour le Développement, BIOBAB Project IRD/ Universidad Mayor de San Andres, CP 9214, La Paz, Bolivia

JEAN YVES BENAÏM, Laboratoire de RCMO, Université de Toulon et du Var, BP 132, F-83957 La Garde Cedex, France

MARC BENEDETTI, Laboratoire de Géochimie et Métallogénie, Université Pierre et Marie Curie, 4 Place Jussieu, F-75000 Paris, France

MARCELO C. BERNARDES, Centro de Energia Nuclear na Agricultura (CENA), Universidade de São Paulo, Avenida Centenário 303, 13416-000 Piracicaba, SP, Brazil

DAVID BINET, Ministère de l'Agiculture, DDAF du Var, BP 122, F-83071 Toulon Cedex, France david.binet@agriculture.gouv.fr

LIGIA BLANCO-BELMONTE, Fundación La Salle de Ciencias Naturales (FLASA), Estación Hidrobiológica de Guayana, Apdo 51, San Félix, Edo. Bolívar, Venezuela bblanco@cantv.net

CHRIS BRADLEY, University of Birmingham, School of Geography and Environmental Sciences, Edgbaston, Birmingham B15 2TT, UK c.bradley@bham.ac.uk

JACQUES BOULÈGUE, IRD, Département Milieux et Environnement, 213 Rue Lafayette, F-75480 Paris Cedex 10, France dme@paris.ird.fr

PLINIO B. CAMARGO, Centro de Energia Nuclear na Agricultura (CENA), Universidade de São Paulo, Avenida Centenário 303, 13416-000 Piracicaba, SP, Brazil

CARLOS EDUARDO CERRI, Centro de Energia Nuclear na Agricultura (CENA), Universidade de São Paulo, Avenida Centenário 303, 13416-000 Piracicaba, SP, Brazil

NOEMÍ CHACÓN, Instituto Venezolano de Investigaciones Científicas (IVIC), Centro de Ecología, Altos de Pipe, Km 11, Caracas, Venezuela nchacon@oikos.ivic.ve

DAVID COLLOMB, 1 Rue de la Fontaine, F-25870 Devecey, France

ASSAD DARWICH, Instituto Nacional de Pesquisas da Amazônia, Avenida André Araújo 2936, 69011 Manaus, AM, Brazil

FELIX DAZA, Fundación La Salle de Ciencias Naturales (FLASA), Estación Hidrobiológica de Guayana, Apdo 51, San Félix, Edo. Bolívar, Venezuela fjdaza@cantv.net

CARLOS EDUARDO VEIGA DE CARVALHO, Laboratório de Ciências Ambientais, Centro de Biociências e Biotecnologia, Universidade Estadual do Norte Fluminense, Avenida Alberto Lamego 2000, Parque Califórnia, 28013-602 Campos dos Goytacazes, RJ, Brazil carvalho@uenf.br

NELDA DEZZEO, Instituto Venezolano de Investigaciones Científicas (IVIC), Centro de Ecología, Altos de Pipe, Km 11, Caracas, Venezuela ndzzeo@oikos.ivic.ve

EDMUNDO C. DRAGO, Instituto Nacional de Limnología, INALI-CONICET, J. Maciá 1933, 3016 Santo Tomé, Santa Fe, Argentina edmundodrago@arnet.com.ar

GLADYS ESCALANTE, Instituto Venezolano de Investigaciones Científicas (IVIC), Centro de Ecología, Altos de Pipe, Km 11, Caracas, Venezuela gescalan@oikos.ivic.ve

INÉS EZCURRA DE DRAGO, Instituto Nacional de Limnología, INALI-CONICET, J. Maciá 1933, 3016 Santo Tomé, Santa Fe, Argentina inesezcurra@arnet.com.ar

EPAMINONDAS S. FERRAZ, Centro de Energia Nuclear na Agricultura (CENA), Universidade de São Paulo, Avenida Centenário 303, 13416-000 Piracicaba, SP, Brazil

DAVID GILVEAR, University of Stirling, Department of Environmental Sciences, Stirling FK9 4LA, UK d.j.gilvear@stir.ac.uk

NIRSON GONZALEZ, Fundación La Salle de Ciencias Naturales (FLASA), Estación Hidrobiológica de Guayana, Apdo 51, San Félix, Edo. Bolívar, Venezuela hidrobiolflasa@cantv.net

ISMAEL GONZALES, Ministerio del Desarrollo Sostenible, Unidad Ordenamiento Territorial, La Paz, Bolivia uot@unete.com

FRANCK GOURDIN, Ministère de l'Agiculture, DDAF Guyane, BP 5002, 97305 Cayenne Cedex, France franck.gourdin@agriculture.gouv.fr

JEAN-LOUP GUYOT, Institut de Recherche pour le Développement, UR069 HiBAm IRD-LMTG, 38 Rue des 36 Ponts, F-31400 Toulouse, France guyot@cict.fr

STEPHEN K. HAMILTON, W. K. Kellogg Biological Station and Department of Zoology, Michigan State University, Hickory Corners, Michigan 49060-9516, USA hamilton@kbs.msu.edu

JÜRGEN KERN, Institute of Agricultural Engineering, Max-Eyth-Allee 100, D-14469 Potsdam, Germany jkern@atb-potsdam.de

CLAUDIA KNAB-VISPO, 1685 Route 5, Canaan, New York 12029, USA knab@students.wisc.edu; *formerly at*: Fundación La Salle de Ciencias Naturales (FLASA), Instituto de Investigaciones Orinoco-Esequibo, Tumeremo, Edo. Bolívar, Venezuela

HEIDI KREIBICH, GFZ GeoForschungsZentrum, Telegrafenberg, D-14473 Potsdam, Germany; *formerly at*: Institute of Agricultural Engineering, Max-Eyth-Allee 100, D-14469 Potsdam, Germany

ALEX V. KRUSCHE, Centro de Energia Nuclear na Agricultura (CENA), Universidade de São Paulo, Avenida Centenário 303, 13416-000 Piracicaba, SP, Brazil

LUCIENE L. S. LARA, Centro de Energia Nuclear na Agricultura (CENA), Universidade de São Paulo, Avenida Centenário 303, 13416-000 Piracicaba, SP, Brazil luciene@cena.usp.br

MERCEDES MARCHESE, Instituto Nacional de Limnología, INALI-CONICET, J. Maciá 1933, 3016 Santo Tomé, Prov. Santa Fe, Argentina mrmarchese@arnet.com.ar

LUIZ A. MARTINELLI, Centro de Energia Nuclear na Agricultura (CENA), Universidade de São Paulo, Avenida Centenário 303, 13416-000 Piracicaba, SP, Brazil zebu@cena.usp.br

LAURENCE MAURICE-BOURGOIN, IRD, French Research Institute for Development, CP 7091 Lago Sul, 71619-970 Brasilia, DF, Brazil lmaurice@unb.br

EMILIO MAYORGA, University of Washington, School of Oceanography, Seattle, Washington 98195-5351, USA emiliomayorga@alum.mit.edu

MICHAEL E. MCCLAIN, Department of Environmental Studies, Florida International University, Miami, Florida 33199, USA mcclainm@fiu.edu

JORGE M. DE MORAES, Centro de Energia Nuclear na Agricultura (CENA), Universidade de São Paulo, Avenida Centenário 303, 13416-000 Piracicaba, SP, Brazil

MARCELO Z. MOREIRA, Centro de Energia Nuclear na Agricultura (CENA), Universidade de São Paulo, Avenida Centenário 303, 13416-000 Piracicaba, SP, Brazil

STÉPHANE MOUNIER, Laboratoire de RCMO, Université de Toulon et du Var, BP 132, F-83957 La Garde Cedex, France mounier@univ-tln.fr

JEAN PIERRE H. B. OMETTO, Centro de Energia Nuclear na Agricultura (CENA), Universidade de São Paulo, Avenida Centenário 303, 13416-000 Piracicaba, SP, Brazil

GEOFFREY PETTS, University of Birmingham, School of Geography and Environmental Sciences, Edgbaston, Birmingham B15 2TT, UK g.e.petts@bham.ac.uk

IRMA QUIROGA, *formerly at:* Universidad Mayor de San Andres, Chemical Institute, CP 330, La Paz, Bolivia

GIOVANNA ROCABADO, Instituto de Ecologìa, Unidad de Limnologìa, CP 10077, La Paz, Bolivia giovanna_rocabado@hotmail.com

JUDITH ROSALES, Universidad Nacional Experimental de Guayana (UNEG), Urbanización Chilemex, Calle Chile, Puerto Ordaz, Edo. Bolívar, Venezuela jrosales@uneg.edu.ve

JOÃO PAULO MACHADO TORRES, Laboratório de Radioisótopos Eduardo Penna Franca, Instituto de Biofísica Carlos Chagas Filho, Centro de Ciências da Saúde, Universidade Federal do Rio de Janeiro, 21941-900 Rio de Janeiro, RJ, Brazil jptorres@biof.ufrj.br

REYNALDO L. VICTORIA, Centro de Energia Nuclear na Agricultura (CENA), Universidade de São Paulo, Avenida Centenário 303, 13416-000 Piracicaba, SP, Brazil

CONRAD VISPO, 1685 Route 5, Canaan, New York 12029, USA knab@students.wisc.edu; *formerly at*: Fundación La Salle de Ciencias Naturales (FLASA), Instituto de Investigaciones Orinoco-Esequibo, Tumeremo, Edo. Bolívar, Venezuela

JEAN-GABRIEL WASSON, CEMAGREF, UR Biologie des Ecosystèmes Aquatiques, Laboratoire d'Hydroécologie Quantitative, 3 bis Quai Chauveau, BP 220, F-69336 Lyon Cedex 09, France jean-gabriel.wasson@cemagref.fr

CHAPTER 1

PROCESSING OF BIOACTIVE ELEMENTS IN THE AMAZON RIVER SYSTEM

Emilio Mayorga & Anthony Aufdenkampe

The Amazon basin drains a vast and heterogeneous region of northern South America, encompassing large areas of seven nations and important population centres such as Manaus, Porto Velho, Iquitos, Cusco, and La Paz (Fig. 1.1). Despite the acceleration of human impact in the form of deforestation and land-use change, mining and oil extraction, dam construction, and urbanization, larger tracts remain relatively pristine, and the main river channels remain largely unaffected by engineering (Lewis *et al.*, 1995; Richey *et al.*, 1997). The Amazon is of undeniable importance for local communities in particular, and Amazonian nations in general; but as the largest river system in the world it also plays a significant continental and global role in modulating climate, biogeochemical cycles, and terrestrial inputs to the ocean (Devol *et al.*, 1994; Meade, 1994; Devol & Hedges, 2001; Marengo & Nobre, 2001).

An understanding of biogeochemical dynamics and controls in river systems is a central requirement of any effort to rationally manage fluvial resources and minimize the impact of human activity on riverine ecosystems. The cycling of bioactive materials (oxygen, carbon, nitrogen, phosphorus, etc.) in rivers, however, is mediated by physical and biological components, such as hydrology, sediment dynamics, and bacterial metabolism. In addition, the linkages between a river and the land it drains are often complex. An integrated approach that crosses disciplinary and landscape boundaries is required to fully elucidate the biogeochemistry of a river system.

In this chapter we review and synthesize current understanding of the biogeochemistry of the Amazon River system. Our goal is to provide a conceptual framework from which watershed managers can evaluate observations, rather than summarize all available data as has been done in the extensive reviews by Guyot (1993) and Lewis *et al.* (1995). In this overview, no new data or analyses are presented nor are observations discussed in extensive detail. Rather, we provide numerous references to the many papers that do so.

We begin by briefly introducing the geography of the Amazon basin and the history of geochemical and related research on the region's rivers. We then describe the principal fractions of bioactive materials carried by rivers—their characteristics, sources and interactions—paying special attention to the role of size in fate and transport. Finally, we illustrate with examples largely from the Amazon mainstem the principal patterns of spatial and temporal variability typical of river systems and discuss how their controls can change regionally and with river scale. Through such a presentation we hope to facilitate the interpretation and use of natural riverine biogeochemical observations in tropical river systems like the Amazon, a task that is of critical importance for evaluating likely impacts of human perturbations. While we refer to relevant research by other groups, we write this chapter from our perspective as members of the CAMREX (**C**arbon in the **AM**azon **R**iver **EX**periment) project, with origins in oceanography that are evident in our view of river processes.

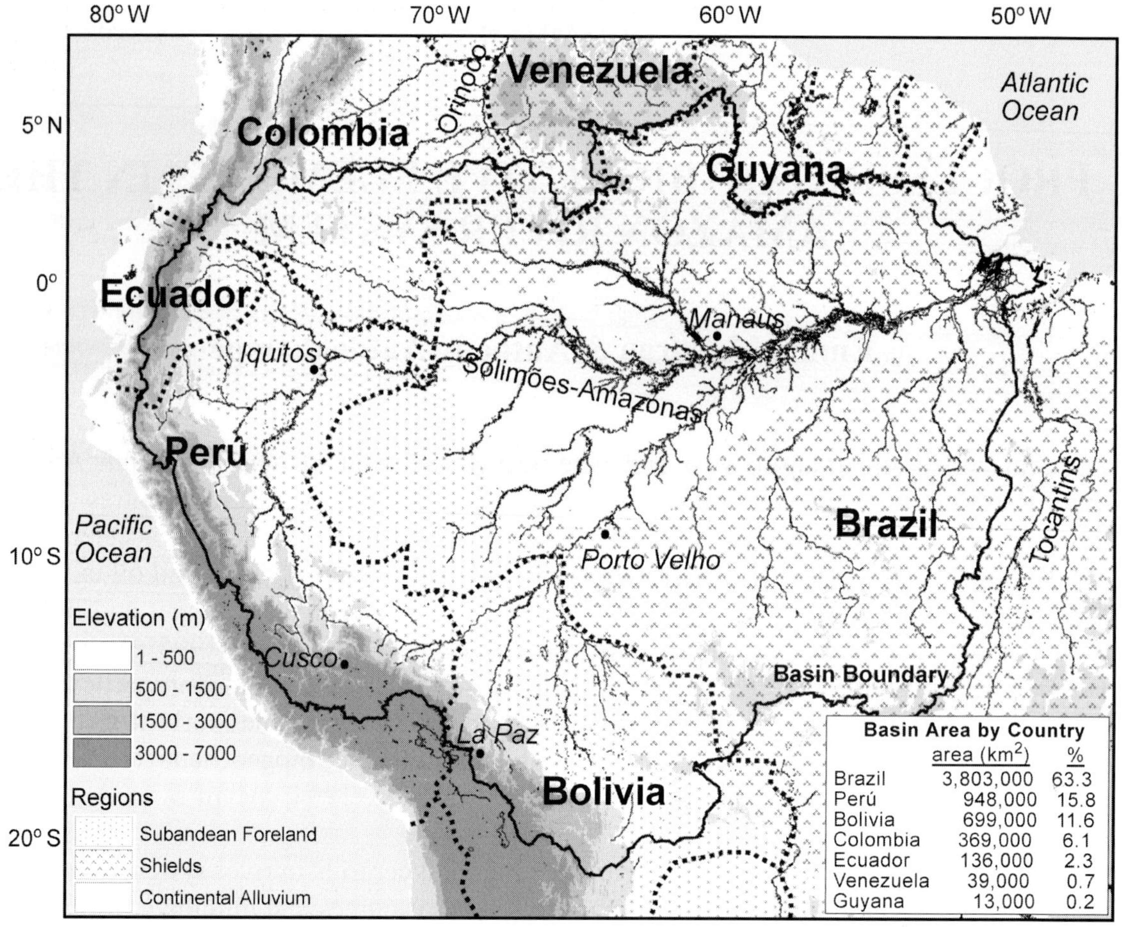

Fig. 1.1 General physiographic features of the Amazon basin. The basin boundary is shown as a thick black line. National boundaries are shown as thick dotted lines. Major cities in the basin are included. The portion of the total basin area encompassed in each country is listed in the inset table; total basin area is 6 007 000 km^2. Also shown are elevation and the main morphostructural regions.

GEOGRAPHICAL AND HISTORICAL PERSPECTIVES OF THE AMAZON

The Amazon River and its basin

The Amazon basin covers a continental-scale region of approximately six million km^2 (not including the Tocantins basin; Figs 1.1 and 1.2). Such a large area will inevitably encompass very dissimilar environments, climates, and ecosystems. The main physiographic features of the basin and the river system have been extensively reviewed elsewhere (Sternberg, 1975; Sioli, 1984b; Lewis *et al.*, 1995), and we provide only a brief outline. The main structural elements include the Precambrian, highly weathered Guayana and Brazilian shields, the Andean mountains to the west, the Andean alluvial foreland, and a central continental alluvium or trough (Fig. 1.1). Soils in the lowlands are generally deep and highly weathered, with prominent occurrences of sandy podzols in the shields. The soils in flood plains and alluvial regions around mainstems draining the Andes originate in the cordillera and are much less weathered. The climate is generally humid, with mean annual rainfall of approximately 2000 mm. However, precipitation in some regions in the northwest and the Andean piedmont can be much larger, whereas large regions in the east, south, and the high Andes can be drier or arid. More specifically, 10–15% of the basin

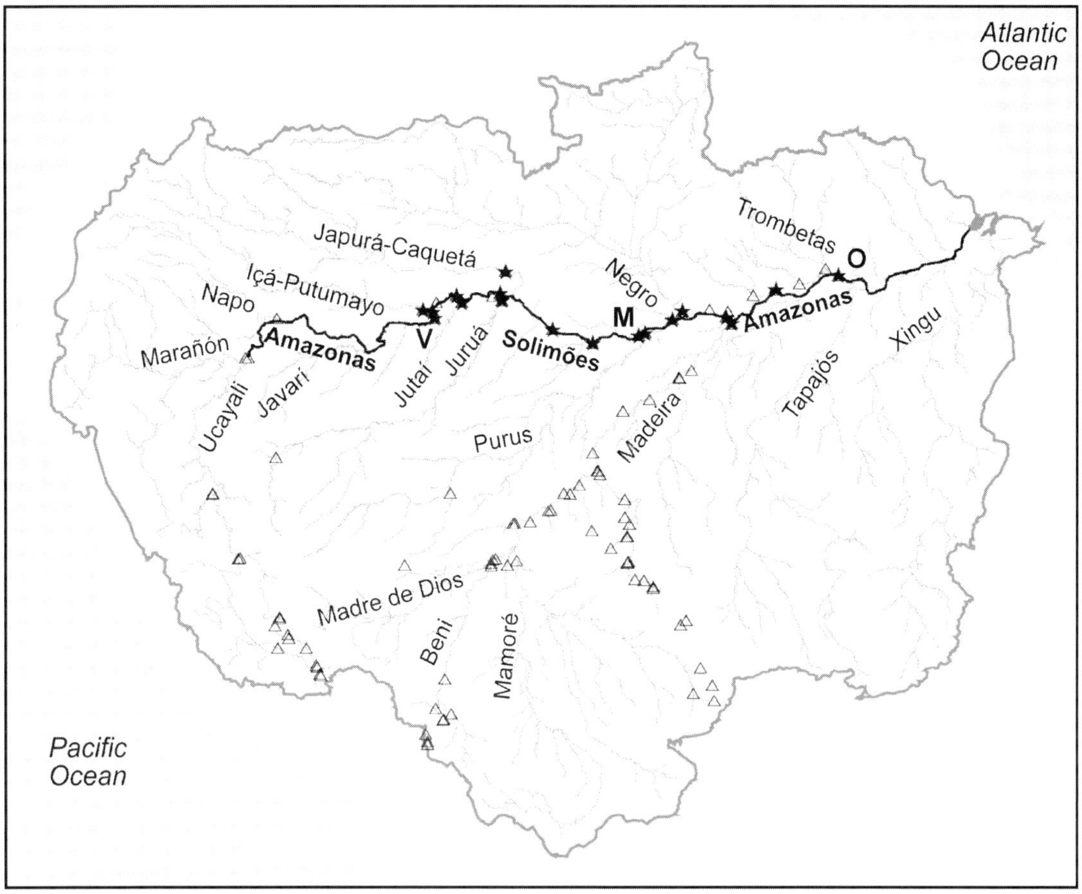

Fig. 1.2 Amazon River network and CAMREX river sampling sites. The names of the mainstem and major tributaries are shown. The Amazon mainstem (thick black line) first begins under the name of "Amazonas" at the confluence of the Ucayali and Marañón in Peru. As it enters Brazil, its name changes to "Solimões", and becomes the "Amazonas" again after the confluence of the Río Negro. The total length of the river is about 6500 km, from its mouth to its source in the Ucayali headwaters south of Cusco. CAMREX mainstem sampling stations are shown as solid stars, from Vargem Grande (V) to Óbidos (O). Eleven mainstem sites and the mouths of the seven major tributaries were sampled through cross-sectional composites during 12 cruises between 1982 and 1991. All other CAMREX sampling sites are shown as open triangles. The site of the Marchantaria–Manacapurú time series is labelled "M".

receives less than 1500 mm of rainfall per year, while 6–10% receives at least 3000 mm of rainfall per year (unpublished data). The length of the dry season in particular has a large role in determining the type of vegetation that may be sustained. The lowlands are primarily covered by evergreen to semi-deciduous rainforest, but large regions have distinct savanna-like vegetation systems, often seasonally inundated in lowlying areas. River discharge generally displays strong seasonality, with changes in stage height of as much as 10 m in the Amazon mainstem driving changes in the extent of inundation within the central lowlands from 100 000 km^2 in November to 350 000 km^2 in May (Richey *et al.*, 2002). Discharge to the ocean is approximately 220 000 m^3 s^{-1}.

History of geochemical research in the Amazon

The variety of environments, massive scale of drainage areas and rivers, and relatively low levels of human impacts, has long attracted geochemical and ecological researchers to the region. Initial investigations centred on large rivers in the Brazilian Amazon, especially the Solimões–Amazonas mainstem, surrounding flood plains and major tributaries, and the region around Manaus (Sioli, 1984b). Early observations of three types of river water based on colour (Wallace, 1853) were later developed extensively and classified by H. Sioli (Furch, 1984; Sioli, 1984a). White-water rivers have a characteristic muddy colour, relatively high concentrations of dissolved solids, and alkaline to neutral pH; their high sediment load originates in Andean or alluvial regions. Examples include the Ríos Solimões and Madeira, and other lowland rivers draining mountainous Andean regions. Black-water rivers are tea-coloured from high concentrations of dissolved organic matter (DOM), have negligible suspended sediment loads and medium transparencies, are very dilute in dissolved ions, and are usually acidic; they typically drain areas of low relief and sandy podzol soils. The Río Negro is the classic example. Finally, clear-water rivers can have high transparencies and are clear or olive-green in colour. Their dissolved load is typically low but varies across systems, and they have a wide range in pH, from acidic to alkaline, and low suspended sediment loads. Examples are the Ríos Tapajos and Xingu. The group led by H. Sioli, K. Furch, W. Junk and associates from the Max-Planck Institut für Limnologie (Plön, Germany) and the Instituto Nacional de Pesquisas da Amazônia (Manaus, Brazil) is responsible for most of the early studies on the geochemistry and ecology of rivers and flood plains in the central Amazon (reviewed in Sioli, 1984b). Others made important contributions to the characterization of inorganic and organic constituents of these rivers of the central basin (Klinge & Ohle, 1964; Williams, 1968), and Gibbs (1967) extended his surveys of river mineralogy and inorganic geochemistry from the lowlands to the Andes.

The cruises of the Research Vessel *Alpha Helix* in 1976–1977 from the Atlantic to Iquitos comprised one of the most influential initiatives in early Amazon research. These expeditions united researchers from diverse fields and nationalities, spawning a substantial number of important publications on sedimentology (Curtis *et al.*, 1979; Meade *et al.*, 1979; Nordin Jr *et al.*, 1980; Sholkovitz & Price, 1980), biogeochemistry (Fisher & Parsley, 1979; Richey *et al.*, 1980; Longinelli & Edmond, 1983; Stallard & Edmond, 1983; Cai *et al.*, 1988), and biology (Fisher, 1979; Wissmar *et al.*, 1981). The Alpha Helix project gave birth to the CAMREX research group, which has continued to unite researchers in Amazon biogeochemical research for the last 20 years. The more than 120 CAMREX publications have focused on understanding physical and biogeochemical dynamics throughout the basin using a large variety of approaches. These include studies of cross-sectionally integrated fluxes of water, sediments and bioactive elements (Meade *et al.*, 1985; Richey *et al.*, 1986; Ferreira *et al.*, 1988; Richey *et al.*, 1990); flood-plain sediment dynamics (Mertes *et al.*, 1996; Dunne *et al.*, 1998); sediment diagenesis in river bars, flood plains and deltas (Victoria *et al.*, 1989; Johnsson & Meade 1990; Martinelli *et al.*, 1993; Keil *et al.*, 1997); nutrient dynamics in rivers, flood plains and *terra firme* streams (Forsberg *et al.*, 1988; Martinelli *et al.*, 1992; McClain *et al.*, 1994; Devol *et al.*, 1995; McClain *et al.*, 1997); biogenic gas production rates and air–water exchange fluxes (Devol *et al.*, 1987; Richey *et al.*, 1988; Devol *et al.*, 1994; Richey *et al.*, 2002); biochemical tracers of organic matter (OM) sources and processes (Ertel *et al.*, 1986; Hedges *et al.*, 1986a; Hedges *et al.*, 1994; Hedges *et al.*, 2000; Aufdenkampe *et al.*, 2001); isotopic tracers of organic matter sources and cycling (Araujo-Lima *et al.*, 1986; Hedges *et al.*, 1986b; Martinelli *et al.*, 1991; Quay *et al.*, 1992; Victoria *et al.*, 1992; Forsberg *et al.*, 1993; Quay *et al.*, 1995; Brandes *et al.*, 1996); isotopic tracers of water cycling (Martinelli *et al.*, 1996); microbiological metabolism of organic matter (Benner *et al.*, 1995; Amon & Benner, 1996a,b); and remote sensing driven models of hydrology, sediment dynamics, and carbon cycling (Mertes *et al.*, 1993).

Other groups have made very substantial contributions to biogeochemical research in the Amazon basin. Several studies have focused on hydrological, nutrient and geochemical dynamics on hillslopes and in small catchments north of Manaus (Franken & Leopoldo, 1984; Brinkmann 1985; Nortcliff & Thornes, 1988; Nortcliff *et al.*, 1990; Forti & Neal

1992; Konhauser *et al.*, 1994). Others have examined the sources and mechanisms yielding high DOM concentrations in the sandy soils of the Río Negro basin (Leenheer, 1980; St John & Anderson, 1982). In the flood plains of the central Amazon, J. Melack and associates have carried out extensive biogeochemical, hydrological, and ecological studies in lakes, small catchments, and the inundated area (Lenz *et al.*, 1986; Lesack & Melack, 1996; Williams & Melack, 1997; Sippel *et al.*, 1998). Studies in other regions of the basin have been primarily focused on terrestrial, hillslope and stream processes. These include the work in the upper Río Negro at San Carlos, Venezuela (Stark & Jordan, 1978; Tiessen *et al.*, 1994); in the Caquetá basin at Araracuara, Colombia (Duivenvoorden & Lips, 1995); and in the Pachitea basin in the Andean piedmont of central Peru (Elsenbeer *et al.*, 1996). In the Peruvian Amazon a group from the University of Turku, Finland, maintained long-term studies of the relationships between geomorphology, river evolution, succession, and flood-plain mineral nutrients (Salo *et al.*, 1986; Puhakka *et al.*, 1992; Kalliola *et al.*, 1993). In the Bolivian headwaters of Río Madeira, from the 1980s the PHICAB (Programa Hidrológico y Climatológico de la Cuenca Amazónica de Bolivia) project conducted extensive research on the hydrology, sediment dynamics, and geochemistry of rivers (Guyot, 1993; Guyot & Wasson, 1994; Guyot *et al.*, 1999). This French-led project has been extended into the Brazilian Amazon, where it is currently active as HiBAm (**Hi**drologia e Geoquímica da **B**acia **Am**azônica) (Guyot *et al.*, 1998; Mounier *et al.*, 1999; Patel *et al.*, 1999; see Chapter 2 of this book). Additional research in the Andean Amazon has been reviewed by McClain *et al.* (1995).

ROLE OF SIZE ON FATE AND TRANSPORT

Three distinct physical size classes of material transported in rivers

Riverine transport of bioactive elements from land to sea forms a major link in the global biogeochemical system. The diversity of matter carried in river water is extraordinary—weathered products of rocks and plants sustain all forms of living organisms. The entire continuum of material size is represented, ranging from individual molecules to boulders and trees. Early on in the study of rivers, materials in water were separated into fractions by size (Williams, 1968; Meybeck, 1982). Coarse particulate, fine particulate, and dissolved fractions were thus operationally defined using sieve pore sizes ranging from 20 to 65 μm to separate particulate fractions, and filter pore sizes ranging from 0.1 to 1.0 μm to isolate dissolved constituents. These exact size cutoffs often differ slightly between research groups, chosen arbitrarily and largely out of convenience. In nature the size distribution of materials across these ranges is actually quite smooth. Despite this, the three size classes—coarse, fine and dissolved—consistently exhibit very distinct transport dynamics, degradation patterns and compositional characteristics. In addition, these three fractions remain compositionally distinct despite interacting with each other on a continual basis (Figs 1.3, 1.4 and 1.5).

Coarse particles

Essentially all the material transported by rivers starts out in the coarse particulate size class. Rock and decaying vegetation are physically and chemically weathered and eroded until they finally reach the river. While the coarse size class includes boulders and dead trees, most riverine studies have focused on sand sizes (defined as 63 μm–2 mm within CAMREX).

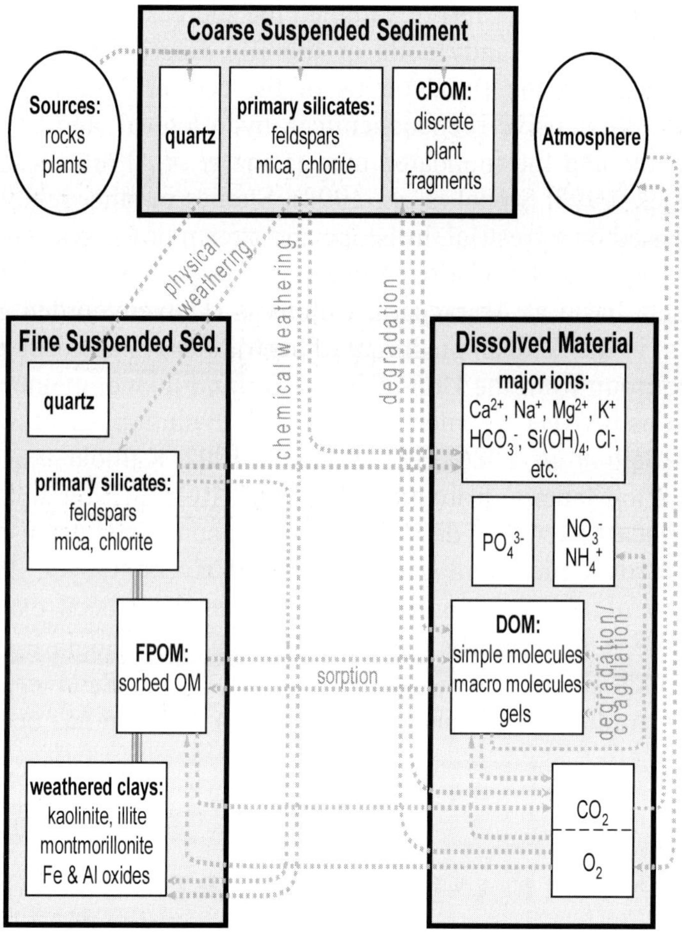

Fig. 1.3 Conceptual model for the composition of the three principal size fractions carried by rivers, and the processes that are responsible for continuous interactions between them. Microbial mediation of these processes is implicit for many of these arrows. Bars connecting boxes within the fine particulate fraction represent the physical associations between mineral and organic components.

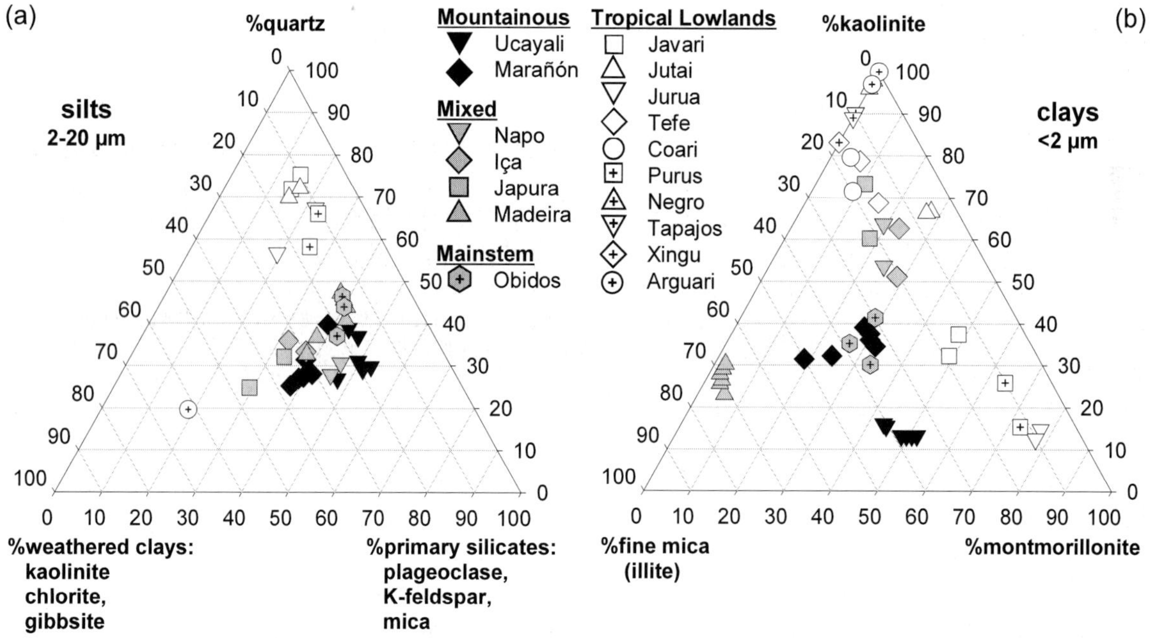

Fig. 1.4 Major mineralogical composition of (a) silt-sized sediments (2–20 µm) and (b) clay-sized sediments (<2 µm) within the Amazon basin. "Mixed" refers to river basins that drain both mountainous and lowland environments. Data from Gibbs (1967).

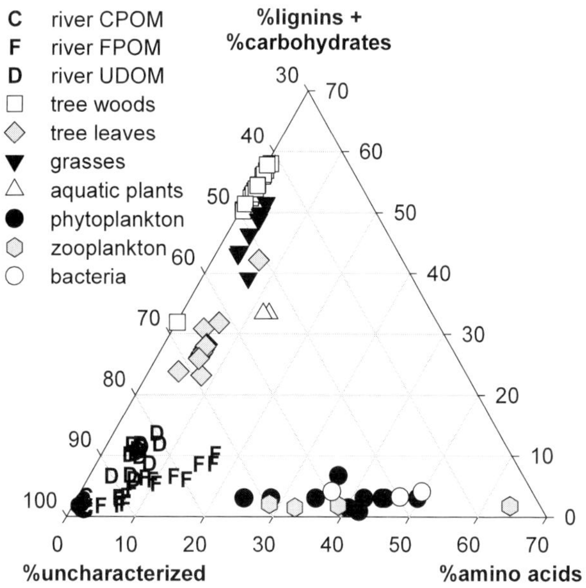

Fig. 1.5 Major biochemical composition of coarse and fine particulate organic matter (OM) and ultrafiltered dissolved OM carried by rivers of the Amazon basin, relative to the composition of source materials. River data from Ertel *et al.* (1986), Hedges *et al.* (1986a), Hedges *et al.* (1994), Hedges *et al.* (2000) and source data from these papers and from Cowie & Hedges (1984, 1992), Goni & Hedges (1992) and unpublished data. Notice that upper and right axes terminate at 70% of the respective biochemicals, due to all samples containing at least 30% of OM not identifiable in these biochemical classes. All DOM compositions fall in the grouping at >95% uncharacterized.

Because these coarse suspended sediments (CSS) settle quickly to the streambed, suspended concentrations are strongly dependent on streamflow velocities (Meade *et al.*, 1985; Devol *et al.*, 1995) and increase substantially with depth in the river (Curtis *et al.*, 1979; Richey *et al.*, 1986). Thus CSS transport is highly episodic or seasonal, with most occurring during flood events. For instance, the mainstem of the Amazon near Manaus has depth-averaged CSS concentrations of 10–20 mg l^{-1} during low water and 45–70 mg l^{-1} at high water (Devol *et al.*, 1995). The bed load, or the CSS travelling very close to, or even as part of, the fluid riverbed, is estimated to be a small fraction of total coarse sediment flux in the Amazon mainstem (Dunne *et al.*, 1998) but it may be important in smaller rivers. As for all the size classes, coarse particulates consist of both inorganic mineral material and organic material derived from the remains of living organisms.

Within the Amazon basin, quartz is the dominant mineral in CSS, forming 25–75% of the total (Gibbs, 1967); other primary silicates such as feldspars, micas, and chlorite comprise the remainder at 2–25% each. These primary minerals in rivers, surrounding soils and aquifers each exhibit distinctive weathering rates and products. Thus through weathering processes, the lithology, relief and climate of a watershed together determine the composition of the finer, secondary minerals such as alumino-silicate clays and iron and aluminium oxides and the composition of major dissolved inorganic ions, such as Ca^{2+}, Na^{+}, Mg^{2+}, HCO_3^{-}, Cl^{-} and Si species (Fig. 1.3) (Stallard & Edmond, 1983, 1987). The composition of CSS as a function of a hydrograph remains unstudied, but hydrodynamic sorting of CSS components would be expected to produce spatial and temporal variations in CSS composition within a reach.

In the Amazon mainstem and other large turbid rivers, organic matter (OM) is a small but important portion of the CSS fraction, comprising only 0.6–3.3% by mass (Richey *et al.*, 1990; Devol *et al.*, 1995). However as CSS concentrations decrease in rivers, OM contributions increase to as high as 10–30% in the clearest waters (Hedges *et al.*, 1994, and unpublished data). In all cases, microscopic studies reveal that most coarse particulate organic matter (CPOM) in rivers resembles partially degraded plant fragments, often retaining visible cell structure (Keil *et al.*, 1997). CPOM is less dense than mineral grains of the same size, hence explaining its higher contribution to CSS in river channels under low flow. Between 10 and 20% of CPOM can be identified biochemically as amino acids, carbohydrates and lignins—the primary molecular building blocks of living organisms—relative to 25–60% within biomass sources (Fig. 1.5) (Hedges *et al.*, 1986a; Hedges *et al.*, 1994; Hedges *et al.*, 2000). Biochemical source indicators, such as carbon to nitrogen ratio and the ratio of cinamyl to vanilyl lignin phenols, all show that Amazon basin CPOM is primarily derived from tree leaves (Fig. 1.6(a)) (Devol & Hedges, 2001). Several other lines of evidence, including stable carbon isotope compositions, confirm these conclusions but also show minor contributions from grasses for certain rivers (data not shown; Quay *et al.*, 1992; Victoria *et al.*, 1992). Biochemical indicators of degradation, such as the contribution of fucose and rhamnose sugars to total carbohydrates and acid to aldehyde ratios in lignin phenols, all support evidence from microscopic studies and major biochemical composition that CPOM is sparingly degraded and rather fresh (Fig 1.6(b)). Radiocarbon analysis of CPOM and low density soil particulate organic matter confirms their recent origin (Hedges *et al.*, 1986b; Trumbore *et al.*, 1995). It is clear that CPOM is actively degrading and

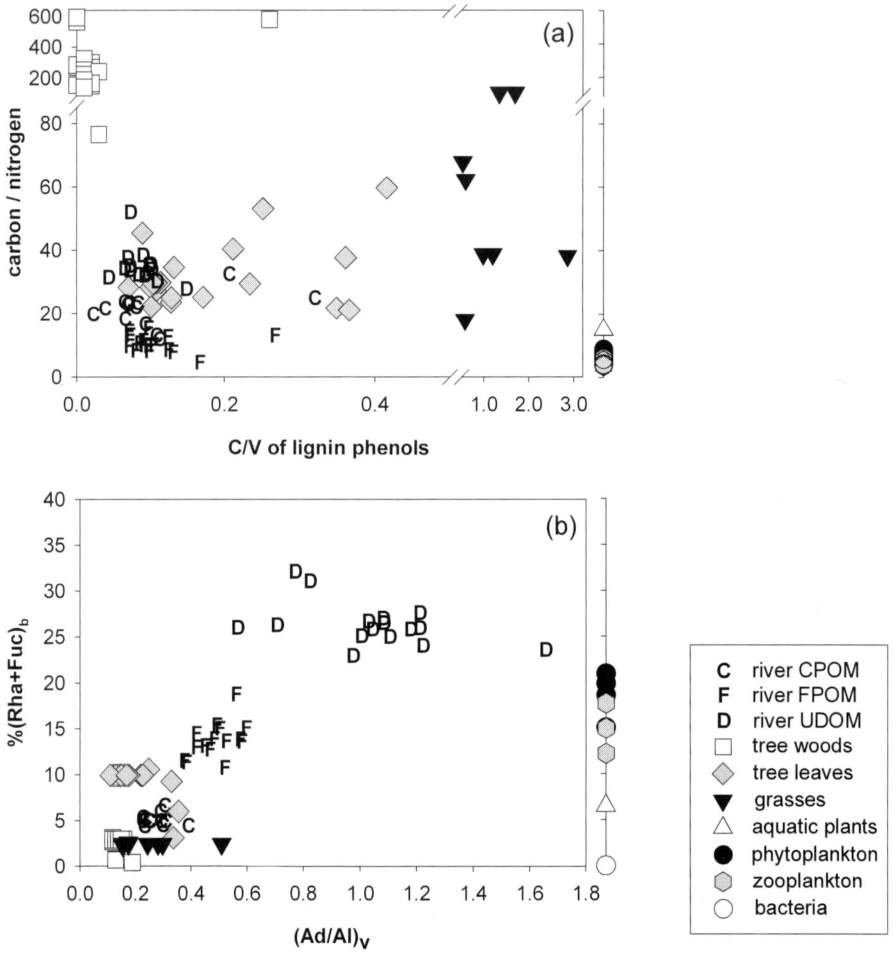

Fig. 1.6 Biochemical parameters commonly used to interpret (a) organic matter sources and (b) organic matter degradation. Riverine coarse and fine particulate OM and ultrafiltered dissolved OM are again presented relative to common source materials. In (a) the ratio of cinamyl to vanilyl lignin phenols (*C/V*) clearly distinguishes between woods, leaves, grasses and all aquatic organisms (which contain no lignin and are thus plotted to the right). In (b), FPOM and UDOM are outside the range of plant sources with respect to both the acid to aldehyde ratio of lignin phenols, $(Ad/Al)_v$, and the percent contribution of sugars rhamnose and fucose to total carbohydrates, $\%(Rha + Fuc)_b$—two parameters that increase with increasing degradation within sediments. Data obtained from the references listed for Fig. 1.5. Average $\%(Rha + Fuc)_b$ values were used for some woods and leaves for which no measurements were made.

leaching, supplying microbes with substrate and releasing dissolved organic and inorganic compounds into the river. These rates have not been directly measured, but are likely to be quite high. Stream and river budgets suggest that CPOM continuously enters the river mostly from bank vegetation and detritus falling directly into the water (McClain *et al.*, 1997).

Coarse suspended river sediments are thus a heterogeneous mixture of sand-sized mineral grains and discrete plant fragments (Fig. 1.3). Both inorganic and organic constituents of CSS are generally fresh and supply the dissolved component (or "pool") as they degrade. CSS deposits on the river bed and in sand bars form a relatively porous sediment buffer or "hyporheic" zone (Boulton *et al.*, 1998) that acts as a biological reactor, supplying the river with its products and forming new riverine environments that can be colonized by vegetation.

Fine particles

Fine suspended sediments (FSS) consist of clays and silts, material between 0.45 and 63 μm in size as defined by CAMREX. Maintained in suspension by the slightest turbulence, FSS is often referred to as the "wash load" by geomorphologists, because once introduced to a stream or river, FSS will generally not settle. Notable environments that often prevent FSS from washing directly out to sea are lakes, reservoirs, and flood plains. For instance, evidence suggests that within the Amazon—with its extensive annually flooded plains—a typical FSS particle passes through flood-plain deposits several times between the Peruvian border and the Atlantic (Martinelli *et al.*, 1993; Dunne *et al.*, 1998). Given the patterns of channel migration, each cycle of flood-plain deposition and re-suspension requires a few thousand years (Mertes *et al.*, 1996). Because FSS is mobilized by erosion events, concentrations are also a function of river discharge, although the tendency for FSS not to be re-deposited results in weaker correlations than those observed for CSS (Devol *et al.*, 1995). The mainstem of the Amazon near Manaus exhibits FSS concentrations of 70–120 mg l^{-1} during low water and annual maximums of 210–340 mg l^{-1} during the rising hydrograph (Devol *et al.*, 1995).

Mineralogy within the FSS fraction can be quite diverse. Silt-size grains (5–63 μm) often retain many of the characteristics of the CSS fraction while at the same time showing signs of being more weathered (Fig. 1.4(a)). Fine quartz is important in more weathered lowland basins whereas other primary silicate minerals dominate in rivers draining the Andes (Gibbs, 1967). Clays and oxides are aggregated with the larger minerals in important quantities. The clay-size fraction (0.45–5 μm) is composed mostly of phylo alumino-silicate clay minerals, which are the weathering products of primary silicates (Figs 1.3, 1.4(b)). As can be seen in Fig. 1.4(a) and (b), mineral diversity within FSS is very large and depends on the geology (source minerals) and climate (weathering rates) of the watershed. As a result, mineralogical compositions within the Amazon are constantly evolving downstream with inputs from tributary watersheds and from weathering during temporary storage in the flood plain (Johnsson & Meade, 1990; Martinelli *et al.*, 1993). All FSS samples do have common characteristics however. This size fraction is generally high in surface area and high in cation and anion exchange capacity. As such, significant quantities of certain inorganic ions (e.g. NH_4^+, PO_4^{3-} and most metals) and organic compounds (e.g. natural organic matter, contaminants, etc.) can potentially be adsorbed to the mineral surfaces of FSS. Thus FSS actively interacts with the dissolved fraction, often acting as a buffer or reservoir for dissolved compounds (Fig. 1.3).

As with the CSS, organic matter contributes only a small fraction of FSS, generally 0.5–2.0% by weight in turbid rivers (Richey *et al.*, 1990; Devol *et al.*, 1995). Unlike CPOM however, fine particulate organic matter (FPOM) is tightly associated with the mineral phase (Fig. 1.3). In the Amazon and other rivers of the world, generally over 90% of the FPOM cannot be physically separated from the mineral material (Keil *et al.*, 1997; Mayer *et al.*, 1998). With low-density material removed, the organic surface loadings of these sediments is consistently 0.5–1.1 mg C m^{-2} of mineral surface area, leading in part to the conclusion that sorption of natural DOM to minerals is the primary pathway in which FPOM is formed (Mayer, 1994; Hedges & Keil, 1995). Evidence suggests however that mineral-associated organic matter does not form uniform coatings over the surfaces, but organic gel-like assemblages that cover <20% of the surfaces (Mayer, 1999; Filius *et al.*, 2000). However, sorbed organic matter affects mineral surface properties significantly by increasing cation exchange capacity and by offering hydrophobic phases into which organic contaminants (e.g. pesticides, petroleum products, etc.) and heavy metals readily sorb (Benedetti *et al.*, 1996). Biochemically, a smaller fraction of FPOM is identifiable as carbohydrates or lignins when compared to CPOM (Fig. 1.5), but often a larger fraction is identifiable as amino acids. It appears that FPOM comes largely from leaf material as does CPOM (Fig. 1.6(a)), but often falls slightly outside of the biochemical range that is possible by mixing biomass alone, as might be expected of diagenetically altered material (Devol & Hedges, 2001). Specific degradation parameters show this clearly (Fig. 1.6(b)), however riverine FPOM is still relatively fresh compared to riverine DOM or deep-sea sediments (Hedges *et al.*, 1994). This appears to be the result of physical protection from microbial attack that is offered by the intimate association of FPOM with mineral surfaces (Keil *et al.*, 1994; Baldock & Skjemstad, 2000; Kaiser & Guggenberger, 2000).

Fine suspended sediment is thus a diverse but tightly associated mixture of minerals and organic matter that interacts strongly with other size fractions (Fig. 1.3). Fine minerals are formed from the weathering of coarse minerals, and FPOM largely results from sorptive interactions with dissolved organics. However, the major characteristics of FPOM most likely form within soils prior to erosion into rivers. Mineralogical and biochemical evidence points to this conclusion (Devol & Hedges, 2001), including carbon-14 dates hundreds to thousands of years old (Hedges *et al.*, 1986b, and unpublished data). However, whereas high relief source basins are the ultimate source for >90% of river-borne FSS (Meade, 1994), stable carbon isotopes show that at Óbidos >60% of the associated FPOM originates in the lowlands (Quay *et al.*, 1992; Hedges *et al.*, 2000). With such high levels of surface exchange and activity, deposited FSS forms highly fertile soils and sediments, often far from the river channel.

Dissolved matter

Just as for the particulate size classes, the dissolved fraction (<0.45 μm as defined by CAMREX) is composed of matter spanning a continuum of sizes, from truly dissolved individual molecules to colloidal minerals and organic gels. The distinguishing physical feature of material in the <0.45 μm size class is that it will stay in suspension indefinitely due to Brownian motion alone. Thus, the transport of material in the dissolved pool is determined by the advection and mixing of the waters that carry it. However, fluxes are not only determined by hydrology but also by processes that transfer material in and out of the dissolved phase, such as degradation, adsorption and coagulation. Perhaps the most important characteristic of dissolved material is that it has the potential to be directly bioavailable. Microbial organisms, plant roots, and many animal tissues will transport dissolved molecules across cellular membranes, both passively and actively. Likewise, contaminants exhibit their highest toxicity when in the dissolved phase. The only way to measure the mass of the entire dissolved pool is by evaporation, however it is much more typical to measure the concentration of individual compounds or compound classes.

The inorganic material within the dissolved fraction includes all major ions, nutrients, dissolved gases, and colloidal minerals. The major inorganic ion composition of the Amazon mainstem is very similar to the world river average, dominated by HCO_3^-, Ca^{2+}, Na^+, $Si(OH)_4$ + $SiO(OH)_3^-$, Cl^-, Mg^{2+} and SO_4^{2-} in order of importance (Stallard & Edmond, 1983). Diversity within the Amazon basin parallels the pattern of other rivers of the world, with a wide range of ionic compositions and concentrations controlled largely by geology and weathering regime with minor inputs from precipitation (Stallard & Edmond, 1983, 1987). Once within the river, these ions mix conservatively, such that conductivity and alkalinity serve as excellent tracers for water sources (Stallard & Edmond, 1983; Devol *et al.*, 1995). Of the bioactive compounds, phosphate shows the least systematic variability, generally 0.4–2.0 μM in turbid rivers as a result of buffering with larger mineral-bound reservoirs in the FSS (Forsberg *et al.*, 1988; Devol *et al.*, 1995). On the other hand, inorganic nitrogen compounds—NO_3^- and NH_4^+—cycle rapidly via remineralization of organic matter and other microbial process such as nitrification and denitrification (McClain *et al.*, 1994; Brandes *et al.*, 1996; Boulton *et al.*, 1998). Dynamics of dissolved gasses, such as O_2 and CO_2, are also largely controlled by the respiration of organic matter because river, flood-plain and lake waters are dominantly heterotrophic (Cole *et al.*, 1994; Cole & Caraco, 2001). These within-river fluxes (more below) are in quasi-steady-state balance with atmospheric gas exchange (Quay *et al.*, 1992; Devol *et al.*, 1995). Waters of the Amazon are characteristically supersaturated in CO_2 with respect to atmospheric equilibrium (commonly by a factor of 20–50) and act as a significant source of CO_2 to the atmosphere (Richey *et al.*, 2002). Thus the dissolved inorganic constituents of river waters are constantly evolving as a result of interactions with non-dissolved phases within the river corridor.

Dissolved organic matter (DOM) exists as a mixture of simple molecules, complex biomacromolecules, their partial degradation products and molecular assemblages or gels. Total concentrations, measured as dissolved organic carbon (DOC), depend strongly upon the mineralogy of the basin, highlighting the importance of sorption to DOM dynamics. Watersheds with soils dominated by quartz sands having low surface area produce black-water rivers with 7–40 mg l^{-1} of DOC, whereas watersheds rich in clays produce surface waters with low DOC (2–7 mg l^{-1}) (Sioli 1984a; Nelson *et al.*, 1993; McClain *et al.*, 1997). To better understand the composition and dynamics of DOM, ultrafiltration techniques have employed membranes with pores as small as 1 nm to separate and concentrate DOM into various size fractions (Hedges *et al.*, 1994; Küchler *et al.*, 1994; Amon & Benner, 1996a,b; Mounier *et al.*, 1999; Patel *et al.*, 1999). Generally, ultrafiltered DOM (UDOM) refers to organic material with molecular weights >1000 g mol^{-1} or daltons (HMW, or high molecular weight DOM in the CAMREX convention), including very high molecular weight subsets variously named VHMW DOM or colloidal organic carbon (COC) by different research groups (1000 daltons is approximately equivalent to a molecule of 1 nm diameter). Inorganic colloids also appear in UDOM fractions, comprising from 5–10% of the total mass in the Amazon lowlands to as much as 80–90% in the high Andes (Hedges *et al.*, 2000).

In Amazon lowland rivers, 70–90% of total DOM can be isolated as UDOM (Hedges *et al.*, 1994), with yields decreasing upstream to 40–50% in the Andes (Hedges *et al.*, 2000). Relative to the two particulate classes, UDOM contains the smallest fraction, 2.5–4.5%, identifiable as major biochemicals (Fig. 1.5), although UDOM also appears to be largely derived from tree leaves (Fig. 1.6(a)). DOM shows the most extensive signs of degradation of all three fractions (Fig. 1.6(b)) (Ertel *et al.*, 1986; Hedges *et al.*, 1986a; Hedges *et al.*, 1994), yet radiocarbon analysis places the average age to be less than decades (Hedges *et al.*, 1986b, and unpublished data). Care must be taken however when interpreting many biochemical trends between DOM and FPOM. It is known that biochemical fractionation of DOM occurs during sorption. Newly sorbed OM tends to be higher in molecular weight, more hydrophobic, more enriched in nitrogen and hydrolysable amino acids, and have different amino acid compositions than the DOM from which it came (Gu *et al.*, 1995; Kaiser & Zech, 1997; Aufdenkampe *et al.*, 2001). Other untested biochemical parameters may also result in part from sorptive fractionation.

Cumulative evidence suggests that DOM is produced largely from the degradation and/or leaching of leaf detritus similar to that in CPOM (Devol & Hedges, 2001). Once in solution, biomacromolecules such as proteins and carbohydrates are easily hydrolysed (at least partially) by exoenzymes for subsequent microbial uptake. As a result degradation tends to decrease both the size and bioavailability of HMW DOM to form the low molecular weight (LMW, ~200–1000 daltons) fraction (Amon & Benner, 1996a). However, as all particulate and dissolved organic carbon fractions degrade, microbial activity and photochemistry can generate a pool of the smallest molecules (<200 daltons)—free amino acids, free sugars, and organic acids such as acetate and citrate (Amon & Benner, 1996b; Moran & Zepp, 1997). Despite the likelihood that this very low molecular weight (VLMW) DOM represents an exceedingly small proportion of total DOM in rivers, these compounds are generally extremely bioavailable and could drive significant biological fluxes. The relative proportions of VHMW, HMW, LMW and VLMW DOM fractions would be expected to evolve down river as a result of their different degradation rates. Coagulation and disassociation of DOM in and out of colloidal gel phases or mineral surfaces complicates these size dynamics significantly however, as many of these processes respond to changes in pH and to ratios of polyvalent to monovalent ions in solution (such as Ca^{2+}/Na^{+}) (Chin *et al.*, 1998; Kaiser, 1998).

The dissolved fraction is particularly characterized by diversity and contrasts (Fig. 1.3). Organics and inorganics exist in both truly dissolved and colloidal phases. This mixture contains the most labile material carried by the river (e.g. NH_4^+, free amino acids, etc.) and also relatively non-labile weathering end-products (e.g. inorganic ions that determine alkalinity, LMW DOM, etc.). Some material is swept away with the water directly to sea, whereas transit times for other components are delayed considerably by interactions with mineral particles.

River metabolism

The CSS, FSS and dissolved fractions are compositionally distinct, yet interact with each other on a continual basis through the processes of degradation and sorption. The dynamics of one fraction cannot be completely understood without considering the dynamics of the other two. Likewise, when considering whole-river processes, such as river metabolism, all three fractions must be considered together because nutrients and substrates (or pollutants) can reside in more than one fraction. A notable omission in the above discussion of river-borne material is microorganisms. Bacterial biomass accounts for only 15–50 μg OC l^{-1} (0.75–2.5 × 10^9 cells l^{-1}) or <5% of total organic carbon (Wissmar *et al.*, 1981; Benner *et al.*, 1995), thus contributing insignificantly to the mass of any fraction. However, the microbial community mediates nearly all biochemical transformations in the natural environment, including the weathering of rocks, and as such the activity of microorganisms is implicit in every discussion of biogeochemistry.

Just as rivers provide microorganisms with a wide array of substrates, nutrients and energy sources, rivers contain within their corridors a diverse set of physicochemical environments within which nearly all biochemical transformations can occur. Most rivers in the Amazon are dominated by heterotrophic respiration over photosynthesis (*R*/*P* ratio > 3) due to low light penetration in turbid rivers and in canopy-covered streams (Quay *et al.*, 1995). Community respiration rates in the Amazon and major tributaries show a wide range, from 0.2–2 µmol C l^{-1} h^{-1} (Richey *et al.*, 1990; Benner *et al.*, 1995; Devol *et al.*, 1995), with no consistent seasonal or river-type patterns. These rates appear to be substrate limited rather than nutrient limited (Benner *et al.*, 1995), further evidence that the bulk of dissolved and particulate OM in rivers is of limited bioavailability but that a small pool of labile compounds fuels the majority of respiration (Amon & Benner, 1996a). Water flow in and out of suboxic and anoxic hyporheic zones in river beds is a likely source of labile fermentation products to the river. Methane gradients from banks to the main river channel can be quite large (Richey *et al.*, 1988; Devol *et al.*, 1994), but quantification of fluxes for other associated fermentation products has not yet been attempted. Lastly, in less turbid waters photochemical reactions can be important in both oxidizing DOM and in releasing VLMW organic compounds from more recalcitrant material for bacterial consumption (Amon & Benner, 1996b; Moran & Zepp, 1997). As light penetration increases, so does photosynthesis. In flood-plain lakes, grassland streams, clear-water rivers and to some extent black-water rivers, respiration to photosynthesis ratios approach one (Quay *et al.*, 1995). Depth integrated gross photosynthesis in these waters is comparable to the more productive regions of the oceans, with rates as high as 160 mmol C m^{-2} day^{-1} (Wissmar *et al.*, 1981; Quay *et al.*, 1995). All of these various biological process are cumulatively important. Concentrations of CO_2 over the Amazon surface exhibit a diurnal cycle with amplitudes of from 70 to 150 ppm (Quay *et al.*, 1989). Thus, to say that large rivers "breath" is not simply metaphorical.

VARIABILITY AND SCALE IN THE RIVER SYSTEM

Importance of natural variability and river scale for biogeochemical processing

The conceptualization of bioactive materials as interacting size classes with distinct properties has led to invaluable insight regarding dominant biogeochemical transformations in riverine environments (Devol & Hedges, 2001). A fuller understanding of biogeochemical dynamics in a river system, however, requires an additional dimension: an examination of the range and patterns of natural variability in concentrations, compositions, and process rates, and of their principal environmental sources and controls (Meyer *et al.*, 1988). In a fluvial system draining a continental-scale basin, natural variability is manifested across all spatial and temporal scales. Most obviously, concentrations and compositions usually vary within a channel cross-section, both laterally and with depth (Richey *et al.*, 1986). This observation is especially relevant for suspended sediments (Curtis *et al.*, 1979), and for highly labile molecules with well-defined sources in the form of discrete inputs or diffusing from the sediment-water interface, such as CH_4 (Richey *et al.*, 1988; Devol *et al.*, 1994). Such variability highlights the need for careful, channel-integrated quantitative sampling in order to obtain representative cross-section averages, especially in larger rivers (Richey *et al.*, 1986).

Regional differences in climate, soils, geomorphology, and vegetation also lead to dramatic contrasts, such as the river water typology already discussed. The progressive downstream growth of rivers that in Amazonia leads to massive channels and vast drainage areas results in not only the mixing of river waters draining dissimilar terrains, but also the creation of geomorphologically distinct riverine landforms and flood plains (Gregory *et al.*,

1991), and longitudinal biogeochemical trends. Likewise, episodic events, the seasonal evolution of climate, and subsequent ecosystem and sediment responses, can result in substantial temporal changes in the biogeochemical functioning of rivers.

Some dominant forms of biogeochemical variability in Amazonian rivers will be described, focusing on the Amazon mainstem and mouths of major tributaries where more data are available and where CAMREX concentrated its efforts in the past. This large-river bias is a common feature of tropical South American research, in contrast with a focus on small catchments in temperate regions (Lewis *et al.*, 1995). River size and system scale are important considerations that will be interwoven in this discussion. We will also emphasize factors such as the interactions of bioactive materials with geomorphology, sediment, and hydrological dynamics, and the sensitivity of different species to reach *vs* basin-scale controls.

Regional heterogeneity, downstream evolution, and large-scale structure

The biogeochemistry of running waters in the Amazon basin is exceedingly diverse. Several examples of this diversity have already been discussed. Spatial heterogeneity can be manifested at multiple scales, from large regional differences between Andean and lowland streams, through downriver longitudinal geochemical trends and interactions with a changing flood-plain environment, down to sub-regional variability and microscale changes. The value of broad classification schemes such as the visual water typology described earlier is diminished when considering subtle sub-regional changes in biogeochemical composition and controls. For example, turbid white-water rivers closer to their Andean headwaters in Peru and Bolivia display substantial contrasts in the sorptive partitioning history and proportion of HMW DOM (Hedges *et al.*, 2000); these differences may be attributable to differences in climate, terrain, and especially lithology and mineralogy (see Fig. 1.4, Ucayali, Marañón, and Madeira rivers). Moreover, smaller rivers at high elevations are often not turbid at all during the dry, low-water season (personal observation; Kvist & Nebel, 2001). Similarly, streams within each of the lowland physiographic domains (Fig. 1.1) can display geochemical composition and dynamics unlike those of the assumed norm, as a result of localized features (Sioli, 1984a; Kvist & Nebel, 2001). Larger rivers tend to average out these small-scale fluctuations and result in the expected water colour type.

The downstream growth of a river yields important biogeochemical consequences, including: a progressively diminishing role of local impacts and inputs compared to upstream influences; mixing of waters from dissimilar regions through the confluence of large tributaries; and creation of particular flood-plain environments through the geomorphological action of the river and its water and sediment regimes (Vannote *et al.*, 1980; Leopold, 1994). Longitudinal trends and "resetting" of the biogeochemistry of the channel often result from these factors. In the Río Beni, Bolivia (Fig. 1.2), such a pattern is seen in the increasingly finer suspended sediment sizes downstream as the river reaches the piedmont area (Guyot *et al.*, 1999); it is also observed in the ^{13}C content of organic matter, which evolves from relatively enriched values in the highlands resulting from an elevation isotopic effect on C3 vegetation (Körner *et al.*, 1988), to a progressive decrease towards the lowlands, followed by a sharp increase in ^{13}C of UDOM that may be a result of input from C4 grasses, and finally a sharp decrease after confluence with the larger Río Madre de Dios (see Fig. 1.6(a) in Hedges *et al.*, 2000).

On a larger scale, CAMREX has investigated changes in concentration, composition, fluxes, and process rates along a 1800-km reach of the Solimões–Amazonas mainstem (Fig. 1.2). While there are often clear differences between longitudinal trends during rising and falling stage (see below), some patterns are persistent (Fig. 1.7). Most biochemical tracers of OM composition, sources, and degradation show relatively insignificant changes downstream, despite the input of very large tributary fluxes and flood-plain exchanges (Hedges *et al.*, 1986a; Hedges *et al.*, 1994); these tracers suggest that organic matter dynamics are operating in an apparent steady-state. ^{13}C of FPOC, on the other hand,

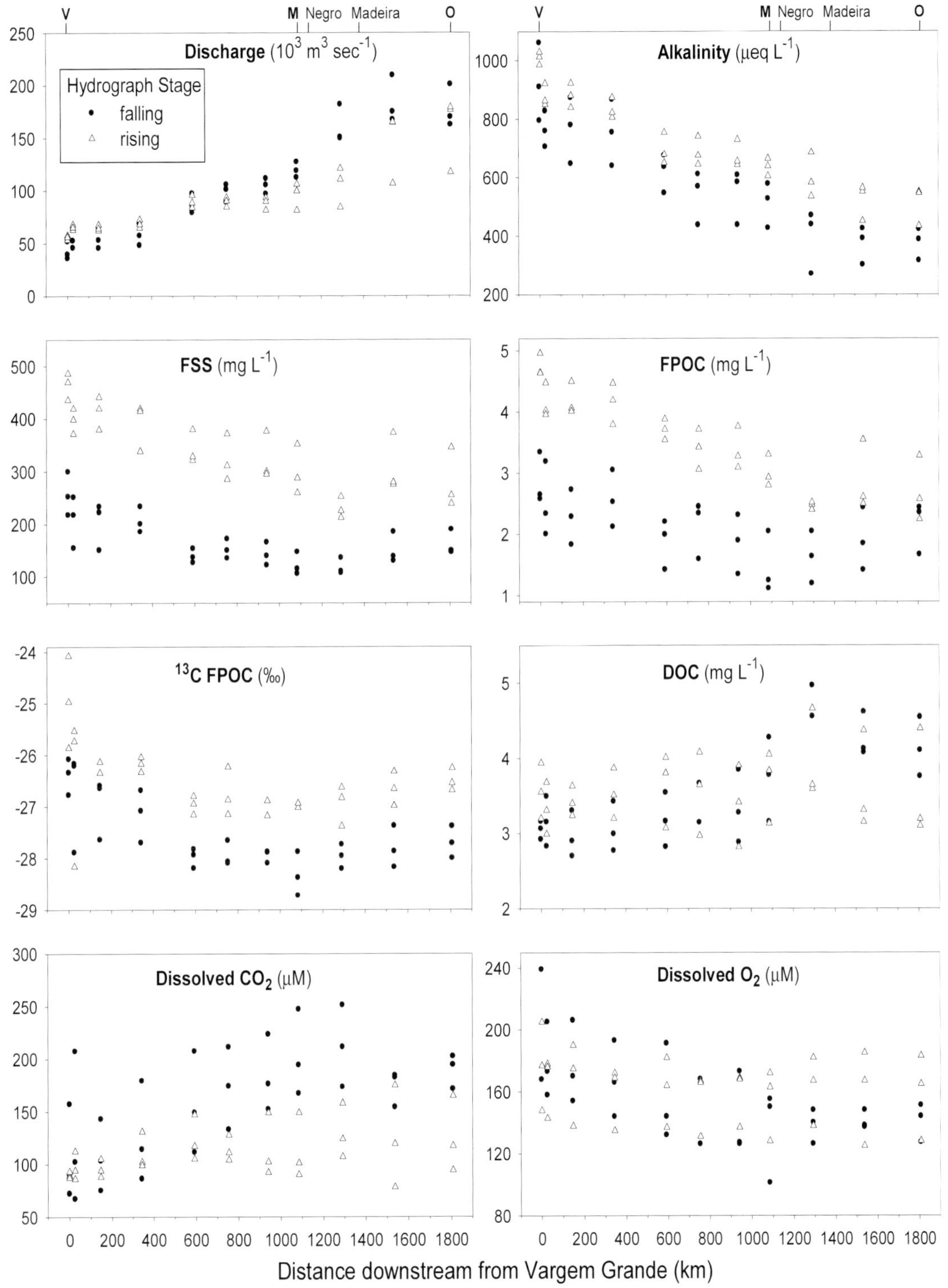

Fig. 1.7 Longitudinal trends along an 1800 km reach in the Solimões–Amazonas mainstem from Vargem Grande to Óbidos in Brazil. The data have been divided into falling (closed circles) and rising (open triangles) stages according to major hydrograph periods, to highlight seasonal differences. Variables shown are river discharge, alkalinity, fine suspended sediment (FSS), fine particulate organic carbon (FPOC), carbon-13 content of FPOC, dissolved organic carbon (DOC), dissolved CO_2 gas, and dissolved oxygen. The sampling station labels at the top of the graph are the same as in Fig. 1.2. Data from Richey *et al.* (1990) and Quay *et al.* (1992).

generally becomes more depleted from Vargem Grande to Manacapurú, reflecting the gradual replacement of Andean FPOC with lowland-derived organic carbon from C3 plants (Quay *et al.*, 1992). Suspended sediment concentrations generally decrease to Manacapurú, though sediment fluxes may follow different patterns and the channel exchanges vast amounts of sediments with the flood plain (Meade *et al.*, 1985; Dunne *et al.*, 1998). Associated particulate species usually track the trends in suspended sediment concentrations (Hedges *et al.*, 1986a; Richey *et al.*, 1990). Conservative dissolved species such as alkalinity, originating primarily in the Andes and alluvial foreland rivers, present primarily a downstream dilution effect from the input of lowland tributaries with low base cation content. For many bioactive dissolved species such as CO_2, on the other hand, the contribution of lowland tributaries and the flood plain is often the opposite, resulting in persistent increases downstream (Richey *et al.*, 1990). DOC concentrations are relatively unchanging in the mainstem, except for substantial increases due to the confluence of the black-water Río Negro. In fact, the input of such a large and distinctive tributary often results in substantial changes in the biogeochemical signature of the mainstem.

Several studies have indicated that the Amazon flood plain is a source of respiration products as well as highly labile organic matter that fuels respiration in the channel, but this input does not appear to affect the composition of bulk river organic matter (Devol & Hedges, 2001). The biogeochemical impact of the flood plain, however, is likely to be variable across the mainstem, given the diversity of geomorphological and ecological functioning of the flood plain in different reaches, driven by large-scale morphotectonic features (Mertes *et al.*, 1996; Dunne *et al.*, 1998). This diverse and changing role is also likely to be important in other flood-plain regions throughout the basin (Puhakka *et al.*, 1992; Guyot, 1993; Kvist & Nebel, 2001).

At a small scale, the extrapolation and regionalization of intensive stream and hillslope studies remains a difficult problem. Few studies of this type have been carried out in the basin (Nortcliff & Thornes, 1988; Elsenbeer *et al.*, 1996; Lesack & Melack 1996; McClain *et al.*, 1997), and their research emphases have varied. A dual challenge exists: defining the appropriate boundaries for regionalization of results from stream studies, and scaling small-stream results to their resulting larger rivers—and conversely, down-scaling from large to small rivers. These issues remain largely unexplored in the Amazon basin, though studies in North American rivers are beginning to point the way forward (Smith *et al.*, 1997; Peterson *et al.*, 2001).

Temporal variability and episodic disturbance

Variability in biogeochemical processing with time can have important ecosystem impacts and must be considered in designing sampling or monitoring systems. It can occur at all temporal scales, from short rainfall events, to regular diurnal cycles and seasonal changes in rainfall; and from interannual climate variability driven by El Niño, to rare episodic disturbance such as earthquakes and mass wasting of hillslopes in the Andes. The consequences in a river are regulated by hydrological regimes and ecosystem and geomorphological responses. In large rivers, the relative timing and magnitude of events in different tributary basins also play a role.

Longitudinal trends along the Amazon mainstem often show marked contrasts between rising and falling stages, resulting from the combined effect of changing relative contributions of tributaries *vs* Andean source waters, local hydraulic conditions, and the contribution of the flood plain in comparison with in-channel transformations (Richey *et al.*, 1990; Devol & Hedges, 2001). Trends in suspended sediment concentration and associated bioactive constituents vary with the hydrograph, showing relatively uniform distributions or lack of clear longitudinal patterns during falling water, but higher concentrations and downstream decreases during rising portions of the hydrograph (Fig. 1.7). These seasonal changes reflect variations in both sources and transport of particulate materials during different stages of the hydrograph, the role of local hydraulic and turbulent conditions as measured by river surface slope, and large-scale changes in sediment dynamics and geomorphological functioning of the flood plain (Meade *et al.*, 1985; Devol *et al.*, 1995; Dunne *et al.*, 1998). However, such well-defined patterns often break down after the confluence of the Ríos Negro and Madeira, the two largest tributaries.

Dissolved conservative species with sources in the Andes, such as alkalinity and Ca^{2+} (Fig. 1.7), show only a dilution effect downstream, where their seasonal variations are controlled largely by the relative contribution and timing of the hydrographs from the headwaters and the major tributaries (Richey *et al.*, 1990). Dissolved bioactive species, on the other hand, are controlled largely by river stage, where stage is a proxy for channel–flood-plain interactions (Richey *et al.*, 1990; Devol *et al.*, 1995). As discussed earlier, oxygen and CO_2 show a local steady-state between transport, respiration, and gas exchange across the river surface. Changing river stage results in variations in the longitudinal patterns of these dissolved gases, with increasing CO_2 super-saturation and decreasing O_2 concentrations downstream observed primarily during falling waters (Fig. 1.7). Lastly, DOC and PO_4^{3-} show little seasonal variability in their upstream–downstream distributions.

A more detailed picture of temporal variability in the mainstem is obtained from a 10-year time series of monthly measurements at the Marchantaria sampling station, in the Río Solimões before its confluence with the Río Negro (Fig. 1.2). Many species present regular, damped variability (Fig. 1.8). This decadal time series facilitated a more rigorous empirical examination of factors controlling the seasonal changes in river constituents (Devol *et al.*, 1995). Sediment and geochemical constituents were regressed against likely controls based on water provenance and hydraulic characteristics: % of water derived from Andean inputs, % derived from "local" water (flood plains and small tributaries feeding the mainstem), river slope, and discharge. Three groups of species with similar controls were found (Table 1.1). Particulate species are controlled by source strength and the hydrodynamics of sediment deposition and remobilization. Conservative dissolved species reflected their sources and dilution effects, and bioactive species vary with the discharge hydrograph. These results are congruent with those gleaned from mainstem cruises.

Variability in streams and mid-sized rivers has been scarcely studied in the Amazon basin. In contrast to large rivers, changes in these systems can occur rapidly in response to storms, often resulting in large and rapid changes in concentrations (Elsenbeer *et al.*, 1996; Lesack & Melack, 1996). In the Bolivian Andes, Guyot (1993) has shown that most annually averaged organic carbon transport occurs in particulate form and is dominated by very high concentrations during a short period of time when FSS concentrations are extremely high. Concentrations of both particulate and dissolved species can vary widely within a few days.

Little is yet known about riverine biogeochemical responses to interannual forcings in the Amazon basin. Richey *et al.* (1989) showed that the mainstem discharge data near Manaus presents substantial variability at frequencies of about 3 years, corresponding to El Niño and La Niña events. Marengo & Nobre (2001) have shown that the impact of ENSO (El Niño Southern Oscillation) events on climate and river hydrographs vary across the basin. Tian *et al.* (1998) and Asner *et al.* (2000) have recently demonstrated via models and remote sensing that carbon cycling in terrestrial ecosystems in the basin also varies with ENSO events. This combined evidence of hydrological and terrestrial biogeochemical responses to interannual climate variability strongly suggests that processing of bioactive materials in the river system will also respond to such forcing. In large rivers, however, this response will be modulated by the relative regional impacts, the influx of tributaries, and the

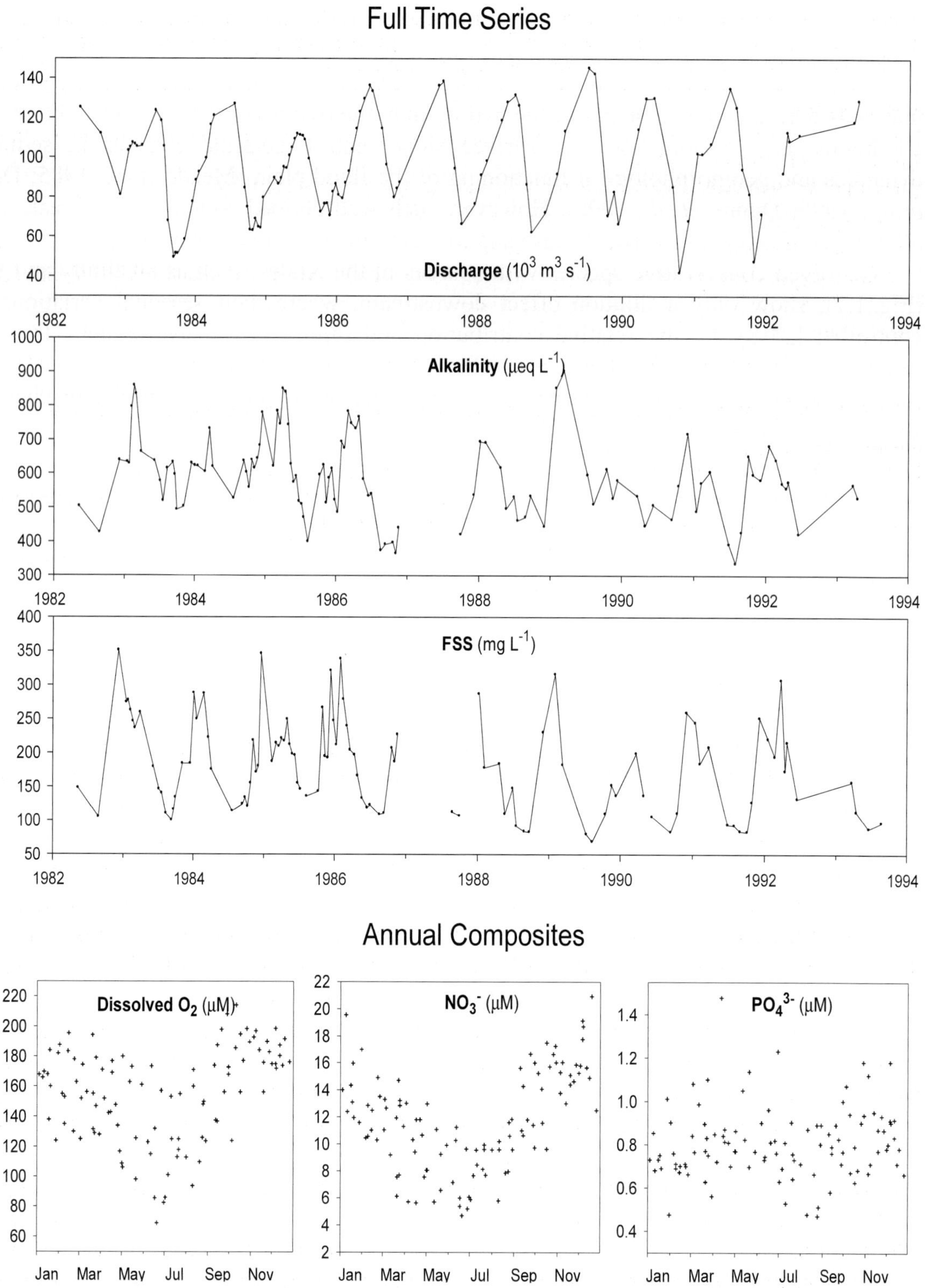

Fig. 1.8 Hydrological and biogeochemical time series from Marchantaria, in the Solimões mainstem near Manaus (see Fig. 1.2). Data were collected at approximately monthly intervals from 1982 through 1993. The complete time series is shown for river discharge, alkalinity, and FSS. Annual composites for dissolved oxygen, nitrate (NO_3^-), and phosphate (PO_4^{3-}) are presented to illustrate the dominant seasonal variability patterns; these time series are created by plotting each value against the day in the year (1–366) when the sample was collected. Data from Devol *et al.* (1995).

relative timing of their hydrographs. Likewise, large episodic disturbance, such as earthquakes and mass wastings in steep Andean slopes, are likely to have substantial and long-lasting biogeochemical impacts on small and mesoscale rivers, as has been shown in small watersheds in Puerto Rico disturbed by hurricanes (Schaefer *et al.*, 2000).

Table 1.1 Hydrological variables controlling seasonal biogeochemical variability at the Marchantaria sampling site, in the Solimões mainstem near Manaus. Biogeochemical parameters are grouped according to the dominant control variable resulting from multiple linear regression.

Andean group

Controlled by the relative contribution of river water from Vargem Grande, serving as a proxy of the Andean drainage of the Amazon mainstem.

Fine suspended sediments (FSS), alkalinity (Alk), calcium (Ca^{2+}), and sulphate (SO_4^{2-})

Stage height group

Controlled by the local stage height at Marchantaria.

Dissolved oxygen (O_2), dissolved carbon dioxide gas (CO_2), nitrate (NO_3^-), pH, sodium (Na^+), chloride (Cl^-), and % carbon by weight in coarse suspended sediments

Local group

Controlled by the relative contribution of river water from small, local tributaries in the central lowlands and the flood plain.

Silicate ($Si(OH)_4$), potassium (K^+), and % carbon and nitrogen by weight in fine suspended sediments

Surface slope group

Controlled by the river surface slope around Marchantaria, as a proxy of local turbulence.

Coarse suspended sediments (CSS), fine particulate organic carbon (FPOC), coarse particulate organic carbon and nitrogen (CPOC and CPON), and total particulate phosphorus (TPP)

CONCLUSIONS

The Amazon River basin—from the Andean Cordillera to the Atlantic Ocean—contains within its border a large variety of river and basin types. Nevertheless, these rivers exhibit remarkable consistencies in the processes that control the fate and transport of bioactive elements. River-borne material resides in three compositionally distinct size fractions that have contrasting transport properties and reaction histories. Coarse particulates are the freshest, degrading to form the fine particulate and dissolved fractions, which continually interact with each other via sorption processes. Organic matter in these fractions contains most of the bioactive material in the river, yet inorganic material is of critical importance due to its ability to complex and protect organic matter from degradation.

Variability in the composition and dynamics of river-borne material from one river (or season) to another is often much less than differences between size fractions within a single river (or season). However, it is often this variability that offers the most insight regarding important processes. Understanding how these processes scale with basin size is of critical importance. The watersheds of streams and small rivers most often represent a single geological and climatic zone. Each small river thus exhibits the narrow range of characteristics typical of its limited basin, yet these rivers respond to episodic forcing (storms, landslides, fires, development) quite quickly and strongly. Large rivers can contain high sub-basin heterogeneity with respect to ecosystems, geology, climate, and timing of weather events. Material carried by these large rivers integrates this heterogeneity, yet the relative influence of each sub-basin is continually changing with time and with distance downstream. Thus the response of large rivers to environmental change is damped relative to its constituent drainages.

The cumulative understanding of natural undisturbed river systems sets the foundation for subsequent work to elucidate human impacts on rivers and to eventually develop sustainable management plans. For example, knowing which size fraction the material of interest (nutrients, pollutants, eroded sediments, etc.) can potentially reside in is a critical first step in evaluating its transport dynamics and bioavailability. The next step might be to evaluate issues of scale. How does the human flux of this material compare to the natural riverine fluxes (transport, degradation, production, etc.) at that location? The answers to this question are likely to change across regions and with the time scale of interest. The evaluation of human impacts on river systems is rarely a simple and straightforward task—the natural world is too variable and complex for textbook solutions to every problem. However, systematic and well planned study of a natural system can yield the information necessary for sound management. Fortunately, a number of such studies are currently in place in the Amazon basin, including CAMREX (http://boto.ocean.washington.edu/camrex/), AARAM (Andean Amazon Rivers Analysis and Management; http://amazonrivers.org/aaram/), and HiBAm (http://www.unb.br/ig/hibam/hibam.htm).

REFERENCES

Amon, R. M. W. & Benner, R. (1996a) Bacterial utilization of different size classes of dissolved organic matter. *Limnology and Oceanography* **41**, 41–51.

Amon, R. M. W. & Benner, R. (1996b) Photochemical and microbial consumption of dissolved organic carbon and dissolved oxygen in the Amazon River system. *Geochimica et Cosmochimica Acta* **60**, 1783–1792.

Araujo-Lima, C. A. R. M., Forsberg, B. R., Victoria, R. & Martinelli, L. (1986) Energy sources for detrivorous fishes in the Amazon. *Science* **234**, 1256–1258.

Asner, G. P., Townsend, A. R. & Braswell, B. H. (2000) Satellite observation of El Niño effects on Amazon forest phenology and productivity. *Geophysical Research Letters* **27**, 981–984.

Aufdenkampe, A. K., Hedges, J. I., Krusche, A. V., Llerena, C. & Richey, J. E. (2001) Sorptive fractionation of dissolved organic nitrogen and amino acids onto sediments within the Amazon basin. *Limnology and Oceanography* **46**(8), 1921–1935.

Baldock, J. A. & Skjemstad, J. O. (2000) Role of the soil matrix and minerals in protecting natural organic materials against biological attack. *Organic Geochemistry* **31**, 697–710.

Benedetti, M. F., Van Riemsdijk, W. H., Koopal, L. K., Kinniburgh, D. G., Gooddy, D. C. & Milne, C. J. (1996) Metal ion binding by natural organic matter: from the model to the field. *Geochimica et Cosmochimica Acta* **60**, 2503–2513.

Benner, R., Opsahl, S., Chin-Leo, G., Richey, J. E. & Forsberg, B. R. (1995) Bacterial carbon metabolism in the Amazon River system. *Limnology and Oceanography* **40**, 1262–1270.

Boulton, A. J., Findlay, S., Marmonier, P., Stanley, E. H. & Valett, H. M. (1998) The functional significance of the hyporheic zone in streams and rivers. *Annual Review of Ecology and Systematics* **29**, 59–81.

Brandes, J. A., McClain, M. E. & Pimentel, T. P. (1996) ^{15}N evidence for the origin and cycling of inorganic nitrogen in a small Amazonian catchment. *Biogeochemistry* **34**, 45–56.

Brinkmann, W. L. F. (1985) Studies on hydrobiogeochemistry of a tropical lowland forest system. *GeoJournal* **11**, 89–101.

Cai, D. L., Tan, F. C. & Edmond, J. M. (1988) Sources and transport of particulate organic carbon in the Amazon river and estuary. *Estuarine, Coastal and Shelf Science* **26**, 1–14.

Chin, W.-C., Verdugo, P. & Orellana, M. (1998) Spontaneous assembly of marine dissolved organic matter into polymer gels. *Nature* **391**, 568–572.

Cole, J. J., Caraco, N. F., Kling, G. W. & Kratz, T. K. (1994) Carbon dioxide supersaturation in the surface waters of lakes. *Science* **256**, 1568–1570.

Cole, J. J. & Caraco, N. F. (2001) Carbon in catchments: Connecting terrestrial carbon losses with aquatic metabolism. *Marine and Freshwater Research* **52**, 101–110.

Cowie, G. L. & Hedges, J. I. (1984) Determination of neutral sugars in plankton, sediments, and wood by capillary gas chromatography of equilibrated isomeric mixtures. *Analytical Chemistry* **56**, 497–504.

Cowie, G. L. & Hedges, J. I. (1992) Sources and reactivities of amino acids in a coastal marine environment. *Limnology and Oceanography* **37**, 703–724.

Curtis, W. F., Meade, R. H., Nordin Jr, C. F., Price, N. B. & Sholkovitz, E. R. (1979) Non-uniform vertical distribution of fine sediment in the Amazon River. *Nature* **280**, 381–383.

Devol, A. H. & Hedges, J. I. (2001) The biogeochemistry of the Amazon River mainstem. In: *The Biogeochemistry of the Amazon Basin and its Role in a Changing World* (ed. by M. E. McClain, R. L. Victoria & J. E. Richey). Oxford University Press, UK.

Devol, A. H., Quay, P. D., Richey, J. E. & Martinelli, L. A. (1987) The role of gas exchange in the inorganic carbon, oxygen, and ^{222}Rn budgets of the Amazon River. *Limnology and Oceanography* **32**, 235–248.

Devol, A. H., Richey, J. E., Forsberg, B. R. & Martinelli, L. A. (1994) Environmental methane in the Amazon river floodplain. In: *Global Wetlands: Old World and New* (ed. by W. J. Mitsch), 151–165. Elsevier, Amsterdam.

Devol, A. H., Forsberg, B. R., Richey, J. E. & Pimentel, T. P. (1995) Seasonal variation in chemical distributions in the Amazon (Solimões) River: a multiyear time series. *Global Biogeochemical Cycles* **9**, 307–328.

Duivenvoorden, J. F. & Lips, J. M. (1995) *A Land-Ecological Study of Soils, Vegetation, and Plant Diversity in Colombian Amazonia.* TROPENBOS Series 12, Backhuys Publishers, Leiden, The Netherlands.

Dunne, T., Mertes, L. A. K., Meade, R. H., Richey, J. E. & Forsberg, B. R. (1998) Exchanges of sediment between the flood plain and channel of the Amazon River in Brazil. *Geological Society of America Bulletin* **110**, 450–467.

Elsenbeer, H., Lack, A. & Cassel, K. (1996) The stormflow chemistry at La Cuenca, western Amazonia. *Interciencia* **21**, 133–139.

Ertel, J. R., Hedges, J. I., Devol, A. H. & Richey, J. E. (1986) Dissolved humic substances of the Amazon River system. *Limnology and Oceanography* **31**, 739–754.

Ferreira, J. R., Devol, A. H., Martinelli, L. A., Forsberg, B. R., Victoria, R. L., Richey, J. E. & Mortatti, J. (1988) Chemical composition of the Madeira River: seasonal trends and total transport. *Mitteilungen der Geologisch Paläontologisches Institut* **66**, 63–75.

Filius, J. D., Lumsdon, D. G., Meeussen, J. C. L. *et al.* (2000) Adsorption of fulvic acid on goethite. *Geochimica et Cosmochimica Acta* **64**, 51–60.

Fisher, T. R. (1979) Plankton and primary production in aquatic systems of the central Amazon basin. *Comparative Biochemistry and Physiology* **62(A)**, 31–38.

Fisher, T. R., Jr & Parsley, P. E. (1979) Amazon lakes: water storage and nutrient stripping by algae. *Limnology and Oceanography* **24**, 547–553.

Forsberg, B. R., Devol, A. H., Richey, J. E., Martinelli, L. A. & dos Santos, H. (1988) Factors controlling nutrient concentrations in Amazon floodplain lakes. *Limnology and Oceanography* **33**, 41–56.

Forsberg, B. R., Araujo-Lima, C. A. R. M., Martinelli, L. A., Victoria, R. L. & Bonassi, J. A. (1993) Autotrophic carbon sources for fish of the Central Amazon. *Ecology* **74**, 643–652.

Forti, N. C. & Neal, C. (1992) Hydrochemical cycles in tropical rainforest: an overview with emphasis on central Amazonia. *Journal of Hydrology* **134**, 103–115.

Franken, W. & Leopoldo, P. R. (1984) Hydrology of catchment areas of central-Amazonian forest streams. In: *The Amazon: Limnology and Landscape Ecology of a Mighty Tropical River and its Basin* (ed. by H. Sioli), 501–519. W. Junk Publishers, Dordrecht, The Netherlands.

Furch, K. (1984) Water chemistry of the Amazon basin: the distribution of chemical elements among freshwaters. In: *The Amazon: Limnology and Landscape Ecology of a Mighty Tropical River and its Basin* (ed. by H. Sioli), 167–199. W. Junk Publishers, Dordrecht, The Netherlands.

Gibbs, R. J. (1967) The geochemistry of the Amazon River System: Part I. The factors that control the salinity and the composition and concentration of the suspended solids. *Geological Society of America Bulletin* **78**, 1203–1232.

Goni, M. A. & Hedges, J. I. (1992) Lignin dimers: Structures, distribution, and potential geochemical applications. *Geochimica et Cosmochimica Acta* **56**, 4025–4043.

Gregory, S. V., Swanson, F. J., McKee, W. A. & Cummins, K. W. (1991) An ecosystem perspective of riparian zones. *BioScience* **41**, 540–551.

Gu, B., Schmitt, J., Chen, Z., Liang, L. & McCarthy, J. F. (1995) Adsorption and desorption of different organic matter fractions on iron oxide. *Geochimica et Cosmochimica Acta* **59**, 219–229.

Guyot, J. L. (1993) *Hydrogeochimie des fleuves de l'Amazonie bolivienne.* Editions de l'ORSTOM (Etudes et Thèses), Paris.

Guyot, J. L. & Wasson, J. G. (1994) Regional pattern of riverine dissolved organic carbon in the Amazon drainage basin of Bolivia. *Limnology and Oceanography* **39**, 452–458.

Guyot, J. L., Filizola, N. & Guimarães, V. (1998) Amazon suspended sediment yield measurements using an Acoustic Doppler Current Profiler (ADCP): first results. In: *Hydrology of the Humid Tropic Environment* (ed. by A. I. Johnson & C. A. Fernández-Jáuregui) (Proc. Kingston, Jamaica, Symp., November 1996), 109–115. IAHS Publ. no. 253.

Guyot, J. L., Jouanneau, J. M. & Wasson, J. G. (1999) Characterisation of river bed and suspended sediments in the Río Madeira drainage basin (Bolivian Amazonia) *Journal of South American Earth Sciences* **12**, 401–410.

Hedges, J. I. & Keil, R. G. (1995) Sedimentary organic matter preservation: an assessment and speculative synthesis. *Marine Chemistry* **49**, 81–115.

Hedges, J. I., Clark, W. A., Quay, P. D., Richey, J. E., Devol, A. H. & Santos, U. d. M. (1986a) Compositions and fluxes for particulate organic material in the Amazon River. *Limnology and Oceanography* **31**, 717–738.

Hedges, J. I., Ertel, J. R., Quay, P. D., Grootes, P. M., Richey, J. E., Devol, A. H., Farwell, G. W., Schmidt, F. W. & Salati, E. (1986b) Organic carbon-14 in the Amazon River system. *Science* **231**, 1129–1131.

Hedges, J. I., Cowie, G. L., Richey, J. E., Quay, P. D., Benner, R., Strom, M. & Forsberg, B. R. (1994) Origins and processing of organic matter in the Amazon River indicated by carbohydrates and amino acids. *Limnology and Oceanography* **39**, 743–761.

Hedges, J. I., Mayorga, E., Tsamakis, E., McClain, M. E., Aufdenkampe, A., Quay, P., Richey, J. E., Benner, R., Opsahl, S., Black, B., Pimentel, T., Quintanilla, J. & Maurice, L. (2000) Organic matter in Bolivian tributaries of the Amazon River: a comparison to the lower mainstem. *Limnology and Oceanography* **45**, 1449–1466.

Johnsson, M. J. & Meade, R. H. (1990) Chemical weathering of fluvial sediments during alluvial storage: the Macuapanim Island point bar, Solimões River, Brazil. *Journal of Sedimentary Petrology* **60**, 827–842.

Kaiser, K. & Zech, W. (1997) Competitive sorption of dissolved organic matter fractions to soils and related mineral phases. *Soil Science Society of America Journal* **61**, 64–69.

Kaiser, K. (1998) Fractionation of dissolved organic matter affected by polyvalent metal cations. *Organic Geochemistry* **28**, 849–854.

Kaiser, K. & Guggenberger, G. (2000) The role of DOM sorption to mineral surfaces in the preservation of organic matter in soils. *Organic Geochemistry* **31**, 711–725.

Kalliola, R., Linna, A., Puhakka, M., Salo, J. & Räsänen, M. (1993) Mineral nutrients in fluvial sediments from the Peruvian Amazon. *Catena* **20**, 333–349.

Keil, R. G., Montluçon, D. B., Prahl, F. G. & Hedges, J. I. (1994) Sorptive preservation of labile organic matter in marine sediments. *Nature* **370**, 549–552.

Keil, R. G., Mayer, L. M., Quay, P. D., Richey, J. E. & Hedges, J. I. (1997) Losses of organic matter from riverine particles in deltas. *Geochimica et Cosmochimica Acta* **61**, 1507–1511.

Klinge, H. & Ohle, W. (1964) Chemical properties of rivers in the Amazonian area in relation to soil conditions. *Verhandlungen—Internationale Vereinigung fur Theoretische und Angewandte Limnologie* **15**, 1067–1076.

Konhauser, K. O., Fyfe, W. S. & Kronberg, B. I. (1994) Multi-element chemistry of some Amazonian waters and soils. *Chemical Geology* **111**, 155–175.

Körner, C., Farquhar, G. D. & Roksandic, Z. (1988) A global survey of carbon isotope discrimination in plants from high altitude. *Oecologia* **74**, 623–632.

Küchler, I. L., Miekeley, N. & Forsberg, B. R. (1994) Molecular mass distributions of dissolved organic carbon and associated metals in waters from Río Negro and Río Solimões. *Science of the Total Environment* **156**, 207–216.

Kvist, L. P. & Nebel, G. (2001) A review of Peruvian flood plain forests: ecosystems, inhabitants and resource use. *Forest Ecology and Management* **150**, 3–26.

Leenheer, J. A. (1980) Origin and nature of humic substances in the waters of the Amazon River basin. *Acta Amazonica* **10**, 513–526.

Lenz, P. H., Melack, J. M., Robertson, B. & Hardy, E. A. (1986) Ammonium and phosphate regeneration by the zooplankton of an Amazon floodplain lake. *Freshwater Biology* **16**, 821–830.

Leopold, L. B. (1994) *A View of the River*. Harvard University Press, Cambridge, Massachusetts, USA.

Lesack, L. F. W. & Melack, J. M. (1996) Mass balance of major solutes in a rainforest catchment in the central Amazon: implications for nutrient budgets in tropical rainforests. *Biogeochemistry* **32**, 115–142.

Lewis, W. M., Hamilton, S. K. & Saunders, J. F. (1995) Rivers of northern South America. In: *River and Stream Ecosystems* (ed. by C. E. Cushing, K. W. Cummins & G. W. Minshall), 219–256. Elsevier, Amsterdam.

Longinelli, A. & Edmond, J. M. (1983) Isotope geochemistry of the Amazon basin: a reconnaissance. *Journal of Geophysical Research* **88**, 3703–3717.

Marengo, J. A. & Nobre, C. A. (2001) General characteristics and variability of climate in the Amazon basin and its links to the global climate system. In: *The Biogeochemistry of the Amazon Basin* (ed. by M. E. McClain, R. L. Victoria & J. E. Richey). Oxford University Press, UK.

Martinelli, L. A., Devol, A. H., Victoria, R. L. & Richey, J. E. (1991) Stable carbon isotopic variation in C_3 and C_4 plants along the Amazon River. *Nature* **353**, 57–59.

Martinelli, L. A., Victoria, R. L., Trivelin, P. C. O., Devol, A. H. & Richey, J. E. (1992) ^{15}N natural abundance in plants of the Amazon River floodplain and potential atmospheric N_2 fixation. *Oecologia* **90**, 591–596.

Martinelli, L. A., Victoria, R. L., Dematte, J. L. I., Richey, J. E. & Devol, A. H. (1993) Chemical and mineralogical composition of Amazon River floodplain sediments, Brazil. *Applied Geochemistry* **8**, 391–402.

Martinelli, L. A., Victoria, R. L., Sternberg, L. S. L., Ribeiro, A. & Moreira, M. Z. (1996) Using stable isotopes to determine sources of evaporated water to the atmosphere in the Amazon basin. *Journal of Hydrology* **183**, 191–204.

Mayer, L. M. (1994) Relationships between mineral surfaces and organic carbon concentrations in soils and sediments. *Chemical Geology* **114**, 347–363.

Mayer, L. M. (1999) Extent of coverage of mineral surfaces by organic matter in marine sediments. *Geochimica et Cosmochimica Acta* **63**, 207–215.

Mayer, L. M., Keil, R. G., Macko, S. A., Joye, S. B., Ruttenberg, K. C. & Aller, R. C. (1998) Importance of suspended particulates in riverine delivery of bioavailable nitrogen to coastal zones. *Global Biogeochemical Cycles* **12**, 573–579.

McClain, M. E., Richey, J. E. & Pimentel, T. P. (1994) Groundwater nitrogen dynamics at the terrestrial-lotic interface of a small catchment in the central Amazon basin. *Biogeochemistry* **27**, 113–127.

McClain, M. E., Richey, J. E. & Victoria, R. L. (1995) Andean contributions to the biogeochemistry of the Amazon river system. *Bulletin de l'Institut Français des Etudes Andine* **24**, 425–437.

McClain, M. E., Richey, J. E., Brandes, J. A. & Pimentel, T. P. (1997) Dissolved organic matter and terrestrial-lotic linkages in the central Amazon basin of Brazil. *Global Biogeochemical Cycles* **11**, 295–311.

Meade, R. H. (1994) Suspended sediments of the modern Amazon and Orinoco Rivers. *Quaternary International* **21**, 29–39.

Meade, R. H., Nordin Jr, C. F., Curtis, W. F., Rodrigues, F. M. C., Vale, C. M. D. & Edmond, J. M. (1979) Sediment loads in the Amazon River. *Nature* **278**, 161–163.

Meade, R. H., Dunne, T., Richey, J. E., Santos, U. d. M. & Salati, E. (1985) Storage and remobilization of suspended sediments in the Lower Amazon River of Brazil. *Science* **228**, 488–490.

Mertes, L. A. K., Smith, M. O. & Adams, J. B. (1993) Estimating suspended sediment concentrations in surface waters of the Amazon River wetlands from Landsat images. *Remote Sensing of the Environment* **43**, 281–301.

Mertes, L. A. K., Dunne, T. & Martinelli, L. A. (1996) Channel–floodplain geomorphology of the Solimões–Amazon River, Brazil. *Geological Society of America Bulletin* **108**, 1089–1107.

Meybeck, M. (1982) Carbon, nitrogen, and phosphorus transport by world rivers. *American Journal of Science* **282**, 401–450.

Meyer, J. L., McDowell, W. H., Bott, T. L., Elwood, J. W., Ishizaki, C., Melack, J. M., Peckarsky, B. L., Peterson, B. J. & Rublee, P. A. (1988) Elemental dynamics in streams. *Journal of the North American Benthological Society* **7**, 410–432.

Moran, M. A. & Zepp, R. G. (1997) Role of photoreactions in the formation of biologically labile compounds from dissolved organic matter. *Limnology and Oceanography* **42**, 1307–1316.

Mounier, S., Braucher, R. & Benaïm, J. Y. (1999) Differentiation of organic matter's properties of the Río Negro basin by crossflow ultra-filtration and UV-spectrofluorescence. *Water Research* **33**, 2363–2373.

Nelson, P. N., Baldock, J. A. & Oades, J. M. (1993) Concentration and composition of dissolved organic carbon in streams in relation to catchment soil properties. *Biogeochemistry* **19**, 27–50.

Nordin Jr, C. F., Meade, R. H., Curtis, W. F., Bósio, N. J. & Landim, P. M. B. (1980) Size distribution of Amazon River bed sediment. *Nature* **286**, 52–53.

Nortcliff, S. & Thornes, J. B. (1988) The dynamics of a tropical floodplain environment with reference to forest ecology. *Journal of Biogeography* **15**, 49–59.

Nortcliff, S., Ross, S. M. & Thornes, J. B. (1990) Soil moisture, runoff and sediment yield from differentially cleared tropical rainforest plots. In: *Vegetation and Erosion: Processes and Environments* (ed. by J. B. Thornes), 419–436. John Wiley & Sons, Chichester, UK.

Patel, N., Mounier, S., Guyot, J. L., Benamou, C. & Benaïm, J. Y. (1999) Fluxes of dissolved and colloidal organic carbon, along the Purus and Amazonas Rivers (Brazil). *Science of the Total Environment* **229**, 53–64.

Peterson, B. J., Wollheim, W. M., Mulholland, P. J., Webster, J. R., Meyer, J. L., Tank, J. L., Martí, E., Bowden, W. B., Valett, H. M., Hershey, A. E., McDowell, W. H., Dodds, W. K., Hamilton, S. K., Gregory, S. V. & Morrall, D. D. (2001) Controls of nitrogen export from watersheds by headwater streams. *Science* **292**, 86–90.

Puhakka, M., Kalliola, R., Rajasilta, M. & Salo, J. (1992) River types, site evolution and successional vegetation patterns in Peruvian Amazonia. *Journal of Biogeography* **19**, 651–665.

Quay, P., King, S., Wilbur, D., Wofsy, S. & Richey, J. (1989) $^{13}C/^{12}C$ of atmospheric CO_2 in the Amazon basin: forest and river sources. *Journal of Geophysical Research* **94**, 18 327–18 336.

Quay, P. D., Wilbur, D. O., Richey, J. E., Hedges, J. I., Devol, A. H. & Victoria, R. L. (1992) Carbon cycling in the Amazon River: implications from the ^{13}C composition of particles and solutes. *Limnology and Oceanography* **37**, 857–871.

Quay, P. D., Wilbur, D. O., Richey, J. E., Devol, A. H., Benner, R. & Forsberg, B. R. (1995) The $^{18}O/^{16}O$ of dissolved oxygen in rivers and lakes in the Amazon basin: a tracer of respiration and photosynthesis. *Limnology and Oceanography* **40**, 718–729.

Richey, J. E., Brock, J. T., Naiman, R. J., Wissmar, R. C. & Stallard, R. F. (1980) Organic carbon: oxidation and transport in the Amazon River. *Science* **207**, 1348–1351.

Richey, J. E., Meade, R. H., Salati, E., Devol, A. H., Nordin, C. F. & dos Santos, U. (1986) Water discharge and suspended sediment concentrations in the Amazon River: 1982–1984. *Water Resources Research* **22**, 756–764.

Richey, J. E., Devol, A. H., Wofsy, S. C., Victoria, R. & Ribeiro, M. N. G. (1988) Biogenic gases and the oxidation and reduction of carbon in the Amazon River and floodplain waters. *Limnology and Oceanography* **33**, 551–561.

Richey, J. E., Nobre, C. & Deser, C. (1989) Amazon River discharge and climate variability: 1903 to 1985. *Science* **246**, 101–103.

Richey, J. E., Hedges, J. I., Devol, A. H. & Quay, P. D. (1990) Biogeochemistry of carbon in the Amazon River. *Limnology and Oceanography* **35**, 352–371.

Richey, J. E., Wilhelm, S. R., McClain, M. E., Victoria, R. L., Melack, J. M. & Araujo-Lima, C. (1997) Organic matter and nutrient dynamics in river corridors of the Amazon basin and their response to anthropogenic change. *Ciência e Cultura* **49**, 98–110.

Richey, J. E., Melack, J. M., Aufdenkampe, A. K., Ballester, M. V. & Hess, L. L. (2002) Outgassing from Amazonian rivers and wetlands as a large tropical source of atmospheric CO_2. *Nature* **416**, 617–620.

Salo, J., Kalliola, R., Häkkinen, I., Mäkinen, Y., Niemelä, P., Puhakka, M. & Coley, P. D. (1986) River dynamics and the diversity of Amazon lowland forest. *Nature* **322**, 254–258.

Schaefer, D. A., McDowell, W. H., Scatena, F. N. & Asbury, C. E. (2000) Effects of hurricane disturbance on stream water concentrations and fluxes in eight tropical watersheds of the Luquillo Experimental Forest, Puerto Rico. *Journal of Tropical Ecology* **16**, 189–207.

Sholkovitz, E. R. & Price, N. B. (1980) The major-element chemistry of suspended matter in the Amazon Estuary. *Geochimica et Cosmochimica Acta* **44**, 163–171.

Sioli, H. (1984a) The Amazon and its main affluents: hydrography, morphology of the river courses, and river types. In: *The Amazon: Limnology and Landscape Ecology of a Mighty Tropical River and its Basin* (ed. by H. Sioli), 127–165. W. Junk Publishers, Dordrecht, The Netherlands.

Sioli, H. (ed.) (1984b) *The Amazon: Limnology and Landscape Ecology of a Mighty Tropical River and its Basin.* W. Junk Publishers, Dordrecht, The Netherlands.

Sippel, S. J., Hamilton, S. K., Melack, J. M. & Novo, E. M. M. (1998) Passive microwave observations of inundation area and the area/stage relation in the Amazon River floodplain. *International Journal of Remote Sensing* **19**, 3055–3074.

Smith, R. A., Schwarz, G. E. & Alexander, R. B. (1997) Regional interpretation of water-quality monitoring data. *Water Resources Research* **33**, 2781–2798.

St John, T. V. & Anderson, A. B. (1982) A re-examination of plant phenolics as a source of tropical black water rivers. *Tropical Ecology* **23**, 151–154.

Stallard, R. F. & Edmond, J. M. (1983) Geochemistry of the Amazon: 2. The influence of geology and weathering environment on the dissolved load. *Journal of Geophysical Research* **88**, 9671–9688.

Stallard, R. F. & Edmond, J. M. (1987) Geochemistry of the Amazon. 3. Weathering chemistry and limits to dissolved inputs. *Journal of Geophysical Research* **92**, 8293–8302.

Stark, N. M. & Jordan, C. F. (1978) Nutrient retention by the root mat of an Amazonian rain forest. *Ecology* **59**, 434–437.

Sternberg, H. O. (1975) *The Amazon River of Brazil*. Franz Steiner Verlag, Wiesbaden, Germany.

Tian, H., Melillo, J. M., Kicklighter, D. W., Mcguire, A. D., Helfrich III, J. V. K., Moore III, B. & Vörösmarty, C. J. (1998) Effect of interannual climate variability on carbon storage in Amazonian ecosystems. *Nature* **396**, 664–667.

Tiessen, H., Chacon, P. & Cuevas, E. (1994) Phosphorus and nitrogen status in soils and vegetation along a toposequence of dystrophic rainforests on the upper Río Negro. *Oecologia* **99**, 145–150.

Trumbore, S. E., Davidson, E. A., Camargo, P. B. D., Nepstad, D. C. & Martinelli, L. A. (1995) Belowground cycling of carbon in forests and pastures of Eastern Amazonia. *Global Biogeochemical Cycles* **9**, 515–528.

Vannote, R. L., Minshall, G. W., Cummings, K. W., Sedell, J. R. & Cushing, C. E. (1980) The river continuum concept. *Canadian Journal of Fisheries and Aquatic Sciences* **37**, 130–137.

Victoria, R., Martinelli, L. A., Richey, J. E., Devol, A. H., Forsberg, B. R. & Ribeiro, M. N. G. (1989) Spatial and temporal variations in soil chemistry on the Amazon floodplain. *GeoJournal* **19**, 45–52.

Victoria, R. L., Martinelli, L. A., Trivelin, P. C. O., Matsui, E., Forsberg, B. R., Richey, J. E. & Devol, A. H. (1992) The use of stable isotopes in studies of nutrient cycling: carbon isotope composition of Amazon varzea sediments. *Biotropica* **24**, 240–249.

Wallace, A. R. (1853) *Narrative of Travels on the Amazon and Río Negro*. Reeve, London.

Williams, M. R. & Melack, J. M. (1997) Solute export from forested and partially deforested catchments in the central Amazon. *Biogeochemistry* **38**, 67–102.

Williams, P. M. (1968) Organic and inorganic constituents of the Amazon River. *Nature* **218**, 937–938.

Wissmar, R. C., Richey, J. E., Stallard, R. F. & Edmond, J. M. (1981) Plankton metabolism and carbon processes in the Amazon River, its tributaries, and floodplain waters, Peru–Brazil, May–June 1977. *Ecology* **62**, 1622–1633.

CHAPTER 2

ORGANIC MATTER SIZE DYNAMICS IN THE AMAZON RIVER

Stéphane Mounier, Marc Benedetti, Jean Yves Benaïm & Jacques Boulègue

Few studies have taken into account the size heterogeneity of natural organic matter in large river systems like the Amazon. This heterogeneity may have a major impact on the fluxes and distribution of carbon during the mixing of river waters with contrasting chemical composition such as occurs in the Amazon system (see Chapter 1). The main objective of this chapter is to present new data on the distribution of carbon in the Amazon River and its dynamics during the mixing of the main tributaries. Through this we wish to illuminate the importance of the colloidal and particulate fractions. The simple extrapolation of discharge *vs* concentration is used in many studies, especially those focused on global change evaluations (Probst *et al.*, 1994; Hedges, 1992; Richey *et al.*, 1990a; Richey *et al.*, 1990b). This is clearly an over simplification and potentially problematic considering settling phenomena and seasonal variation. Already in estuary environments this type of extrapolation falls apart and corrections must be applied for flocculation and settling (Ludwing *et al.*, 1996). Due to its distinct types of water, the Amazon does not behave conservatively and is subject to pronounced mixing phenomena (Küchler *et al.*, 1994). Several studies have been done on the particulate (Probst *et al.*, 1994) and dissolved (Devol *et al.*, 1995) material carried by thc Amazon. This study examines the role of the colloidal phase and determines if the Amazon behaves conservatively between the cities of Manaus (03°04.40′S, 60°15.74′W) and Óbidos (01°54.72′S, 55°32.60′W). Particularly, major elements like Ca, Mg, and Na, and trace metals like Ba or Sr and organic matter were studied. The Amazon River is the single most important fluvial input to the world ocean. It contributes 7.3–8.7% of the carbon riverine input (Probst *et al.*, 1994). The Río Negro, Río Solimões and the Río Madeira are its main tributaries. Within the mixing zones of these chemically distinct rivers, processes occur that affect the carbon inputs. The Río Solimões corresponds to a typical white-water river and the Río Negro inputs typical black-water river. This work presents studies on the mixing area between black water from the Río Negro river (coloured, acidic, mineral poor) and the white water from Río Solimões and Río Madeira (carbonated pH, sediment charged, mineral rich). The study focuses more precisely on the behaviour of organic carbon.

At least two processes are highly influenced by particle size: hydrological transport and sorption of dissolved elements onto particle surfaces. Variations in hydrological transport as a function of particle size significantly complicate mass flux calculations, as sampling techniques must integrate fluxes as a function of depth and velocity (Richey *et al.*, 1990a). Differing properties of sorption with changing particle size significantly influence the transport and fate of a whole range of surface reactive elements and compounds (Benaïm & Mounier, 1998). Thus again techniques for quantifying fluxes and processes must take care to accurately reflect different particle size classes.

Before carbon fluxes and processing can be calculated as a function of time and runoff, preliminary hydrological budgets must be made and controlled for losses or gains. This is the main topic of the next section and is based on previously published literature as well as recent measurements made in the HiBAm project ((Hidrologia e Geoquímica da Bacia Amazônica), an international research project, based at the Geosciences Institute of the University of Brasilia, on the hydrology and geochemistry of the Amazon basin. Before discussing fluxes of carbon, however, we need to check the discharge recovery at Óbidos, the most downstream station on the river. For this, the discharge measured for the Río Negro, the Río Madeira and the Río Solimões were summed and compared to the Río Amazon at Óbidos. The results showed a calculated discharge varying from measured by –5 to 20% according to the maximum 10% error measurement. The variation in carbon fluxes did not vary linearly with gain of water. For example in September, the calculated flow at Óbidos was 3% more than measured flow, while the carbon flux was about 40% less than expected.

EXPERIMENTS AND RESULTS

Seven sampling campaigns were carried out between 1994 and 1996 and results from these campaigns permit an estimation of the annual loss of organic carbon during mixing events. Annual values are extrapolated from individual campaigns by assuming linear correlations between experimental and missing data. Sampling sites are located at Manacapurù for the Río Solimões, Paricatuba for the Río Negro, 20 km up river from the mouth of the Río Madeira and at Óbidos, 496 km further downstream. All organic carbon concentrations were measured in the RCMO laboratory (France) using the Dorhmann DC80 wet oxidation method (10% reproducibility). Discharge was measured by an acoustic Doppler current profiler (ADCP) (10% reproducibility) operated by the DNAEE/ORSTOM team (Brazil). The size distribution was determined by sequential tangential ultra filtration (STUF) and fluxes were calculated for dissolved, colloidal and particulate carbon as the product of concentration and measured discharge.

On-site filtrations were done with a Sartorius UltraSart Mini tangential filtration system which separated material into 0.22 μm, 100 and 5 kD (kilo dalton) size classes. All filters had a surface area of 0.1 m^2 and were composed of cellulose acetate, polysulfon and cellulose triacetate, respectively. On-site filtration is necessary to avoid an enrichment of the particulate and colloidal fractions by aggregation of organic compound during storage (De Vitre *et al.*, 1987). Fifty litres of water were collected from a depth of 20 cm in the middle of each station in a pre-washed high density polypropylene carbon. Three cartridges were used to reach 0.3 m^2 of filtering surface. The first cut off is about 0.22 μm (Sartorius, Sartocon Mini, SM 303 123 07 01W, cellulose acetate). The first 5 litres of sample was filtered and discarded (Buesseler *et al.*, 1996), and the permeate, which passed through the 0.22 μm pore size, was used to rinse the second flask. When steady state was obtained the V_i volume was noted. A concentration factor of about 10 was achieved with the 45 litres of sample. The concentrated volume, labelled $V_{c,0.22\mu m}$, was about 4.5 litres and the permeate volume, labelled $V_{p,0.22\mu m}$, was about 40.5 litres. For each of these volumes, 250 ml were subsampled, poisoned with 100 μl of NaN_3 (1M) and stored at 4°C in ice.

The second step in the filtration process utilized the permeate from the first (i.e. $V_{p,0.22\mu m}$) and the 100-kD cut-off membrane (Sartorius, Sartocon Mini, SM 303 146 69 01E, polysulfon). After stabilization the concentrated $V_{p,0.22\mu m}$ volume was noted as well as $V_{p,100kD}$ and $V_{c,100kD}$. The third filtration step used a 5-kD cut-off membrane (Sartorius, Sartocon Mini, SM 303 145 29 01E, triacetate of cellulose). At the end of the sequential tangential filtration run seven fractions were collected for each sampling site: one raw fraction, V_0, without filtration or treatment, three concentrated fractions C_1, C_2 and C_3 for the $V_{c,0.22\mu m}$, $V_{c,100kD}$ and $V_{c,5kD}$ volumes, respectively, and three permeate samples P_1, P_2

Table 2.1 Values obtained for the station from the TOC of the concentrated fraction. Units are in µmole of carbon (µM C) per litre and discharge in $m^3 s^{-1}$ was measured by ADCP. TOC is the bulk measured value, POC, COC_g and COC_f are calculated values from STUF data.

	Dates	TOC (µM C)	POC (µM C)	COC_g (µM C)	COC_f (µM C)	DOC (µM C)	Discharge ($m^3 s^{-1}$)
Río Negro	March 1995	558	84	74	116	143	7 240
	June 1995	577†	-	-	-	-	50 865*
	July 1996	583	292	84	36	83	64 380
	Sept. 1994	621	103	11	295	168	49 620
	Oct. 1996	554	318	71	70	122	32 990
	Nov. 1995	465	103	42	74	149	12 311
Río Solimões	March 1995	663	3	0	197	230	84 670
	June 1995	460†	-	-	-	-	127 005*
	July 1996	392	59	1	2	283	120 110
	Sept. 1994	330	99	0	117	463	104 000
	Oct. 1996	206	26	16	67	120	55 940
	Nov. 1995	385	5	3	31	312	52 477
Río Madeira	March 1995	295	0	0	166	483	37 440
	June 1995	247	52	27	79	127	19 760
	July 1996	200	19	2	5	158	13 370
	Sept. 1994	173	5	7	54	117	5 940
	Oct. 1996	171	8	11	9	124	7 410
	Nov. 1995	178	8	3	24	177	5 132
Río Amazonas (Óbidos)	March 1995	251	4	1	111	123	157 380
	June 1995	478	124	61	202	104	196 700
	July 1996	408	72	41	2	267	189 400
	Sept. 1994	337	67	13	243	135	155 000
	Oct. 1996	300	52	48	55	158	109 440
	Nov. 1995	232	15	38	36	153	81 090

* Value extrapolated from another sampling campaign.
† Weighted mean for extrapolation.

and P_3 for the $V_{p,0.22\mu m}$, $V_{p,100kD}$ and $V_{p,5kD}$ volumes, respectively. Samples C_1, C_2 and C_3 contained particulate, coarse colloidal and fine colloidal concentrated fractions, respectively. The P_3 permeate was considered to be the dissolved fraction which passed through the 5-kD-pore-size membrane. All the permeate fractions were not concentrated (Dai *et al.*, 1995; Guo & Santschi, 1996; Reitmeyer *et al.*, 1996).

The results are shown in Table 2.1. Mass balances were calculated for each step in the filtration process (Buffle *et al.*, 1982; De Vitre *et al.*, 1987; Hoffmann *et al.*, 1981; Reitmeyer *et al.*, 1996), making it possible to compare the initial concentration (C_0) with the calculated recovered concentration (C_{calc}). Mass recoveries with less than 10% of standard deviation were considered acceptable (Hoffmann *et al.*, 1981; Reitmeyer *et al.*, 1996). Some authors even accept to 20% of standard deviation (Buesseler *et al.*, 1996).

NEW INSIGHTS ON AMAZON DISCHARGE CHEMISTRY

Carbon fluxes

Recoveries for carbon fluxes were calculated by subtracting the sum of upstream inputs from the value measured at Óbidos. Positive values indicated an undetected discharge contribution (such as from the flood plain or small tributaries) or gains, while negative values implied water retention or losses (Fig. 2.1). As expected the most important carbon inputs were made by the Río Negro, the Río Solimões and the Río Madeira. This was due to their high discharge and small inputs from other tributaries such as the Río Trombetas, which contributes less than 1% of the carbon fluxes in the Amazon. Moreover, the distribution pattern depended mainly on the Río Negro in September and on the Solimões in March. The Río Madeira was responsible for a relatively small portion of the total organic carbon discharge. The Río Negro contributed less in March due to its low-water period and the damming effects of higher water in the Río Solimões. Carbon inputs varied during the sampling period from 287 to 1189 kg s^{-1}. Carbon flux recovery calculations showed that there were losses of organic carbon between the Negro/Solimões confluence and the Óbidos station.

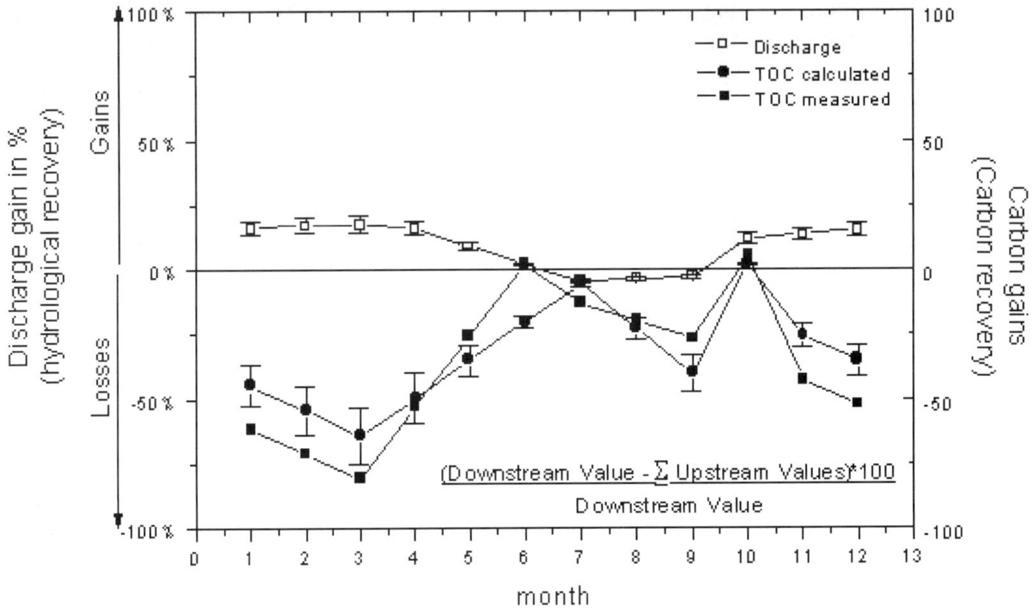

Fig. 2.1 Organic carbon loss and gain between Manaus and Óbidos stations.

Metal ions and carbon

A strong linear correlation was observed between major and trace elements and organic carbon content, implying a physical association between these forms. For example, a strong correlation was observed between Ca and TOC in black-water and white-water rivers (Fig. 2.2). In this type of correlation the most important feature is the ordinate. If its value is small, then only a small quantity of the element is "free" relative to organic matter. Inversely if it is high then the element is more associated with the mineral phase of the size fraction. More research is required to improve the interpretation of this type of correlation, but it seems to provide useful information on the role of the organic matter in particular size fractions (dissolved, colloidal and particulate) in the transport of dissolved elements (Fig. 2.5).

Another type of relationship is that shown by barium (Fig. 2.3), which is strongly associated with organic matter from the Río Negro but not with organic matter in white-water rivers. Based on additional results obtained for other metals (not show here), it appears that organic matter from the Río Negro had more affinity for trace metals than organic matter from the white-water rivers. This also explains why the fate of metals in the mixing zone of black and white waters is complex. The prediction of metal fluxes at Óbidos based on the inputs from Manaus could not be done correctly if information was not provided on the physico-chemical transformations linked to the organic matter size classes.

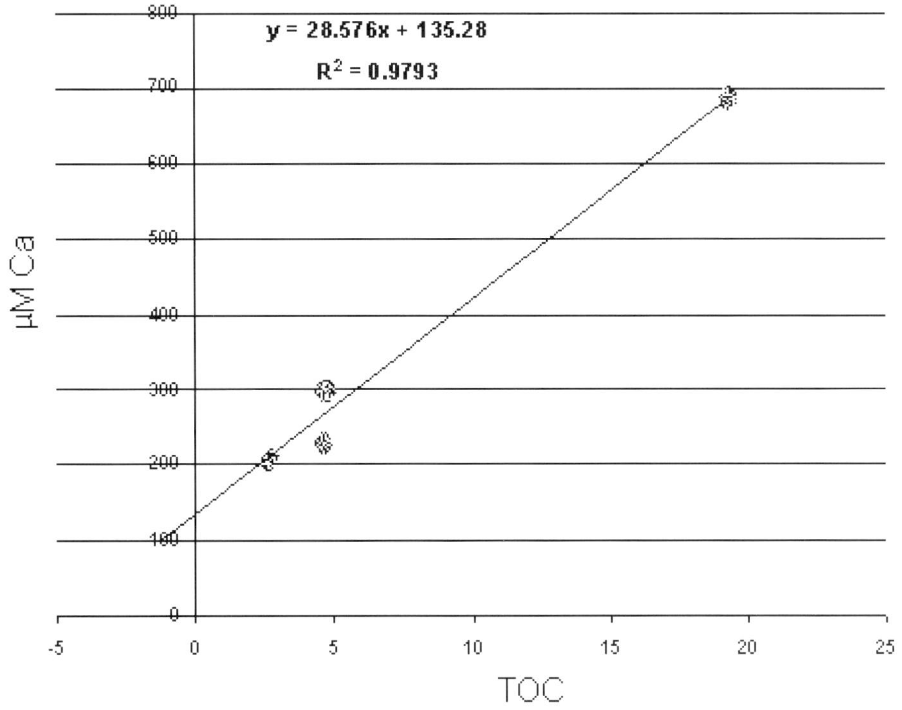

Fig. 2.2 Organic carbon correlation with calcium for white and black river water.

Carbon size dynamics

Significant losses of organic matter were calculated between Manaus and Óbidos from January to March and in October (Fig. 2.4). The maximum loss occurred during October 1994, with approximately 340 kg s^{-1} of carbon "missing" at Óbidos. During the high water level of the Amazon River little carbon seems to be deposited after the Río Negro and the Río Solimões join the Amazon. In October, when the Río Negro water levels drop, there is an important colloidal gain of 150 kg s^{-1} and simultaneously a dramatic dissolved organic carbon loss of 400 kg s^{-1}. We hypothesize that this may be a result of flocculation and coagulation phenomena within the mixing zone. Dissolved (from the Río Solimões) and particulate carbon are converted into colloidal carbon that is deposited before Óbidos. At the beginning of the year, the loss of organic carbon is significant in the dissolved and colloidal size fractions.After March, the loss of carbon seems to be insignificant.

The Río Negro seems to play an important role in element transfers, not by its proper inputs, which range from 20% to 40%, but by the particular forms of organic matter which it carries. Organic matter in this river is mostly colloidal. Thus the mixing with Solimões waters leads to a coagulation reaction and flocculation or sinking of elements to the river bottom. By this process 30–45% of the carbon input by the Río Negro and Río Solimões during the low-water period accumulates in the reach of the mainstem between Manaus and Santarèm.

Results show that at all stations except the Río Madeira, concentrations of element are higher during the low-water period. Moreover, spatial variations in the concentrations of some elements are very rapid over short distances along the river (e.g. iron and aluminium in the Río Solimões). Calculations of fluxes show some losses that could be explained by flocculation or precipitation of these elements. This illustrates that long distance extrapolations of fluxes, for example from Santarèm to the estuary, are likely to be incorrect unless physico-chemical interaction are well known.

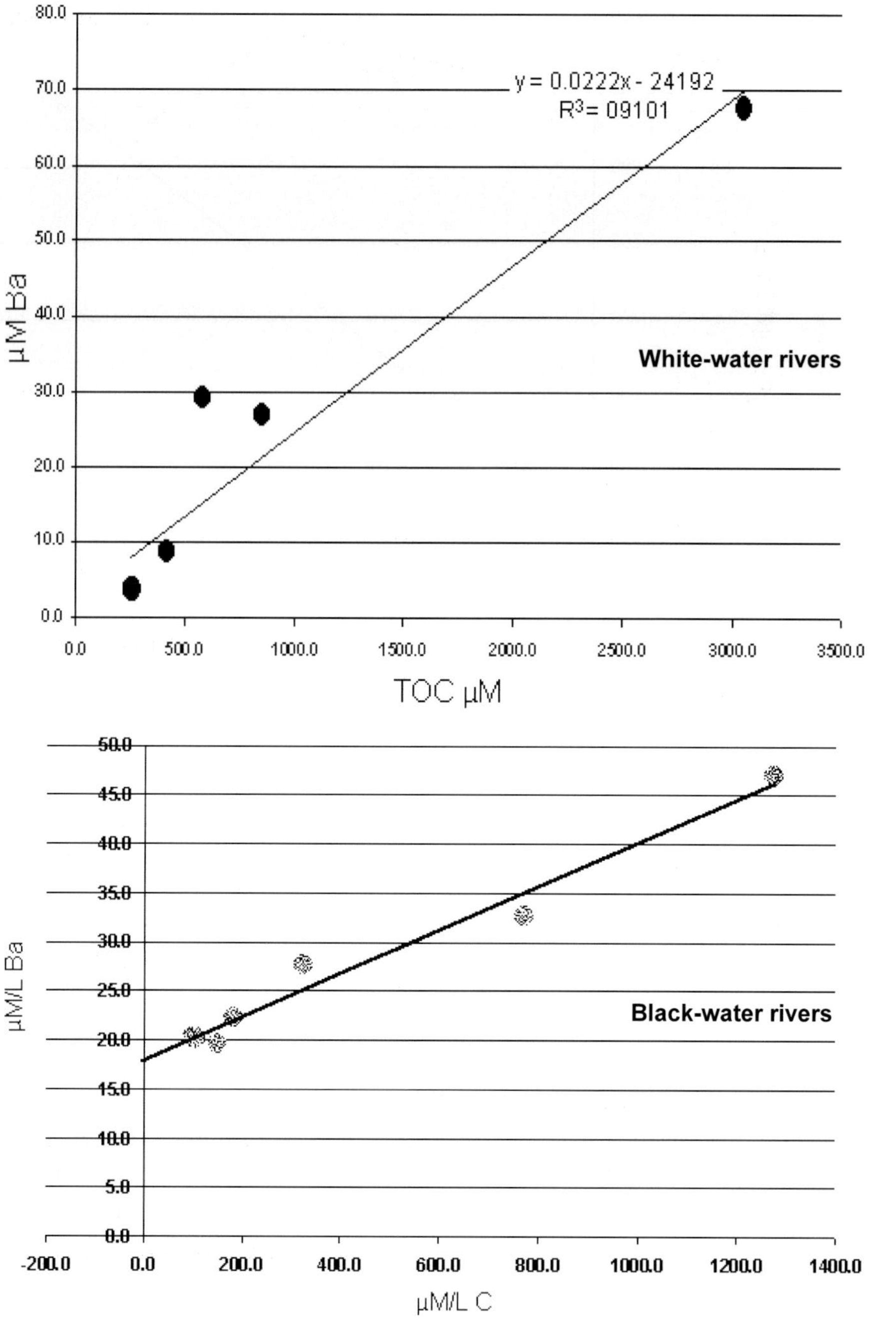

Fig. 2.3 Organic carbon correlation with barium for white and black river water.

Understanding size speciation variation

Figure 2.4 presents different mass balances between Manaus and Óbidos for several sampling dates. Over the course of the year there was a continuous loss of carbon. The maximum losses were observed in March and September, which correspond to low- and high-water periods, respectively. During rising water, losses of carbon focused on the dissolved rather than colloidal fraction (Fig. 2.4(d) and (e)). During this period less particulate organic matter is imported by rivers. Because respiration is not sufficient to explain this loss, we hypothesize that flocculation followed by coagulation explains the dissolved organic carbon (DOC) losses.

$$DOC \rightarrow COC \rightarrow POC$$

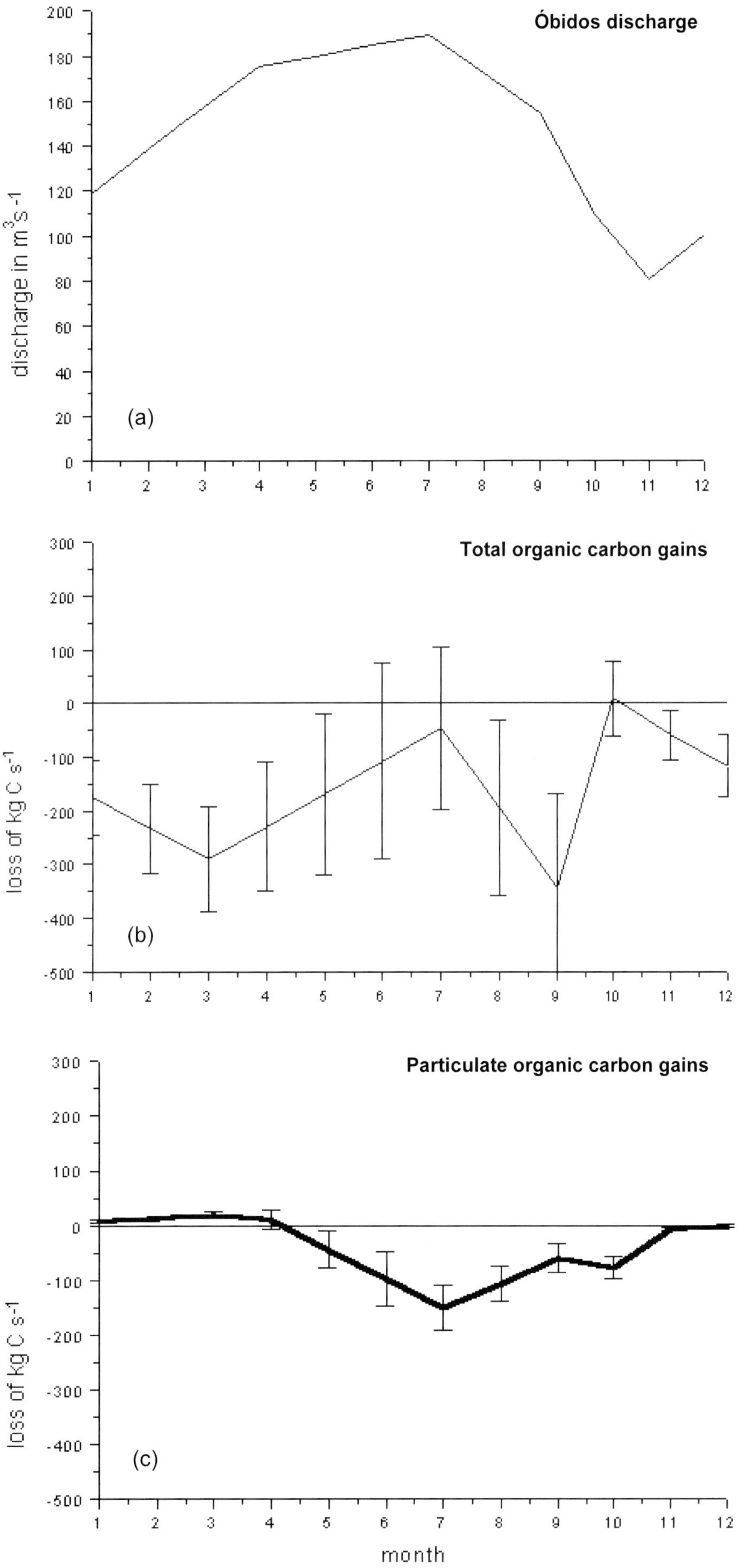

Fig. 2.4 Loss and gain for all carbon fraction sizes.

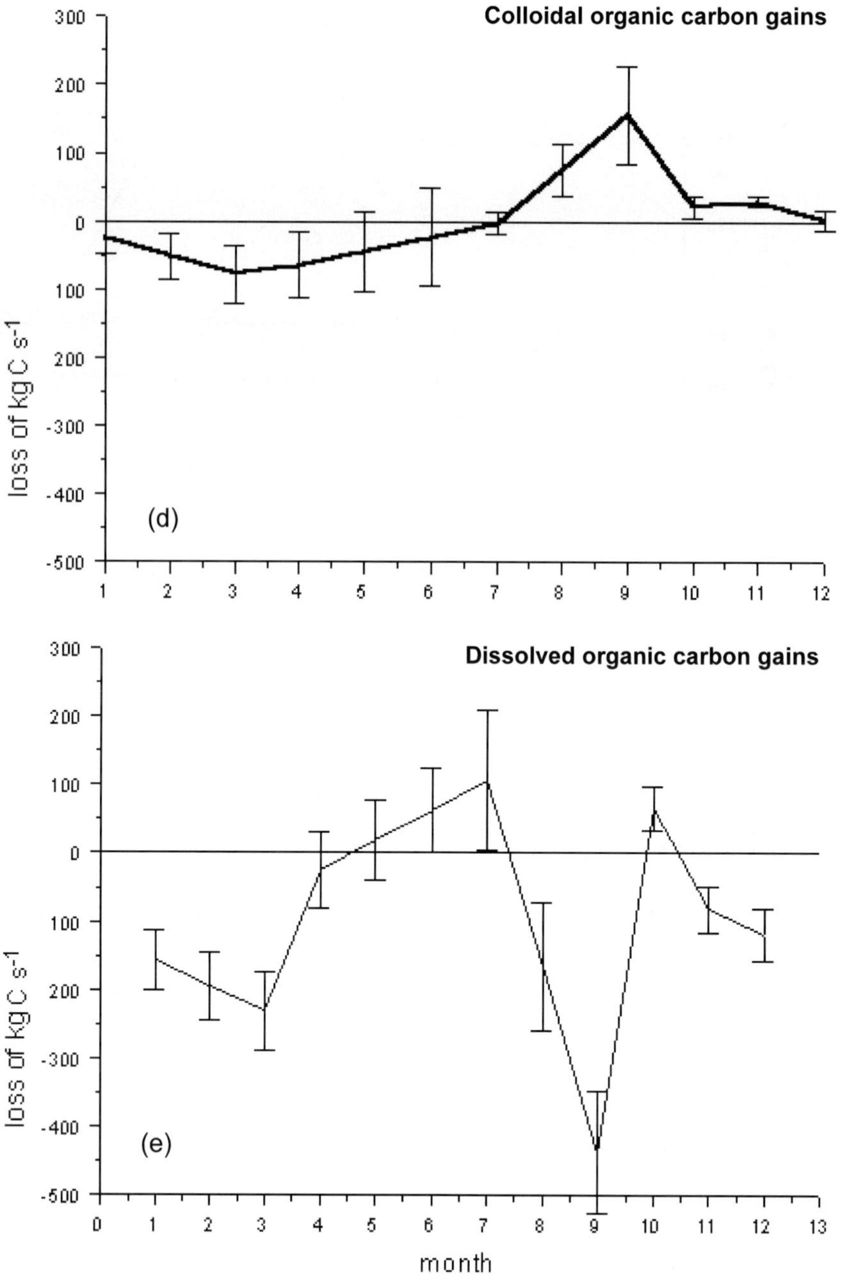

Fig. 2.4 *continued*

where COC is colloidal organic carbon and POC is particulate organic carbon. Conversely before July there was an increase in DOC. The inverse reaction is applied to explain this increase.

POC → COC → DOC

Just after the high-water period (November), a break occurs: DOC (–400 kg s^{-1}) and POC (–60 kg s^{-1}) decreased and COC increased by 150 kg s^{-1}. These variations show an unusual colloidal fraction behaviour. This could be explained if the association/dissociation processes are as follows:

POC + DOC → COC

POC → COC ← DOC

During the rising water period, inputs of both DOC and COC are important. Flocculation of DOC around colloidal particles could explain the loss of DOC and the small

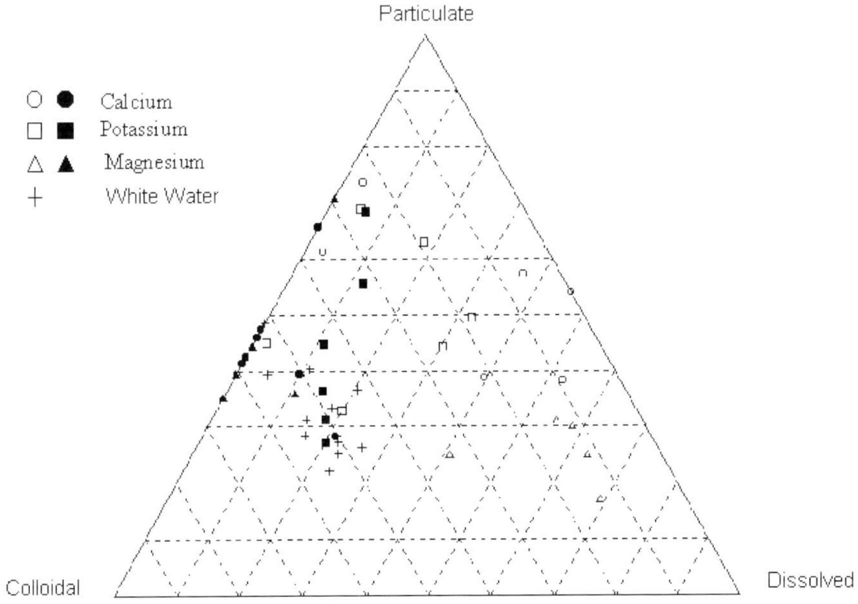

Fig. 2.5 Major elements size speciation for white- and black-water rivers.

gain in POC. At the end of the high-water period, DOC became the dominant input of organic carbon while POC diminished. In conclusion, the loss in organic carbon for this section seems not to be due simply to the loss of POC by sinking. DOC appears to be lost by the way of POC formation during mixing, accentuated by the two very different types of water. Mixing seems to result in simple flocculation. The increase in COC during the falling water period suggests that the flocculation process is reversible. The principal reaction may be COC formation by DOC flocculation and continued flocculation to form POC, which then sinks. A more intensive short-term observation period of the mixing zone might provide a better understanding of the exact mechanism.

Estimation of carbon fluxes to the estuary

Values obtained during this project allow for an evaluation of the quantity of carbon transported by the Amazon to its estuary. Our estimations are presented in Table 2.2, along with a comparison with values reported by other authors. Our values are similar to all but those obtained by Richey (1980).

Table 2.2 Estimation of organic carbon fluxes for the Amazon at Óbidos station.

Reference	Year	Month	Flux (Tg year^{-1})
Richey (1980)	1977	-	60
Richey *et al.* (1990a)	1982–1984	-	36
this study	1994–1995	-	20
this study	Min.	10	6.6
this study	Max.	06	38

CONCLUSIONS

Our results demonstrate that the Río Negro is distinct among Amazon tributaries and plays a regulatory role in determining mainstem organic matter fluxes. Its contribution of organic carbon accounts for only 20–40% of mainstem fluxes, but it contributes a higher proportion within the colloidal size fraction. Coagulation and flocculation within the mixing zone of the Negro and Solimões rivers is hypothesized to promote sedimentation of organic matter in the reach between Manaus and Óbidos. Through mass balance calculations coupled with detailed size fraction analyses, we estimate organic carbon losses of 30–40% along this reach during the low-water period. This amounts to 20×10^6 t $year^{-1}$ of losses made up of the following size fractions: 42% DOC, 40% COC, and 18% POC. There are also losses of metals bound to organic matter.

Organic material deposited along this reach must be removed through bed-load transport or biological decomposition. An improved understanding of the fate of sedimented organic matter will require additional research.

Acknowledgements This work was part of the PEGI program (environmental intertropical geosphere programme) which is supported by IRD/CNRS INSU. The authors are also indebted to ANEEL for the preparation of the sampling campaigns in Brazil.

REFERENCES

Benaïm, J. Y. & Mounier, S. (1998) Metal transport by organic carbon in the Amazon basin. *Croatica Chemica Acta* **71**(2), 405–419.

Buesseler, K. O., Bauer, J. E., Chen, R. F., Eglinton, T. I., Gustafsson, O., Landing, W., Mopper, K., Moran, S. B., Santschi, P. H., Vernon-Clark, R. & Wells, M. L. (1996) An intercomparison of cross-flow techniques used for sampling marine colloids: overview and organic carbon results. *Marine Chemistry* **55**, 1–31.

Buffle, J., Deladoey, P., Zumstein, J. & Haerdi, W. (1982) Analysis and characterization of natural organic matter in fresh waters. I. Study of analytical techniques. *Schweizerische Zeitschrift für Hydrologie* **44**(2), 325–362.

Dai, M., Martin, J. M. & Cauwet, G. (1995) The significant role of colloids in the transport and transformation of organic carbon and associated trace metals (Cd, Cu, and Ni) in the Rhône delta (France). *Marine Chemistry* **51**, 159–175.

De Vitre, R. R., Bulard, F., Bernard, C. & Buffle, J. (1987) A novel *in situ* cascade ultrafiltration unit specifically designed for field studies of anoxic waters. *International Journal of Environmental Analytical Chemistry* **31**, 145–163.

Devol, A. H., Forsberg, B. R., Richey, J. E. & Pimentel, T. P. (1995) Seasonal variation in chemical distributions in the Amazon (Solimões) River: a multiyear time series. *Global Biogeochemical Cycle* **9**(3), 307–328.

Guo, L. & Santschi, P. H. (1996) A critical evaluation of the cross-flow ultrafiltration technique for sampling colloidal organic carbon in seawater. *Marine Chemistry* **55**, 113–127.

Hedges, J. I. (1992) Global biogeochemical cycles: progress and problems. *Marine Chemistry* **39**, 67–93.

Hoffmann, M. R., Yost, E. C., Eisenreich, S. J. & Maier, W. J. (1981) Characterisation of soluble and colloidal-phase metal complexes in river water by ultrafiltration. A mass-balance approach. *Environmental Science and Technology* **15**(6), 655.

Küchler, I. L., Miekeley, N. & Forsberg, B. R. (1994) Molecular mass distributions of the dissolved organic carbon and associated metals in waters from Rio Negro and Rio Solimões. *The Science of the Total Environment* **156**, 207–216.

Ludwing, W., Amiotte-Suchet, P. & Probst, J. L. (1996) River discharge of carbon to the world's oceans: determining local inputs of alkalinity and of dissolved and particulate organic carbon. *Comptes Rendus de l'Académie des Sciences* Série IIa **323**, 1007–1014.

Probst, J. L., Mortatti, J. & Tardy, Y. (1994) Carbon river fluxes and weathering CO_2 consumption in the Congo and Amazon river basins. *Applied Geochemistry* **9**, 1–13.

Reitmeyer, R., Powell, R. T., Landing, W. M. & Measures, C. I. (1996) Colloidal aluminum and iron in seawater: an intercomparison between various cross flow ultrafiltration systems. *Marine Chemistry* **55**, 75–91.

Richey, J. E.(1980) Organic carbon: oxidation and transport in the Amazon River. *Science* **207**, 1348–1351.

Richey, J. E., Hedges, J. I., Devol, A. H., Quay, P. D., Victoria, R., Martinelli, L. & Forsberg B. R. (1990a) Biogeochemistry of carbon in the Amazon River. *Limnology and Oceanography* **35**(2), 352–371.

Richey, J. E., Victoria, R. L., Salati, E. & Forsberg, B. R. (1990b) The biogeochemistry of a major river system: the Amazon case study. In: *Biogeochemistry of Major World Rivers.* John Wiley, Chichester, UK.

CHAPTER 3

NITROGEN DYNAMICS ON THE AMAZON FLOOD PLAIN IN RELATION TO THE FLOOD PULSE OF THE SOLIMÕES RIVER

Jürgen Kern, Heidi Kreibich & Assad Darwich

The flood plain of the mainstem Amazon River is closely linked to the river channel. In Brazil alone about 150 000 km^2 are periodically inundated (Sippel *et al.*, 1992) and blanketed by suspended solids carried in from the river channel. The river itself reflects physico-chemical and biological processes in the catchment, and the flood plain is an important part of the whole. Due to the flood pulse of the Amazon River there are predictable oscillations in water level and nutrient supply. In the central Amazon at Manaus the mean water level fluctuation is 9.9 m (1902–1993) and leads to a regular shift between aquatic and terrestrial environments on the flood plain, with high water levels dominating in the middle of the year and low water levels in October–November (Fig. 3.1). High biomass production, rapid nutrient turnover and unique adaptations of plants and animals are characteristic in this aquatic–terrestrial transition zone as described by Junk *et al.* (1989), Piedade *et al.* (1991), Adis (1992) and Waldhoff *et al.* (1998) and make the flood plain important in the trophic ecology of the mainstem Amazon River. Many commercially important Amazonian fish are specially adapted to food and feeding habits of the flood-plain forest (Saint-Paul, 1992).

The high productivity of the flood plain (referred to locally as the *várzea*) requires a constant supply of nutrients, but this supply varies according to the flood pulse of the river. No macronutrient is more affected by the flood pulse than nitrogen, which may be enriched by N_2 fixation or lost by denitrification. Nitrogen limitation may be compensated for by various kinds of N_2 fixation which have already been described for the *várzea* by Salati *et al.* (1982), Martinelli *et al.* (1992), Doyle & Fisher (1994) and Kern & Darwich (1997). Little information exists on the opposing process of denitrification. Because most habitats in the Amazon basin, such as surface waters, groundwaters and soils, are characterized by low nitrogen contents, considerable nitrogen losses by denitrification have not been expected. Within the aquatic–terrestrial transition zone of the Solimões River (as the Amazon is called above its confluence with the Negro River) it has been shown, however, that denitrification occurs at high rates even during the low-water period (Kern *et al.*, 1996).

Plant production may be enhanced by N_2 fixation as has been studied primarily in developing countries where there is an economic incentive to minimize the use of nitrogen fertilizers (Campbell *et al.*, 1995; Döbereiner *et al.*, 1995; Kadiata *et al.*, 1996). To what extent fixation of atmospheric nitrogen contributes to soil fertility of the Amazonian lowland and how much it may counteract nitrogen loss due to river export has not yet been quantified. More information is needed to better understand nitrogen processes such as biological N_2 fixation and denitrification in the Solimões River and its flood plain.

To gain more insight into the nitrogen dynamics of the central Amazon flood plain the pathways of nitrogen input and output were measured in various ecotopes of one *várzea* lake on Marchantaria Island, Lake Camaleão. This lake is only influenced by the white water of the Solimões River and is subject of very little anthropogenic disturbance. Figure 3.2

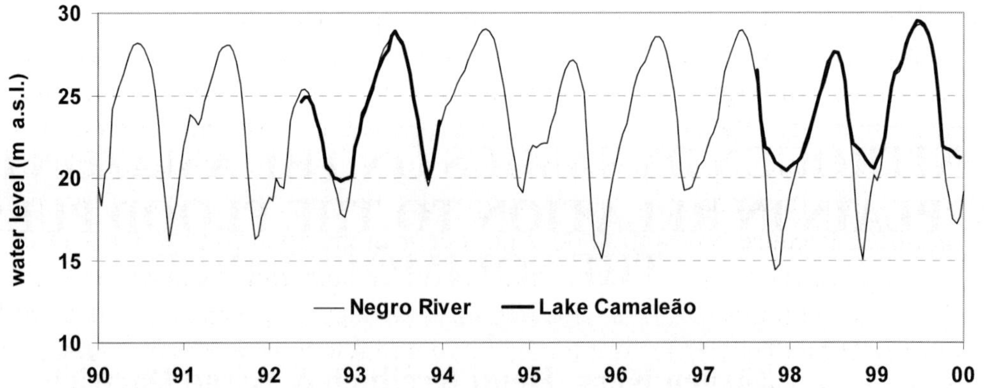

Fig. 3.1 Water level fluctuations of the Negro River at Manaus and of Lake Camaleão between 1990 and 2000.

illustrates a typical profile of the aquatic–terrestrial transition zone of the *várzea* with flooded and exposed sediments. Flood-plain forest grows at the highest elevational boundaries above 21 m a.s.l. A significant amount of nitrogen reaches the flood plain in the dissolved and particulate load of the Solimões River during floods and N_2 fixation contributes less than 20% to the total input as calculated by Kern & Darwich (1997). Their study, however, did not include the flood-plain forest, where a considerable nitrogen input may be assumed due to the abundance of legumes which may form symbioses with N_2-fixing bacteria.

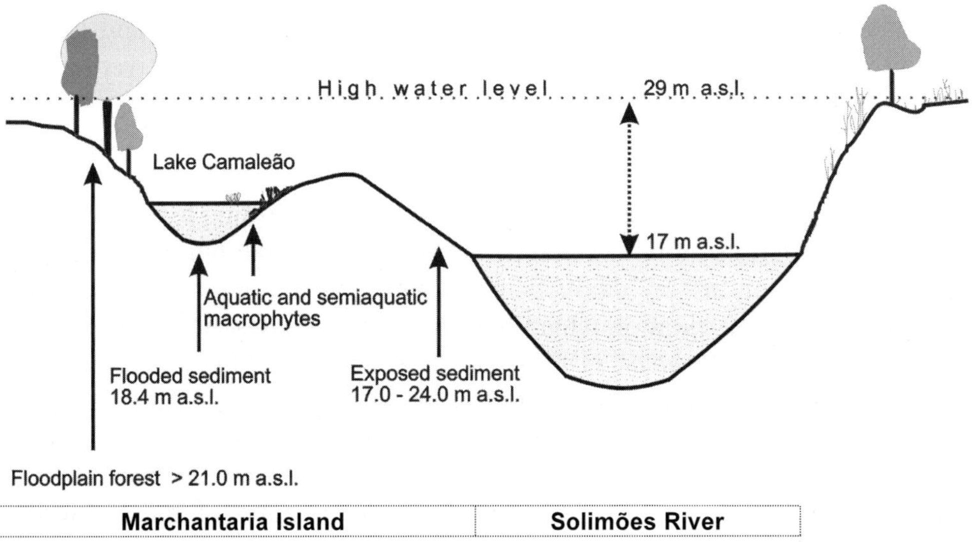

Fig. 3.2 Cross-section of a typical Amazon flood plain with the aquatic–terrestrial transition zone during the low-water period.

HYDROCHEMISTRY OF THE SOLIMÕES RIVER AND LAKE CAMALEÃO

The hydrochemical composition of the Solimões River reflects physical and chemical weathering processes in the Andes and sub-Andean lowlands and the interaction of the river with its flood plain. Large amounts of metals are transported as fine-grained suspended solids (5.5% Fe, 1.8% Ca, 1.2% Mg, 1.0% Ca) in the river and deposited on the flood plain (Irion *et al.*, 1997). This has a fertilizing effect and supports high biomass production rates and the accumulation of carbon and nitrogen fixed from the atmosphere. To date, transfers of anthropogenically derived nitrogen from terrestrial to aquatic systems through soil leaching and through direct waste-water discharge plays a small role in the Amazon basin (Howarth *et al.*, 1996). In the case of nitrogen, natural biochemical processes of N_2 fixation and denitrification result in gains and losses of nitrogen. Gaseous nitrogen exchange within the river system seems to be mostly restricted to the flood plain with its changing hydrological conditions because N_2 fixation does not occur in the open river water with its high turbidity and currents. Also, open water denitrification is negligible because, despite relatively high NO_3^- concentrations, the river channel does not support the anoxic environment required for denitrification (Table 3.1). Within the flood plain both processes are running as indicated by enhanced concentrations of NH_4^+ and organic nitrogen (N_{org}) and reduced concentrations of NO_3^- in Lake Camaleão compared to the Solimões River. These concentration patterns point to a rapid uptake and storage of nitrogen by plants, particularly during rising water, and also to intensive denitrification (Kern *et al.*, 1998; Villar *et al.*, 1998).

Despite predictable flood events, the extent of inundation may differ from one year to another. For example the extremely low river level during high water in May/ June 1992 (Fig. 3.1), did not result in overbank flow at Marchantaria Island. There was an accumulation of NH_4^+, which was then exported by outflowing water from the flood plain to the river. High river concentrations of NH_4^+ during receding water in 1992 reflect that exceptional situation (Fig. 3.3). The differing hydrochemical properties between the flood-plain lakes and the river channel derive primarily from the supply of oxygen, which greatly depends on oxygen consumption during decomposition of organic matter and carbon turnover. Studies by Furch (1999) and Kern & Darwich (1997) showed that concentrations of dissolved organic nitrogen, NH_4^+ and NO_2^- were much higher in the flood plain as

Table 3.1 Hydrochemical characteristics of surface water from Solimões River and Lake Camaleão during two different hydrological cycles from November 1981 to October 1982 (Furch, 1999) and from December 1992 to November 1993. Each study ran over a 12 month period when the river level changed between 11 and 12 m.

		1981/82: Solimões	Lake Camaleão: entrance	centre	1992/93: Solimões	Lake Camaleão: entrance	centre
K_{25} ($\mu S\ cm^{-1}$)	Min–max	59.1–96.9	60.8–426	60.8–685	61.0–80.0	61.3–298	60.9–450
	Mean	**74.1**	**121.1**	**176.4**	**68.3**	**113.9**	**143.9**
	SD	11.6	105.3	195.4	5.30	83.9	133.3
pH	Min–max	6.20–7.38	6.20–7.81	6.14–7.66	5.95–8.12	6.02–7.73	5.92–7.84
	Mean	**6.79**	**6.76**	**6.66**	**6.55**	**6.67**	**6.54**
	SD						
O_2 ($mg\ l^{-1}$)	Min–max		0–5.6	0–6.7	1.3–7.5	1.6–7.7	0.8–6.8
	Mean		**2.3**	**2.0**	**4.7**	**4.0**	**3.5**
	SD		1.7	2.0	1.7	1.9	1.6
N_{org} ($\mu mol\ l^{-1}$)	Min–max				0.3–24.3	0–45.3	1.1–54.5
	Mean				**10.19**	**15.35**	**20.17**
	SD				6.98	11.41	14.36
NH_4 ($\mu mol\ l^{-1}$)	Min–max	1.33–9.89	1.93–40.64	1.09–54.71	0.14–2.43	0.52–43.14	0.39–46.93
	Mean	**5.32**	**11.79**	**11.71**	**0.86**	**4.41**	**7.09**
	SD	2.31	13.29	16.21	0.45	9.02	11.73
NO_2 ($\mu mol\ l^{-1}$)	Min–max	0.1–0.36	0.03–30.57	0.04–3.31	0–0.17	0–4.98	0–16.0
	Mean	**0.16**	**2.81**	**0.41**	**0.06**	**0.45**	**1.12**
	SD	0.11	8.76	0.91	0.06	1.22	3.40
NO_3 ($\mu mol\ l^{-1}$)	Min–max	2.36–10.37	0.89–45.64	0.35–8.29	2.6–14.9	1.6–20.7	0.5–20.0
	Mean	**6.84**	**7.34**	**2.64**	**7.99**	**6.18**	**5.05**
	SD	2.57	12.21	2.75	3.32	4.29	4.05

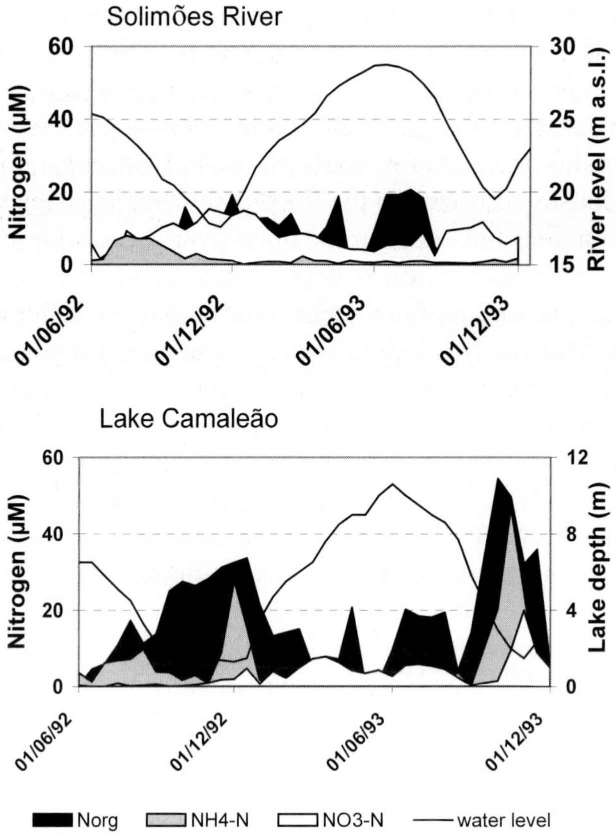

Fig. 3.3 Seasonal variability of nitrogen components in a non-cumulative pattern. N_{org} derived from $N_{tot} - (NH_4 + NO_3)$.

compared to the river channel, where NO_3^- is the predominant nitrogen compound (Table 3.1). This means that the flood plain acts as a sink for NO_3^- and as a source for other dissolved nitrogen forms due to the water flow. During receding water *várzea* lakes, as represented by Lake Camaleão, export more N_{org} and NH_4^+ than NO_3^- to the river, which may then be oxidized in the oxic environment of the river. This is supported by the findings of Brandes *et al.* (1996) who suggested that remineralization of N_{org} within the channel is the major source of dissolved inorganic nitrogen to streams of the central Amazon. During low-water periods in 1992 and 1993, N_{org} began to increase in the lake water, followed by NH_4^+ and NO_3^-, reflecting leaching and seepage from flood-plain forests on high elevational boundaries, as well as sedimentary re-suspension and mineralization within the lake (Fig. 3.3).

When nitrogen compounds reach the river, nitrification seems to proceed rapidly because river NO_3^- concentrations are twice as high during low river levels, when the river is fed by N_{org} and NH_4^+. Other bioactive compounds such as O_2 and CO_2 may also be controlled by inchannel processes linked to river stage (Devol *et al.*, 1995). The more stable part of N_{org} subsequently returns to the flood plain with the rising river water which on average has a total suspended solids (TSS) concentration of 80 mg l^{-1} (Mertes, 1994). A considerable amount of nitrogen is transferred to the flood plain and may reach up to 150 kg N ha^{-1} $year^{-1}$. This does not, however, meet the requirement of forest vegetation which has been estimated at 292 kg N ha^{-1} $year^{-1}$ (Furch, 1997). This is remarkable because nitrogen is the only macronutrient which is obviously limited with regard to particulate matter. Furthermore, the deposited nitrogen is not readily available and is therefore primarily buried in the sediments of *várzea* lakes. Some nitrogen compounds may undergo further microbial degradation and become part of internal nitrogen recycling.

The content of nitrogen adsorbed to riverine TSS is only about 1‰, which means that the export of particulate nitrogen by the Amazon to the Atlantic Ocean is about 1.2×10^6 t $year^{-1}$, assuming 1.2×10^9 t year^{-1} of sediment transport (Martinelli *et al.*, 1989). This is about one third of the total nitrogen flux (3.3×10^6 t year^{-1}; Howarth *et al.*, 1996). A significant loss of nitrogen from the Amazon basin results from the downstream transport of riverine NO_3^-. Mean annual NO_3^- concentrations of 8 µM in the Solimões River near Manaus are decreased to 2 µM at the Amazon estuary about 1600 km downstream, indicating a relatively low export to the Atlantic Ocean (Fig. 3.4). This seems to be typical for a couple of tropical rivers, where high temperatures may contribute to high microbiological activities and thus to a rapid nitrogen turnover. A much better predictor of river NO_3^- concentrations, however, is the human population density in a basin, as reported by Peierls *et al.* (1991). The higher the human influence on waste-water loading, agriculture and deforestation in the basin, the higher the NO_3^- export as reflected by the rivers Thames, Rhine and Danube. Although the Amazon River does not seem to be affected by an overload of nitrogen during the near future, the consequences of ongoing deforestation of the Amazon catchment should be taken as a serious problem at least for upland soils. One example is the export of total nitrogen and total phosphorous which was two times and eight times higher in surface runoff of partially deforested areas, when compared to forested areas (Williams & Melack, 1997).

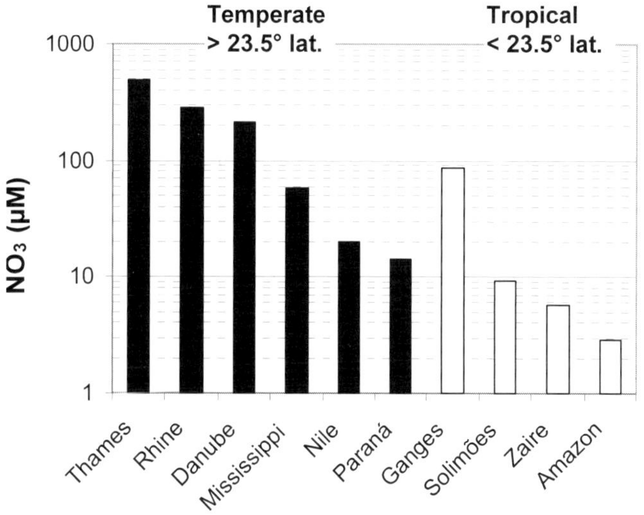

Fig. 3.4 Nitrate export by large rivers of the world (after Peierls *et al.*, 1991).

SEASONAL NITROGEN PROCESSES IN FLOODED AND EXPOSED SEDIMENTS

The Amazon flood plain is built up by deposited clay minerals, such as montmorillonite and illite, that play an important role in the storage of nutrients due to their high capacities for adsorbing NH_4^+ and protecting it from leaching and microbial oxidation. Sediments may be flooded or exposed according to the river stage and due to the elevational boundary within the flood plain. At low elevational boundaries we find flooded sediments which are covered by water for the whole year, a situation which is favoured when water bodies are hydrologically isolated from the river channel during low-water periods. A typical succession of *várzea* lakes can be observed as a result of abiotic processes like sedimentation of mineral matter, which are carried by the river and as a result of lake internal biomass production and accumulation of organic matter. Evidence for such a continuous process is given by increasing water depth of Lake Camaleão during the last decade (Fig. 3.1).

By this process lake water has become less concentrated during low water as reflected by the decrease in electrical conductivity (Table 3.1). The overlying water becomes a diffusion barrier for oxygen, which is needed for nitrification and consecutive denitrification in the sediment. Consequently in *várzea* lake sediments almost no NO_3^- is detectable, in contrast to high concentrations of NH_4^+ resulting from high production and decomposition of biomass during high water (Junk & Piedade, 1993). High NH_4^+ concentrations in the sediment interstitial waters suppress the nitrogenase system which probably is the reason for the lack of N_2 fixation throughout the year (Fig. 3.5). Due to the NH_4^+ gradient between the water column and sediment and due to re-suspension, diffusion and bioturbation, NH_4^+ may be released from the sediment. During low-water periods such as in 1992 and 1993

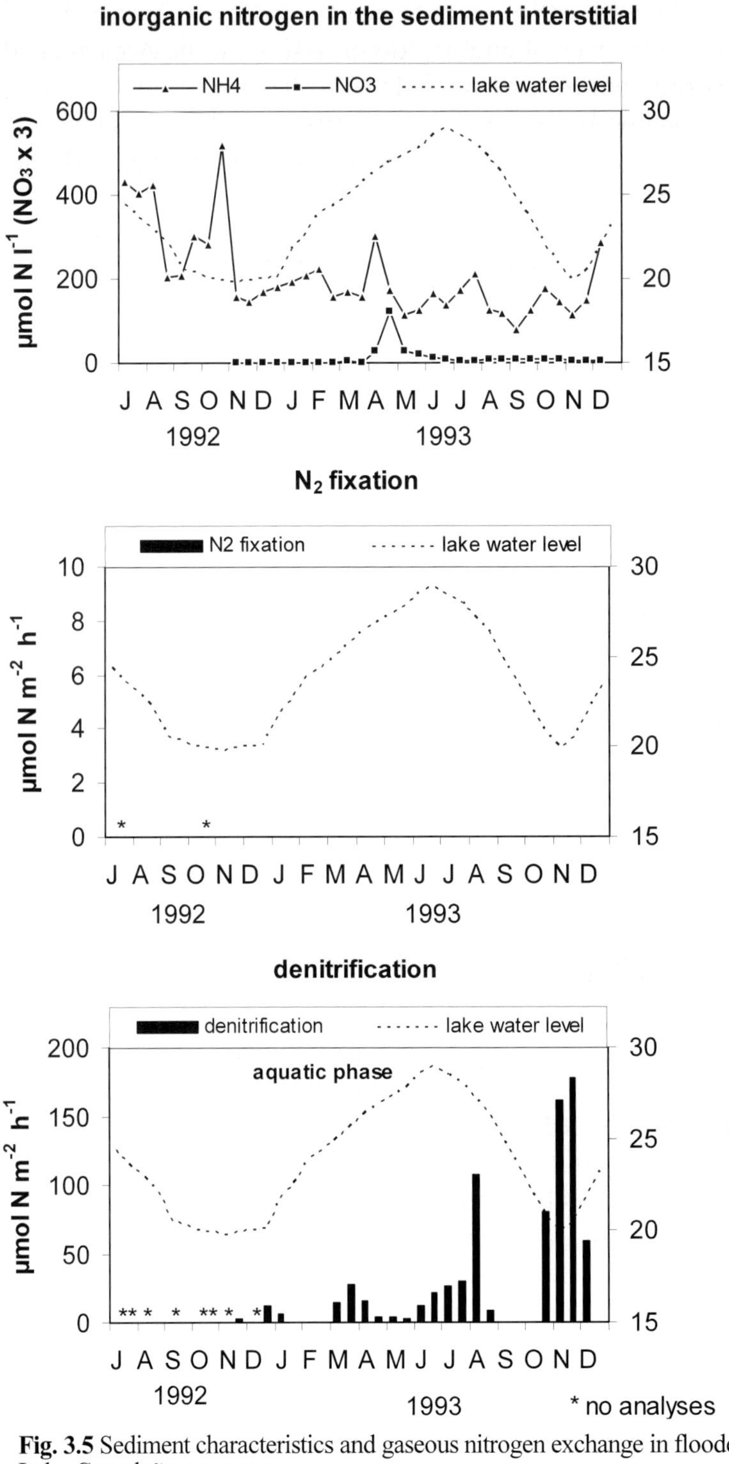

Fig. 3.5 Sediment characteristics and gaseous nitrogen exchange in flooded sediments (0–5 cm) of Lake Camaleão.

seepage from exposed sediments and NH_4^+ release from flooded sediments into the lake water are the initial steps for nitrification. With a sufficient supply of NO_3^-, denitrification can occur in flooded lake sediments, which may reach high rates of up to 200 µmol m^{-2} h^{-1} during low-water periods corresponding to maximum NO_3^- concentrations in the overlying water (Fig. 3.3). High NO_3^- concentrations may be explained by nitrification, which is initiated by increasing O_2 input into the diminishing water column.

Another situation may occur during exceptional hydrological situations like those observed at Lake Camaleão in April 1993. In that month overbank flow was increasing and different current situations led to a stagnation of water flow before lake water became the same as river water. A remarkable sediment–internal nitrification reaction seemed to occur due to high concentrations of NH_4^+ mixing into an aerated water supply. This resulted in 10 times higher NO_3^- concentrations in the sediment interstitial waters compared to the overlying water body. This event can be considered as unusual, because the inflow of river water is generally the main source of NO_3^- for *várzea* lakes.

Sediments are occasionally exposed to air depending on their elevation relative to the changing water level. As time after exposure increases, both N_2 fixation and denitrification increase. First nitrogen-fixing cyanobacteria are established as long as water content is sufficient to allow algal cells to survive. Although processes of N_2 fixation and

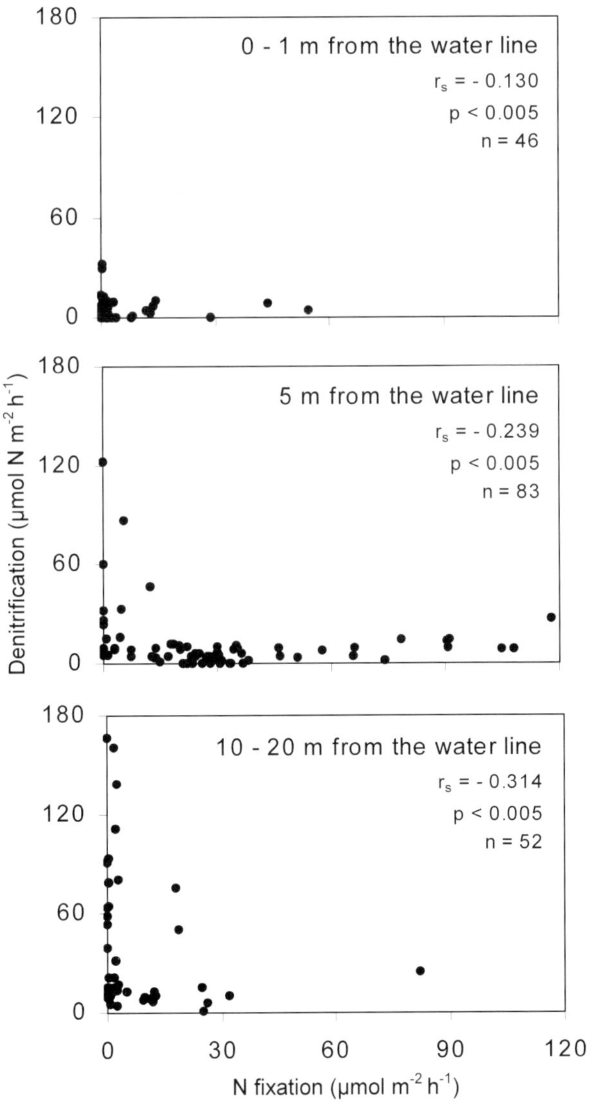

Fig. 3.6 N_2 fixation and denitrification in exposed sediments without vegetation cover: 2 September 1992–1 December 1993.

denitrification may run simultaneously, negative correlations suggest that they do not (Fig. 3.6). The highest rates of N_2 fixation occur 5 m from the water line, while maximum denitrification rates were measured 10–20 m from the water line of *várzea* lakes. This means that N_2 fixation only plays a role after sediments are exposed. Denitrification may reach the same range as in flooded sediments. However, it plays a greater role in exposed sediments, particularly in sediments, which are vegetated during the terrestrial phase. In these ecotopes large amounts of organic material deriving from *Echinochloa polystachya* and other semiaquatic macrophytes support denitrification (Kern & Darwich, 1997).

In contrast to permanently flooded sediment, the soil of the flood-plain forest on higher elevational boundaries is less homogeneous and interlaced with roots. This may explain the

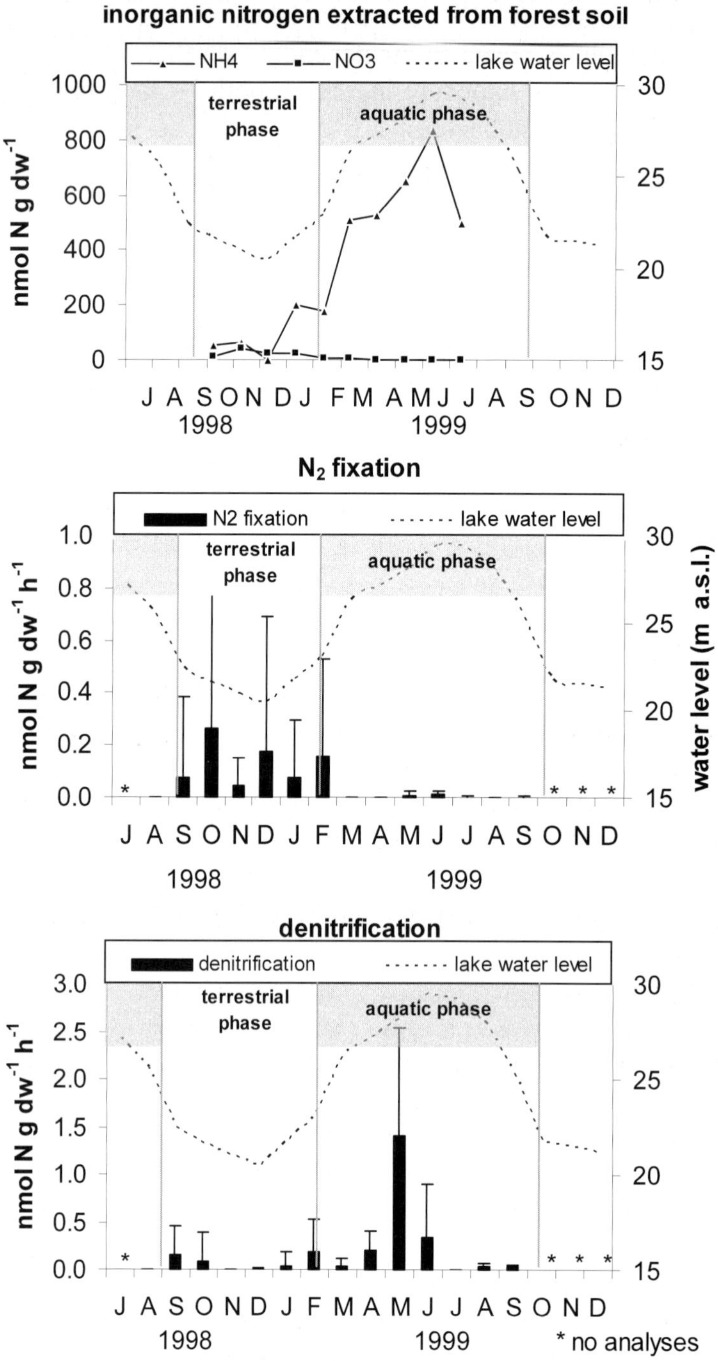

Fig. 3.7 Soil characteristics and gaseous nitrogen exchange in semi-terrestrial sediments of the flood-plain forest at Lake Camaleão.

much higher variability of both processes within a few square centimetres of forest ground. Mean N_2 fixation in the soil varied between 0.04 and 0.26 nmol N g dw^{-1} h^{-1} (dw = dry weight) during the terrestrial phase (Fig. 3.7). As with flooded sediments N_2 fixation in flooded forest soils can be neglected during the aquatic phase. Mean denitrification of sediments within the forest varied between 0 and 1.40 nmol N g dw^{-1} h^{-1}. Reaction rates are highest during flooding and transition periods (Fig. 3.7) due to the availability of oxygen which primarily regulates denitrification within the forest. High denitrification rates during the transition periods are caused by small spatial changes between oxic and anoxic conditions. Upon exposure to air NH_4^+ is oxidized, producing nitrate that feeds denitrification. A strong relationship between denitrification and soil water content is in accordance with experiments by Bollmann & Conrad (1998) and field measurements by Kilian & Werner (1996). During the terrestrial phase, when soil water content decreases below 35%, there is little denitrification. The highest rates of denitrification occurred in May 1999, possibly due to increasing oxygen and nitrate concentrations in the water column resulting from the overbank flow of the river. This event took place under similar hydrological conditions as described for flooded sediments, when the water level was between 26 and 27 m a.s.l., suggesting that hydrology has a significant influence on oxygen supply and consequently on microbial activity during flooding.

ROLE OF N_2 FIXATION FOR THE AMAZON FLOOD PLAIN

In many tropical forests nitrogen is in relative excess compared to temperate forests, suggesting that tropical forests have more open nitrogen cycles (Martinelli *et al.*, 1999). This however, only applies in part to the *várzea*. By constructing a lake input–export budget for total nitrogen, Kern & Darwich (1997) showed that the nitrogen demand of macrophytes and woody plants growing at Lake Camaleão is not met by the inflow of river water. Therefore, it was assumed that either lake–internal nitrogen fluxes are much higher than the nitrogen flux between the lake and the river channel or N_2 fixation contributes a substantial amount to the nitrogen budget of the *várzea*.

Intensive studies in the *várzea* revealed a substantial contribution of N_2 fixation, particularly in the aquatic phase. Nitrogen fixation by cyanobacteria within the periphyton of adventitious roots of semiaquatic grasses was found to be most important. During the terrestrial phase the settlement of N_2-fixing pioneer organisms is obvious. Particularly near the water line and in hollows of exposed sediments free-living cyanobacteria form mats. Among the trees, the pioneer species *Salix humboldtiana*, *Cecropia latiloba* and *Pseudobombax munguba* do not give clear indications of N_2 fixation, in contrast to legume trees (Kern *et al.*, 2000).

The amount of nitrogen fixed by Brazilian native legumes may be significant as reported by Sprent *et al.* (1996) and Martinelli *et al.* (1999). The importance of legume N_2 fixation to nitrogen requirements in the *várzea* forest was studied on Marchantaria Island by comparison of nitrogen stable isotopic composition of tree leaves. Results are expressed in $\delta^{15}N$ with low values indicating N_2 fixation. Low foliar $\delta^{15}N$ between –2 and 5 is obtained from woody legumes and relatively high values from non-legume trees (Table 3.2). Although $\delta^{15}N$ values of both groups overlap, the groups are significantly different ($p < 0.05$) in contrast to their leaf nitrogen content. There seems to be no correlation between N_2 enrichment in leaves and N_2 fixation, with the exception of *Albizia multiflora* ($R^2 = 0.53$) with the highest nitrogen content and *Dalbergia riparia* with the lowest $\delta^{15}N$ ($R^2 = 0.33$). As expected for legumes *Dalbergia riparia*, *Campsiandra comosa*, *Dalbergia inundata*, *Mimosa pigra*, *Zygia inaequalis*, *Albizia multiflora*, and *Pterocarpus amazonum* have $\delta^{15}N$

Table 3.2 N_2 fixation within the flood-plain forest indicated by foliar $\delta^{15}N$. Calculation of nitrogen derived from atmosphere (Ndfa) according to Shearer & Kohl (1988).

	$\delta^{15}N$	% Nitrogen	*n*	% Ndfa
Woody legumes				
Dalbergia riparia	−1.15 ± 1.22	2.53 ± 0.43	6	92 ± 19
Campsiandra comosa	−0.01 ± 1.22	2.23 ± 0.28	4	74 ± 13
Dalbergia inundata	0.30 ± 1.20	2.70 ± 0.54	4	69 ± 18
Mimosa pigra	0.69 ± 0.68	3.50 ± 0.63	4	63 ± 10
Albizia multiflora	1.11 ± 1.69	4.19 ± 1.87	7	57 ± 26
Zygia inaequalis	1.22 ± 1.04	2.59 ± 0.29	6	55 ± 16
Pterocarpus amazonum	2.79 ± 0.57	2.74 ± 0.16	3	31 ± 9
Macrolobium acaciifolium	4.90 ± 0.92	2.03 ± 0.15	3	
Woody non-legumes				
Tabebuia barbata	3.11 ± 2.01	2.13 ± 0.18	3	
Nectandra amazonum	3.51 ± 0.95	2.15 ± 0.07	3	
Cecropia latiloba	3.97 ± 0.56	3.09 ± 0.14	3	
Vitex cymosa	4.93 ± 1.46	1.62 ± 0.10	3	
Pseudobombax munguba	5.36 ± 0.34	1.83 ± 0.08	3	
Crateva benthami	7.87 ± 1.62	2.90 ± 0.23	3	

Mean ± standard deviation; *n* = number of individual plants studied.

values significantly lower than non-legumes, indicating N_2 fixation. Obviously dominant lianas such as *Dalbergia riparia* and *Dalbergia inundata* play a key role in N_2 fixation in the *várzea* forest. The same may be said for *Mimosa pigra*, a small liana which grows all along the lake bank. *Zygia inaequalis*, a relatively small tree (7–10 m) grows in the understory of the higher elevational boundaries of the forest. *Campsiandra comosa*, *Albizia multiflora*, and *Pterocarpus amazonum* are high trees (15–20 m), forming part of the tallest canopy growing at the middle and higher elevational boundaries of the forest. The roots of all these species are nodulated, with the exception of *Macrolobium acaciifolium* whose $\delta^{15}N$ does not differ from non-legumes.

Mean $\delta^{15}N$ values of non-legume trees range between 3 and 8. There is also a high intraspecific and interspecific variability in $\delta^{15}N$, reflecting both species and site effects. Some of the differences may relate to other strategies of nitrogen uptake as shown for 81% of tree species in the *várzea* which have vesicular–arbuscular mycorrhizal symbioses. Roots of *Tabebuia barbata*, for instance, show a high density of intercellular hyphae and vesicles, while *Crateva benthamii* has no mycorrhiza (Meyer, 1991), so that it may be suggested that *Tabebuia* can more easily compete for nitrogen derived from legume trees. For the nodulated legumes under study, the nitrogen extent which derived from N_2 fixation was calculated according to Shearer & Kohl (1988). Relatively high standard deviations reflect the intraspecific variability in N_2 fixation activity (Table 3.2), which depends on the micro-symbiont, plant condition and growth stage, as well as on soil properties such as water content, temperature, and nutrient availability (Williams *et al.*, 1988; Carranca *et al.*, 1999). Maximal nitrogen contents derived from the atmosphere were found in the lianas, suggesting that N_2 fixation is an important nitrogen source for these high productive plants. However, legume trees are also able to fix significant amounts of nitrogen as reflected by low $\delta^{15}N$. Because foliar $\delta^{15}N$ is suggested to be related to the residence time of the whole ecosystem nitrogen (Handley *et al.*, 1999), the large number of woody plants under study with low $\delta^{15}N$ points to rapid nitrogen turnover within the Amazon flood plain. It may be concluded therefore, that apart from the nitrogen input from the Solimões River, N_2 fixation seems to be the most powerful mechanism for the nitrogen enrichment in the periodically inundated *várzea* forest.

LAND-USE STRATEGIES ON THE AMAZON FLOOD PLAIN

The Amazon flood plain, which is fed by the white water of the Amazon River is one of the world's most productive ecosystems. Small-scale agriculture and shifting cultivation have been practiced for centuries by indigenous peoples, providing subsistence income while conserving soil, water and forest resources (Coomes & Burt, 1997). The annual nutrient input from the river confers a high agricultural potential to the white-water flood plain. The white water is rich in major solutes but not in nitrogen, which may be limiting for biomass production as described above. Therefore, development of more sophisticated, productive and sustainable management concepts requires a better understanding of the nitrogen dynamics within the seasonally flooded soils. Background temporal variability in soil organic carbon, total nitrogen and available mineral nutrient pools exceeds effects of different land-use treatments, suggesting that soil nutrient pools in this environment are well buffered against the effects of cultivation (Zarin *et al.*, 1998). But physical limitations imposed by flooding and fluvial erosion restrict the potential of agricultural intensification. Annual flooding makes the *várzea* very sensitive to physical compaction and erosion associated with mechanized agriculture, ranching and indiscriminate timber harvesting practices. Therefore, the protection of the area requires a population density and consumption that remains within the limits of human carrying capacity. Increasing pressure on resources is already visible and has been documented several times during the past (Fearnside, 1986, 1997; Etchart, 1997; Kolk, 1998). In recent years the use of resources in the flood plain has led to conflicts between fishing, agriculture and forestry. Besides laws, an active participation of local people is needed to resolve conflicts and to promote a sustainable use of the *várzea*.

In developed countries there has been increasing concern about environmental problems associated with fertilizer use during the last two decades. Therefore alternative nitrogen sources such as N_2 fixation increased in importance and are often applied today in pasture legumes and in rotating crop systems. Grain legumes and forage legumes also play an important role in tropical agriculture as reviewed by Giller & Wilson (1991). To what extent this can also be realized for woody legumes in the Amazon flood plain has to be investigated by future research. Despite the difficulties due to periodical flooding, the development of agroforestry systems that support the cultivation of legumes may open new perspectives for sustainable agriculture in the central Amazon by improving soil fertility, reducing erosion and transferring the fertilizing effects to associated crops. By taking full advantage of the biological potential of the flood plain, the use of nutrient resources transported along the Amazon River will become much more efficient.

REFERENCES

Adis, J. (1992) How to survive six months in a flooded soil: strategies in chilopoda and symphyla from central Amazonian flood plains. *Studies on Neotropical Fauna and Environment* **27**, 117–129.

Bollmann, A. & Conrad, R. (1998) Influence of O_2 availability on NO and N_2O release by nitrification and denitrification in soils. *Global Change Biology* **4**, 387–396.

Brandes, J. A., McClain, M. E. & Pimentel, T. P. (1996) ^{15}N evidence for the origin and cycling of inorganic nitrogen in a small Amazonian catchment. *Biogeochemistry* **34**, 45–56.

Campbell, C. A., Myers, R. J. K. & Curtin, D. (1995) Managing nitrogen for sustainable crop production. *Fertilizer Research* **42**, 277–296.

Carranca, C., Devarennes, A. & Rolston, D. E. (1999) Biological nitrogen fixation estimated by ^{15}N dilution, natural ^{15}N abundance, and N difference techniques in a subterranean clover–grass sward under Mediterranean conditions. *European Journal of Agronomy* **10**, 81–89.

Coomes, O. T. & Burt, G. J. (1997) Indigenous market-oriented agroforestry: dissecting local diversity in western Amazonia. *Agroforestry Systems* **37**, 27–44.

Devol, A. H., Forsberg, B. R., Richey, J. E. & Pimentel, T. P. (1995) Seasonal variation in chemical distributions in the Amazon (Solimões) River: a multiyear time series. *Global Biogeochemical Cycles* **9**, 307–328.

Döbereiner, J., Urquiaga, S. & Boddey, R. M. (1995) Alternatives for nitrogen nutrition of crops in tropical agriculture. *Fertilizer Research* **42**, 339–346.

Doyle, R. D. & Fisher, T. R. (1994) Nitrogen fixation by periphyton and plankton on the Amazon floodplain at Lake Calado. *Biogeochemistry* **26**, 41–66.

Etchart, G. (1997) Sustainable resource management in the Brazilian Amazon: the case of the community of Tiningu. *Coastal Management* **25**, 205–226.

Fearnside, P. M. (1986) *Human Carrying Capacity of the Brazilian Rainforest*. Columbia University Press, New York.

Fearnside, P. M. (1997) Human carrying capacity estimation in Brazilian Amazonia as a basis for sustainable development. *Environmental Conservation* **24**(3), 271–282.

Furch, K. (1997) Chemistry of várzea and igapó soils and nutrient inventory of their floodplain forests. In: *The Central Amazon Floodplain: Ecology of a Pulsing System* (ed. by W. J. Junk), 47–67. Ecological Studies Publ. no. 126, Springer-Verlag, Berlin.

Furch, K. (1999) Zur Biogeochemie eines charakteristischen Überschwemmungsgebietes Zentralamazoniens, der Várzea auf der Ilha de Marchantaria nahe Manaus, Brasilien. *Habilitationsschrift*. Universität Hamburg.

Giller, K. E. & Wilson, K. J. (1991) *Nitrogen Fixation in Tropical Cropping Systems*. CAB International, Wallingford, Oxfordshire, UK.

Handley, L. L., Austin, A. T., Robinson, D., Scrimgeour, C. M., Raven, J. A., Heaton, T. H. E., Schmidt, S. & Stewart, G. R. (1999) The ^{15}N natural abundance ($\delta^{15}N$) of ecosystem samples reflects measures of water availability. *Australian Journal of Plant Physiology* **26**, 185–199.

Howarth, R. W., Billen, G., Swaney, D., Townsend, A., Jaworski, N., Lajtha, K., Downing, J. A., Elmgren, R., Caraco, N., Jordan, T., Berendse, F., Freney, J., Kudeyarov, V., Murdoch, P. & Zhu, Z. L. (1996) Regional nitrogen budgets and riverine N & P fluxes for the drainages to the North Atlantic Ocean: natural and human influences. *Biogeochemistry* **35**, 75–139.

Irion, G., Junk, W. J. & Mello de, J. A. S. N (1997) The large central Amazonian river flood plains near Manaus: geological, climatological, hydrological and geomorphological aspects. In: *The Central Amazon Floodplain: Ecology of a Pulsing System* (ed. by W. J. Junk), 23–46. Ecological Studies Publ. no. 126, Springer-Verlag, Berlin.

Junk, W. J. & Piedade, M. T. F. (1993) Biomass and primary-production of herbaceous plant communities in the Amazon floodplain. *Hydrobiologia* **263**, 155–162.

Junk, W. J., Bayley, P. B. & Sparks, R. E. (1989) The flood pulse concept in river-floodplain systems. In: *Proceedings of the International Large River Symposium* (ed. by D. P. Dodge), 110–127. Canadian Special Publication on Fisheries and Aquatic Science no. 106, NRC Research Press, Ottawa.

Kadiata, B. D., Mulongoy, K. & Isirimah, N. O. (1996) Time course of biological nitrogen fixation, nitrogen absorption and biomass accumulation in three woody legumes. *Biological Agriculture and Horticulture* **13**, 253–266.

Kern, J. & Darwich, A. (1997) Nitrogen turnover in the várzea. In: *The Central Amazon Floodplain: Ecology of a Pulsing System* (ed. by W. J. Junk), 119–135. Ecological Studies Publ. no. 126, Springer-Verlag, Berlin.

Kern, J., Darwich, A., Furch, K. & Junk, W. J. (1996) Seasonal denitrification in flooded and exposed sediments from the Amazon floodplain at Lago Camaleão. *Microbial Ecology* **32**, 47–57.

Kern, J., Darwich, A. & Furch, K. (1998) The contribution of gaseous nitrogen flux in the nitrogen budget on the Amazon floodplain at Lago Camaleao. In: *Proceedings of the International Association of Theoretical and Applied Limnology*, vol. 26, 926–928. Schweizerbart'sche Verlagsbuchhandlung, Stuttgart.

Kern, J., Darwich, A. & Förstel, H. (2000) Studies on the role of N_2 fixation in the floodplain forest in the central Amazon. In: *Proceedings of the International Association of Theoretical and Applied Limnology*, vol. 27, 610–614. Schweizerbart'sche Verlagsbuchhandlung, Stuttgart.

Kilian, S. & Werner, D. (1996) Enhanced denitrification in plots of N_2-fixing faba beans compared to plots of a non-fixing legume and non-legumes. *Biology and Fertility of Soils* **21**, 77–83.

Kolk, A. (1998) From conflict to cooperation: international policies to protect the Brazilian Amazon. *World Development* **26**, 1481–1493.

Martinelli, L. A., Victoria, R. L., Devol, A. H., Richey, J. E. & Forsberg, B. R. (1989) Suspended sediment load in the Amazon basin: an overview. *GeoJournal* **19.4**, 381–389.

Martinelli, L. A., Victoria, R. L., Trivelin, P. C. O., Devol, A. H. & Richey, J. E. (1992) ^{15}N natural abundance in plants of the Amazon River floodplain and potential atmospheric N_2 fixation. *Oecologia* **90**, 591–596.

Martinelli, L. A., Piccolo, M. C., Townsend, A. R., Vitousek, P. M., Cuevas, E., McDowell, W., Robertson, G. P., Santos, O. C. & Treseder, K. (1999) Nitrogen stable isotopic composition of leaves and soil: tropical versus temperate forests. *Biogeochemistry* **46**, 45–65.

Mertes, L. A. K. (1994) Rates of flood-plain sedimentation on the central Amazon River. *Geology* **22**, 171–174.

Meyer, U. (1991) Feinwurzelsysteme und Mykorrhizatypen als Anpassungsmechanismen in zentralamazonischen Überschwemmungswäldern—Igapó und Várzea. PhD Thesis, Universität Hohenheim.

Peierls, B. L., Caraco, N. F., Pace, L. & Cole, J. (1991) Human influence on river nitrogen. *Nature* **350**, 386–387.

Piedade, M. T. F., Junk, W. J. & Long, S. P. (1991) The productivity of the C_4 grass *Echinochloa polystachya* on the Amazon floodplain. *Ecology* **72**, 1456–1463.

Saint-Paul, U. (1992) Status of aquaculture in Latin America. *Journal of Applied Ichthyology* **8**, 21–39.

Salati, E., Sylvester-Bradley, R. & Victoria, R. L. (1982) Regional gains and losses of nitrogen in the Amazon basin. *Plant and Soil* **67**, 367–376.

Shearer, G. & Kohl, D. H. (1988) Natural ^{15}N abundance as a method of estimation the contribution of biologically fixed nitrogen to N_2-fixing systems: potential for non-legumes. *Plant and Soil* **110**, 317–327.

Sippel, S. J., Hamilton, S. K. & Melack, J. M. (1992) Inundation area an morphometry of lakes on the Amazon River floodplain, Brazil. *Archiv für Hydrobiologie* **123**, 385–400.

Sprent, J. I., Geoghegan, I. E., Whitty, P. W. & James, E. K. (1996) Natural abundance of ^{15}N and ^{13}C in nodulated legumes and other plants in the cerrado and neighbouring regions of Brazil. *Oecologia* **105**, 440–446.

Villar, C. A., Cabo, D. L., Vaithiy Anathan, P. & Bonetto, C. (1998) River–floodplain interactions: nutrient concentrations in the Lower Parana River. *Archiv für Hydrobiologie* **142**, 433–450.

Waldhoff, D., Junk, W. J. & Furch, B. (1998) Responses of three central Amazonian tree species to drought and flooding under controlled conditions. *International Journal of Ecology and Environmental Sciences* **24**, 237–252.

Williams, M. R. & Melack, J. M. (1997) Solute export from forested and partially deforested catchments in the central Amazon. *Biogeochemistry* **38**, 67–102.

Williams, W. M., Hoh, C. H., Lenz, F. & Broughton, W. J. (1988) Rhizobia in tropical legumes: environmental factors and the reduction of nitrogen. *Soil Biology and Biochemistry* **20**(5), 667–675.

Zarin, D. J., Duchesne, A. L. & Hiraoka, M. (1998) Shifting cultivation on the tidal floodplain of Amazonia: impacts on soil nutrient status. *Agroforestry Systems* **41**, 307–311.

CHAPTER 4

TOTAL MERCURY DISTRIBUTION AND IMPORTANCE OF THE BIOMAGNIFICATION PROCESS IN RIVERS OF THE BOLIVIAN AMAZON

Laurence Maurice-Bourgoin & Irma Quiroga

Mercury contamination in humans and edible fish and its distribution in waters and soils are an environmental problem of increasing concern in the Amazon basin. Mercury (Hg) released from gold-mining activities was commonly held responsible for the Hg contamination of aquatic ecosystems (Malm *et al.*, 1990; Nriagu *et al.*, 1992; Pfeiffer *et al.*, 1993; Malm *et al.*, 1995).

But recent studies have shown the importance of pre-anthropogenic sources in the elevated mercury concentrations measured in the superficial mineral horizons of remote forested oxisols (Roulet & Lucotte, 1995). Considering the relative influence of anthropogenic Hg, Roulet *et al.* (1999a) calculated, for example, that more than 97% of the Hg accumulated at the soil surface of the Tapajós basin is pre-anthropogenic. In Andean tributaries of the Amazon River, the importance of the soil erosion process during the rainy season in the mercury content of freshwaters has been clearly demonstrated (Maurice-Bourgoin *et al.*, 2000b). In the Coroico River, an Andean river, the total Hg (THg) concentrations range from few ng l-1 in the dry season to 169 ng l-1 at high stage and the Hg associated with the sediment particles represents 49 and 92% of the THg.

In the case of the Madeira River basin, there are two main sources of mercury: (a) natural emissions, as long-term low-level atmospheric inputs from natural processes such as volcanism, and important terrestrial inputs due to the erosion of contaminated soils and to the weathering of mercury-bearing rocks during the rainy season. These soils may have become contaminated through natural crustal degasification and dissolution of Hg, and (b) from anthropogenic emissions during gold-mining activities. The environmental impact of these activities is due more to the deforestation of the exploited areas and the release of thousands of tonnes of sediment in exploited rivers, than to the direct release of Hg used during the last stages of the gold extraction. We have estimated that in Bolivia (Maurice-Bourgoin *et al.*, 2000a), more than 330 t of mercury have been released into the environment since 1952 (assuming a mean Au:Hg ratio of 1:5). In 1997, more than 60 000 Bolivian people were directly involved in gold mining, as employees of some 1200 mining companies or members of around 300 local cooperatives. Most of the gold-mining activities in the upper Beni River basin take place in the rivers Tipuani, Mapiri and K'aka (Fig. 4.1). Here, approximately 200 cooperatives extract between 5 and 10 kg of gold per month during recent years and 100 kg per month during the best years (1993–1994). Hence it may be surmised that 250–500 kg of mercury is used each year and that 50–70% of this is directly released into the environment via rivers, soils and atmosphere. Depending upon the mining company, mercury is not always recovered and 5–45% of the total mercury used (Malm *et al.*, 1990; Maurice-Bourgoin *et al.*, 1999) can be released into the river itself. Finally, mercury passes into the atmosphere by open-air burning of the amalgam. The simplicity of this technique explains its widespread use in small-scale mining in developing countries. Rough field conditions, absence of monitoring and available data precludes accurate estimation of the gold extracted, and then the mercury released.

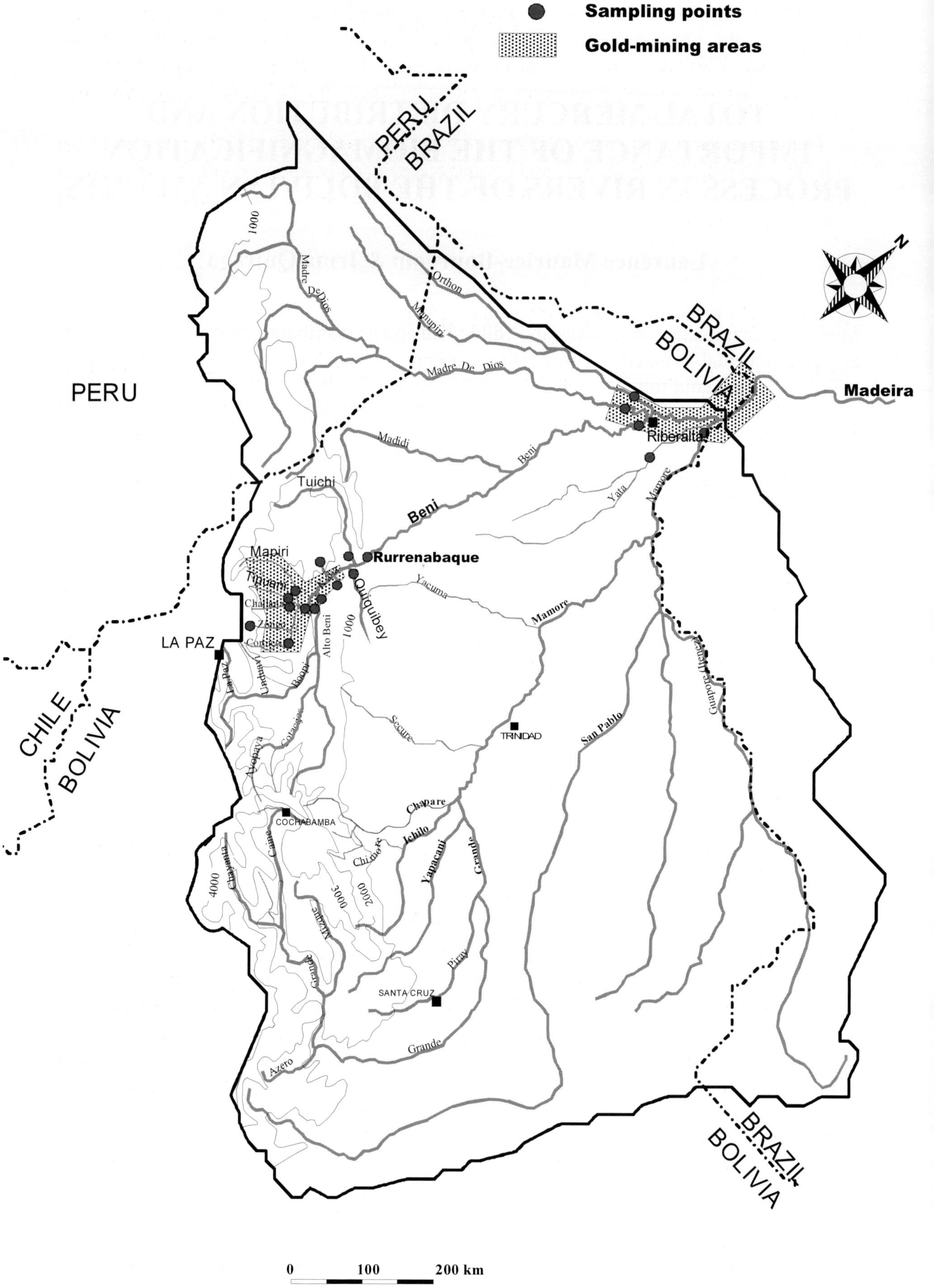

Fig. 4.1 Hydrological map of the Bolivian Amazon basin showing the gold-mining areas and the sampling points.

Based on the results we obtained from the suspended matter and the soils of the main tributaries of the Amazon River, in the Madeira (Maurice-Bourgoin *et al.*, 1997), Tapajós (Roulet *et al.*, 1998) and Negro (Silva-Forsberg *et al.*, 1999) basins, we suggest that there are preferential storage basins depending on their climatology (wind and rain), hydrology (presence of flood plain) and soil type.

The objectives of this study are to analyse the Hg distribution in surface waters of the Bolivian Madeira basin (Fig. 4.1) and to identify contaminated areas and the fish species which present a major contamination risk. Some results of the impact on riparian populations are also presented.

DESCRIPTION OF THE STUDY AREA

The Bolivian Amazon basin (Fig. 4.1) represents the Andean headwaters of the Madeira River, one of the four most important tributaries of the Amazon River. In Bolivia, its drainage basin is 0.9×10^6 km^2, with 25% in the Andes, 27% on the Brazilian shield and 48% in the plain. The Andean cordillera uplift led to subsidence of the shield causing the formation of a subsidence "deep" on the Andean piedmont, where sediments easily accumulate. The study of the biotic compartments contamination is located in the upper Beni River in Rurrenabaque, at the edge of the Bolivian Andean piedmont. Heights above sea level range from 6400 m at the Zongo headwaters to 200 m at Rurrenabaque and 115 m at the Brazilian border. Andean tributaries of the Beni and Mamore rivers drain both semiarid areas of high altitude and areas of tropical humid forest of the piedmont. Andean rivers export an average of 500×10^6 t of sediments per year, and half of this is transported by the Beni River in Rurrenabaque (Guyot, 1993).

CONCENTRATIONS AND DISTRIBUTION OF MERCURY SPECIES IN SURFACE WATERS

Recent improvements of sampling, storage and analytical methods have improved the detection limits of the total Hg concentration (THg) in natural waters and have generated a growing database of aqueous methylmercury (MeHg) concentrations that were not available a decade ago. The ultra-clean techniques (Ahlers *et al.*, 1990) used for sampling and storage compiled with the extreme sensitivity of fluorescence atomic spectrometry now allows the publication of reliable data on Hg concentrations in the Amazon basin (Vinson *et al.*, 1997; Maurice-Bourgoin *et al.*, 1997, 1999, 2001, 2002; Roulet *et al.*, 1998) and then to correct the high values published around 1990 (compiled by Lacerda and Salomons in Pfeiffer *et al.*, 1989, 1991). Total Hg concentrations in surface waters of the Amazon basin range from 1 to 30 ng l^{-1}; higher values can be measured directly at gold-mining sites (Telmer *et al.*, 1999) or during the first floods in the Andean tributaries characterized by a very high suspended solids content (Maurice-Bourgoin *et al.*, 2000b).

Methylmercury which is the prevalent Hg species in animal flesh, usually constitutes only a minor fraction of THg in natural waters; its proportion varies widely from <1% to >30% of THg. The proportion of MeHg appears to increase with the concentration of organic matter and to decrease with the content in mineral particles. It is typically 0.5–5% in turbid rivers, 1–10% in clear-water drainage lakes, 3–30% in most humic and some eutrophic lakes and streams (Meili, 1997). Around wetlands, over flooded soils and in the bottom of stratified lakes, these ranges are higher than the 1–2% observed in precipitation, indicating significant terrestrial or aquatic sources of MeHg, notably in environments rich in organic matter.

The transport, the speciation and the transformation of Hg in most freshwaters are largely controlled by the cycling of organic matter. But, in environments poor in organic matter, for example in high turbid waters (in white-water rivers of the Madeira basin), oxyhydroxides of Fe, Al and Mn, reduced sulphur compounds, and clay minerals are important ligands for heavy metals. Soil profiles (Porcella *et al.*, 1995; Roulet & Lucotte, 1995; Roulet *et al.*, 1998) suggest that the mass-related affinity of Hg for Fe oxyhydroxides alone is over 10-fold higher than for organic matter alone.

METHODOLOGY

Water samples were collected using decontaminated Teflon bottles and stored in polyethylene bags, at 4°C, until filtration. All handling operations were performed using "ultra-clean" techniques (Ahlers *et al.*, 1990; Nolting & De Jong, 1994; Gaudet *et al.*, 1995), including a portable laminar flow hood to avoid contamination. Water samples were filtered between 1 and 6 h after sampling on pre-washed (5% v/v distilled HNO_3) and pre-burned membranes (Whatman QM/A). The dissolved fractions were kept in Teflon flasks, and immediately stabilized with distilled HCl (5%).

Water samples were analysed in an aqueous matrix decontaminated by argon bubbling and acidified with H_2SO_4 (1% v/v). Particulate mercury retained on the filters was solubilized using decontaminated and acidified seawater. Samples were treated with ultra sound for 40 minutes, to strip particles from the membranes. Organic mercury complexes were broken down by addition of 50 μl of $KMnO_4$ (6‰) (Quémerais & Cossa, 1995).

Atomic fluorescence spectrophotometry (Merlin PSA) was used for the detection of dissolved and particulate mercury in water samples. Total mercury was reduced to

Table 4.1 Mercury concentrations and specific fluxes in surface waters in the dissolved (DHg) and particulate (PHg) fractions of the Madeira River and its tributaries, from the Andean headwaters to its confluence with the Amazon; values obtained at the end of the dry season.

	Area of the drainage basin (km^2)	Suspended solids (mg l^{-1})	DHg (ng l^{-1})	PHg (ng g^{-1})	THg (ng l^{-1})	Partition coefficient log K_d	Total Hg flux (mg day^{-1} km^{-2})
Glacial water							
Zongo	2	2.0–2.5	1.95–2.40	121–145	2.24–2.57	4.70–4.87	5.70–6.55
Andean tributaries exploited for their alluvial gold							
Tipuani	1 400	188–338	1.80–3.80	5–7	3.11–5.37	3.09–3.59	5.98–8.67
Mapiri	10 100	32–135	1.30–6.15	7–26	2.25–6.99	3.63–3.73	1.07–5.52
K'aka	18 800	62–95	2.90–5.45	7–50	3.50–6.08	3.27–3.34	3.74–6.49
Edge of the Andean piedmont							
Beni (Rurrenabaque)	67 500	126–233	3.60–6.55	7–61	7.22–19.09	3.04–5.06	9.47–9.52
Amazonian plain							
Beni (Riberalta)	243 000	167–537	2.55–6.03	2–8	3.77–7.64	2.95–3.10	2.57
Madeira (Porto Velho)	954 285	122	5.80	10	7.00	3.23	7.33
Madeira (Manicore)	1 123 670	149	8.50	13	10.43	3.18	11.76
Madeira (Borba)	1 328 581	47	8.60	20	9.51	3.36	12.22
Confluence with the Amazon							
Madeira (Foz Amazonas)	1 370 000	33–38	5.55–10.40	140–209	9.49–10.86	3.12	4.27–3.01

elemental mercury by addition of $SnCl_2$, vaporized by argon bubbling, and transported by an argon current to a gold trap and detected by atomic fluorescence spectrophotometry at 253.7 nm. Analyses were conducted in triplicate. Reproducibility was 0.1–2% and accuracy was 5 pg. The reactive blanks were in the same range as the limit of detection fixed by the argon blank (5 pg Hg). The acid blanks averaged 8 pg l^{-1}, which represents a contribution of 0.1% to the aliquots of 10 ml. The accuracy of the particulate mercury analysis was limited by the standard deviation of membrane blanks, which reached 25 pg Hg.

TOTAL Hg DISTRIBUTION IN RIVERS OF THE BOLIVIAN AMAZON

During the low water period, total mercury concentrations measured in surface waters of the Madeira River basin varied from 2.24 to 2.57 ng l^{-1} in glacial waters of the Zongo River and from 9.49 to 10.86 ng l^{-1} at its confluence with the Amazon River (Table 4.1). During the dry season, in waters of the Andean rivers exploited for their alluvial gold, total mercury concentrations ranged from 2.25 to 6.99 ng l^{-1}. We can note an increasing of the mercury concentration in surface waters from the Andean tributaries of the Madeira River to its confluence with the Amazon. But in the dry season, the highest mercury concentrations and fluxes were not found in the rivers where the mining activities take place but at the outlet of the Andean sub-basins exploited for their alluvial gold, 200 km downstream (Fig. 4.2). The low quantity of mercury directly released into the streams, the local hydrological regimes and the dynamic of the sediment transport can explain this finding. The elemental form of Hg does not contaminate directly the rivers exploited for their alluvial gold but can travel for hundreds of kilometres before it is deposited in different drainage basins to those of its emission. At the Andean piedmont where high concentrations of mercury have been measured, we can observe (a) high values of precipitation in Bolivia, (b) a subsidence zone

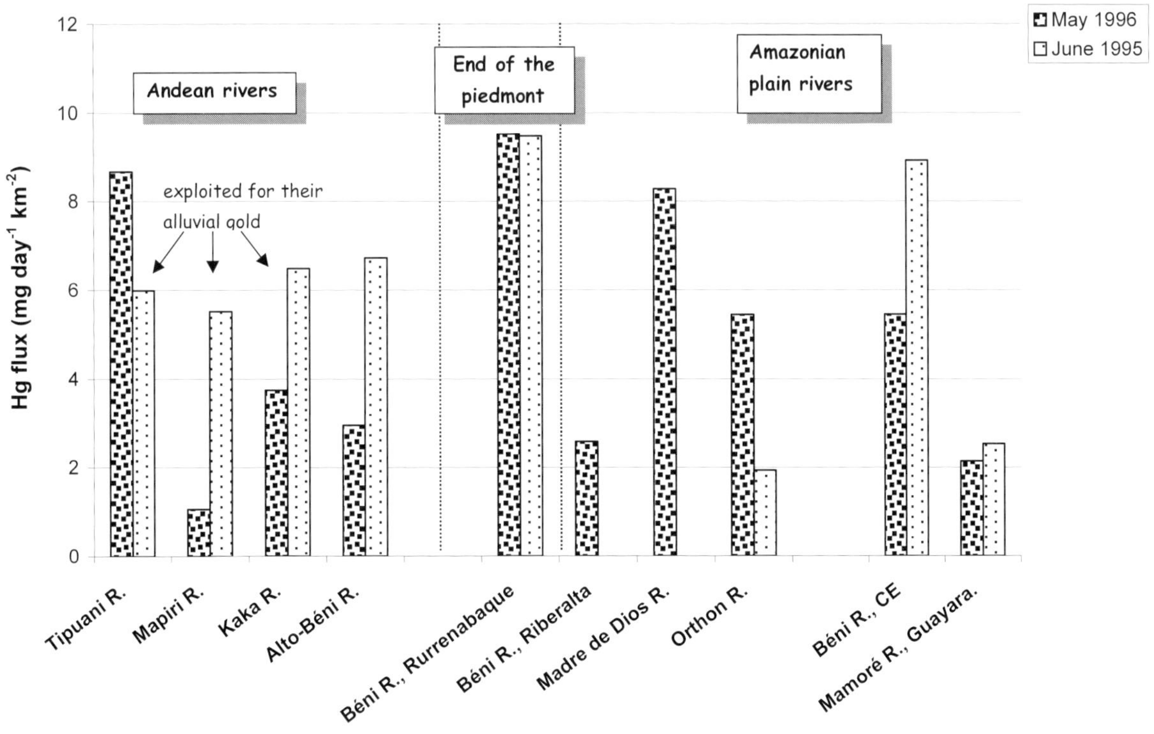

Fig. 4.2 Total mercury fluxes normalized by the drainage basin area along the Andean tributaries of the Madeira River, during the dry season.

where particles transported from the Andes are deposited (c) drainage basins characterized by soils enriched in iron oxyhydroxides (Tuichi and Quiquibey basins) and by natural Hg from rocks like black shales. During the wet season, from November to March, particles are transported from the Andean sub-basins, characterized by steep slopes and by high erosion rates enhanced by recent agriculture practices, to the Amazonian plain (Maurice-Bourgoin *et al.*, 2000b). The high adsorption capacity of mercury especially on fine particles (clays), and the stability of its carbon binders are the reasons why, in any hydrosystem, most of the mercury is transported in particulate matter (Benes & Havlik, 1979; Rae & Aston, 1982; Langston, 1982; Maurice-Bourgoin *et al.*, 1999). This occurs mainly during the rainy season, when soil erosion is most significant. In the white waters of the Andean rivers, dissolved organic carbon concentrations are reduced but on the other hand, in the high water season, 3–15% of the total suspended solids (TSS) in the Beni River at Rurrenabaque consist of clays (Guyot, 1993). The values of particulate mercury published here are quite low (Table 4.1) because of the sampling period; since during the dry season, deposition is much more evident than erosion. Guyot (1993) and Maurice-Bourgoin *et al.* (2001) estimated, from two different methodologies, that around 40% of the suspended particles are deposited in the Amazonian plain of the Beni River.

In agreement with our results, in the Tapajós River, Roulet *et al.* (1998) found that the Hg content in the water column is independent of gold-mining activities upstream.

PARTITIONING OF Hg BETWEEN DISSOLVED AND PARTICULATE FRACTIONS IN SURFACE WATERS

Dissolved mercury concentrations increase from upstream to downstream, with mean values ranging from 2 to 10 ng l^{-1} in glacial waters and at the confluence of the Madeira River with the Amazon, respectively (Table 4.1). In contrast to the downstream waters, we measured low particulate mercury values in Andean tributaries exploited for their alluvial gold. This can be explained by the very important burden of old and recent alluvial sediments removed and thrown into the rivers during the gold extraction processes, increasing the concentration of total suspended solids in the waters, which can dilute the mercury concentration of the suspended particles.

The affinity of a given compound for the particulate phase is described by the distribution coefficient K_d defined as: $K_d = Cp/Cd$, where Cp is the particulate Hg content (PHg) in ng kg^{-1} and Cd is the dissolved Hg concentration (DHg) in ng l^{-1}. Higher K_ds suggest a stronger affinity for the particulate fraction. In the white-water rivers of the Bolivian Amazonian basin, characterized by low organic carbon content, log K_d for Hg ranges from 2.95 to 5.06. The lowest log K_d values are calculated for the Madeira River downstream. A negative regression of log K_d for Hg with suspended solids in two Andean rivers (Fig. 4.3) suggests a greater affinity of suspended particulate matter for Hg at low suspended solids concentrations. This could also be due to the "particle concentration effect" (PCE) which is caused by metals associated with colloidal particles but counted with the dissolved fraction (Benoit, 1995). The PCE could be caused by a decrease in surface area and surface complexation sites due to re-suspension of larger particles under high suspended solids concentrations.

It appears that on the one hand, during the dry season mercury emitted by gold-mining activities in Bolivia, at the headwaters of the Amazon basin, does not directly contaminate the exploited rivers. On the other hand, mercury concentrations measured in suspended matter indicate that mercury is preferentially released in the hydrological system during the rainy season and is then transported adsorbed on fine particles in abundance in the Andean

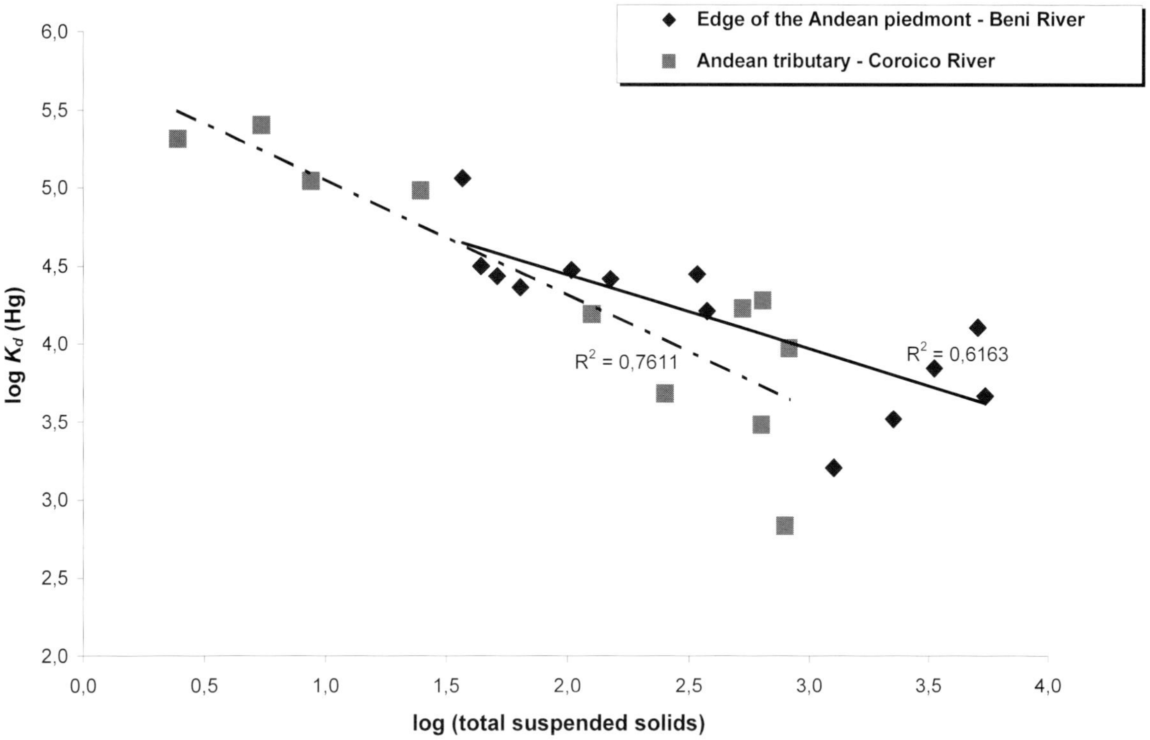

Fig. 4.3 Relationship between the distribution coefficient for Hg in waters of two Bolivian tributaries of the Madeira River and their total suspended solids content.

rivers. Its contamination affects drainage basins downstream characterized by high precipitation, high sedimentation rates, and by soils and sediments enriched in iron or manganese oxyhydroxides.

Even if the hydrodynamic conditions in this area favour the transport downstream and the dispersion of mercury in freshwater, mercury pollution could still become a problem in the Bolivian Amazonian basin because of fish consumption by indigenous riparian communities as a primary protein source.

BIOMAGNIFICATION AND BIOACCUMULATION OF Hg IN FISH

Methodology

Fish specimens of known origin, genera and habitat, were obtained from fishermen or from local markets at Rurrenabaque. The edible portions (flesh) were separated and frozen immediately in liquid nitrogen and kept frozen until analysis.

Fish samples were processed using the method of Agemian & Cheaner (1978). All samples were then neutralized with hydroxylamine (NH_2OH,HCl 12 g l^{-1}).

Cold-vapour generation coupled with atomic absorption spectrophotometry, CV–AAS, (Perkin Elmer 3110), was used for total mercury determination in fish and hair samples. Methylmercury in muscle tissue of fish was determined by chromatography, ECD-GC, after dithizone-benzene extraction (Akagi *et al.*, 1995; Kehrig *et al.*, 1998). Instruments were calibrated each day of the analysis.

Reproducibility (0.5–7%) of the mercury determination by atomic absorption spectrophotometry was checked in triplicate and by intercalibration with standard samples supplied by the Swedish Food Administration. The detection limit for fish was 0.060 μg g^{-1}.

RESULTS OF TOTAL Hg AND METHYLMERCURY IN FISH OF THE BENI, TUICHI AND QUIQUIBEY RIVERS

Between November 1996 and October 1999, 92 fish specimens belonging to 24 different species were analysed, having both different diets and habitats. All these species are consumed by the local population. The distribution of mercury in fish depends mainly on their dietary habits. The feeding habitats were determined for each species from the stomach content analysis and literature review (Mendes dos Santos *et al.*, 1984; Lowe-McConnell, 1987). The concentrations of Hg (Table 4.2) found in non carnivorous fish range from 0.009 $\mu g\ g^{-1}$ (wet weight) in a herbivorous species, Tambaqui (*Colossoma macropomum*) to 0.193 $\mu g\ g^{-1}$ (wet weight) in an omnivorous species, Giro (*Oxydoras* sp.). In piscivorous fish, total mercury concentrations varied from 0.217 $\mu g\ g^{-1}$ (wet weight) for Pintado (*Pseudoplatystoma fasciatum*), up to 2.304 $\mu g\ g^{-1}$ (wet weight) for Surubi (*Pseudoplatystoma tigrinum*) with an average for 10 specimens of 0.986 μg Hg g^{-1}. The last results indicate that the mercury concentration can exceed the WHO safety limit (WHO, 1976) by more than 4-fold (Fig. 4.4). The methylmercury content in omnivorous and piscivorous fish ranges respectively from 71 to 94% of THg and from 86 to 100% of THg; the highest proportion has been measured in large predatory fish (*Pseudoplastytoma tigrinum*) at high trophic level. No significant correlation, for the same species, between total mercury concentration and weight was observed; for example, for the Surubi (*P. tigrinum*), a piscivorous species, we can observe (Fig. 4.4) an increasing tendency of total Hg content with weight but the number of individuals is insufficient to be significant. These species, *Pseudoplastystoma tigrinum* and *fasciatum,* can migrate along the river for several hundred kilometres, especially during the high water season for reproduction. Between high and low waters, their dietary habits may change but most piscivorous species eat fish in abundance in the river, mud feeders, detritivorous or other carnivorous fish. For a species with the same trophic level, the mercury content could be dependant on many factors, including age, sex and fat content.

BIOAVAILABILITY AND BIOCONCENTRATION

The large range of MeHg observed in freshwater ecosystems reflects the combined natural variation in the input and the transformation of Hg and in the bioavailability of MeHg. A commonly used tool to describe bioavailability is the bioconcentration factor (BCF), which is defined by the concentration ratio between the tissue of an aquatic organism (ng kg^{-1} wet weight) and the surrounding water (ng l^{-1}). The BCF corresponds to the volume of water that contains the same amount of a substance as a given unit of biomass.

In the Beni River, at the edge of the Bolivian Andean piedmont (Rurrenabaque), the BCF for omnivorous fish at the end of the dry season, calculated from the dissolved Hg concentration in waters, ranges from 5190 to 130 000 l (kg organic wet weight)$^{-1}$.

This ratio is lower than that calculated for freshwater lake systems, rich in organic matter, which is on the order of 10^6 l (kg organic wet weight)$^{-1}$ and which can correspond to a direct uptake of MeHg dissolved in waters. It appears that the BCF should be strongly related to the organic carbon content of waters. For example, in French Guyana, in the Leblond River, the BCF observed for the piscivorous species *Hoplias aimara* reaches 150 000 for a ratio calculated from the DHg concentration in waters and 15×10^6 for calculations made using the MeHg concentration in waters (Boudou *et al.*, 2000). In another clear-water river, the Tapajós, with a higher dissolved organic carbon content than the Beni River, data published by Roulet et al. (1998) permit us to estimate a BCF range for omnivorous fish

Table 4.2 Total mercury concentrations in fish captured in the Beni River and its piedmont affluents, from November 1996 to December 1998.

Sample origin	Family	Scientific name (common name)	*n*	Dietary habits	Weight range (g)	Hg concentration range (µg g^{-1} wet weight)	Mean Hg (µg g^{-1} wet weight)	Standard deviation range
Beni River (Rurrenabaque area)	Characidae	*Hydrolicus* sp. (Cachorro)	5	P	2500–3900	0.534–1.485	0.989	0.050–0.010
		Brycon sp. (Llorona)	1	O	830		0.034	0.0002
	Curimatidae	*Prochilodus nigricans* (Sabalo)	8	M	120–1830	0.020–0.169	0.078	0.001–0.005
	Serrasalmidae	*Piaractus brachypomum* (Pacu)	3	H	2400–4200	0.022–0.147	0.094	0.001–0.018
		Colossoma macropomum (Tambaqui)	6	H	510–4200	0.010–0.145	0.094	0.002–0.011
		Mylossoma duriventre (Jatara)	7	H	160–1175	0.010–0.086	0.042	0.007
		Pygocentrus nattereri (Palometa)	2	P	1170–1200	1.206–1.233	1.219	0.040–0.047
	Cetopsidae	*Hemicetopsis* sp. (Pez ciego)	1	P			1.370	0.094
	Doradidae	*Oxydoras* sp. (Giro)	1	O	4200		0.193	0.006
	Loricariidae	*Pterygoplichthys multiradiatus* (Zapato)	3	M	415–610	0.028–0.176	0.021	0.001–0.002
	Pimelodidae	*Pimelodus* sp. (Bagre)	2	P–O		1.224–1.819	1.522	0.106–0.130
		Practocephalus sp. (Coronel)	1	P	6900		1.089	0.140
		Brachyplatystoma flavicans (Dorado)	6	P	1190–16000	0.290–1.476	1.021	0.020–0.029
		Pseudoplastystoma fasciatum (Pintado)	5	P	200–10400	0.217–0.579	0.374	0.011–0.042
		Pseudoplastystoma tigrinum (Surubi)	10	P	1300–16300	0.456–2.304	0.986	0.016–0.133
		Pimelodus maculatus blochii (Griso)	3	P–O	125–260	0.100–0.468	0.236	0.011
Quiquibey River	Characidae	*Triportheus angulatus* (Panete)	3	O	100–150	0.098–0.182	0.137	0.008–0.010
	Serrasalmidae	*Colossoma macropomum* (Tambaqui)	1	H	6400		0.021	0.001
		Mylossoma duriventre (Jatara)	1	H	700		0.075	0.006
		Serrasalmus spilopleura (Piraña)	1	P–O	120		0.049	0.007
	Pimelodidae	*Pimelodus* sp. (Bagre)	1	P–O			0.125	0.010
	Curimatidae	*Prochilodus nigricans* (Sabalo)	1	M			0.039	0.003
Tuichi River	Serrasalmidae	*Colossoma macropomum* (Tambaqui)	1	H	4500		0.009	0.001
		Mylossoma duriventre (Jatara)	1	H	1000		0.034	0.002
	Curimatidae	*Prochilodus nigricans* (Sabalo)	1	M	2500–3900	0.534–1.485	0.055	0.002

H: herbivorous; M: mud feeder; O: omnivorous; P: piscivorous.

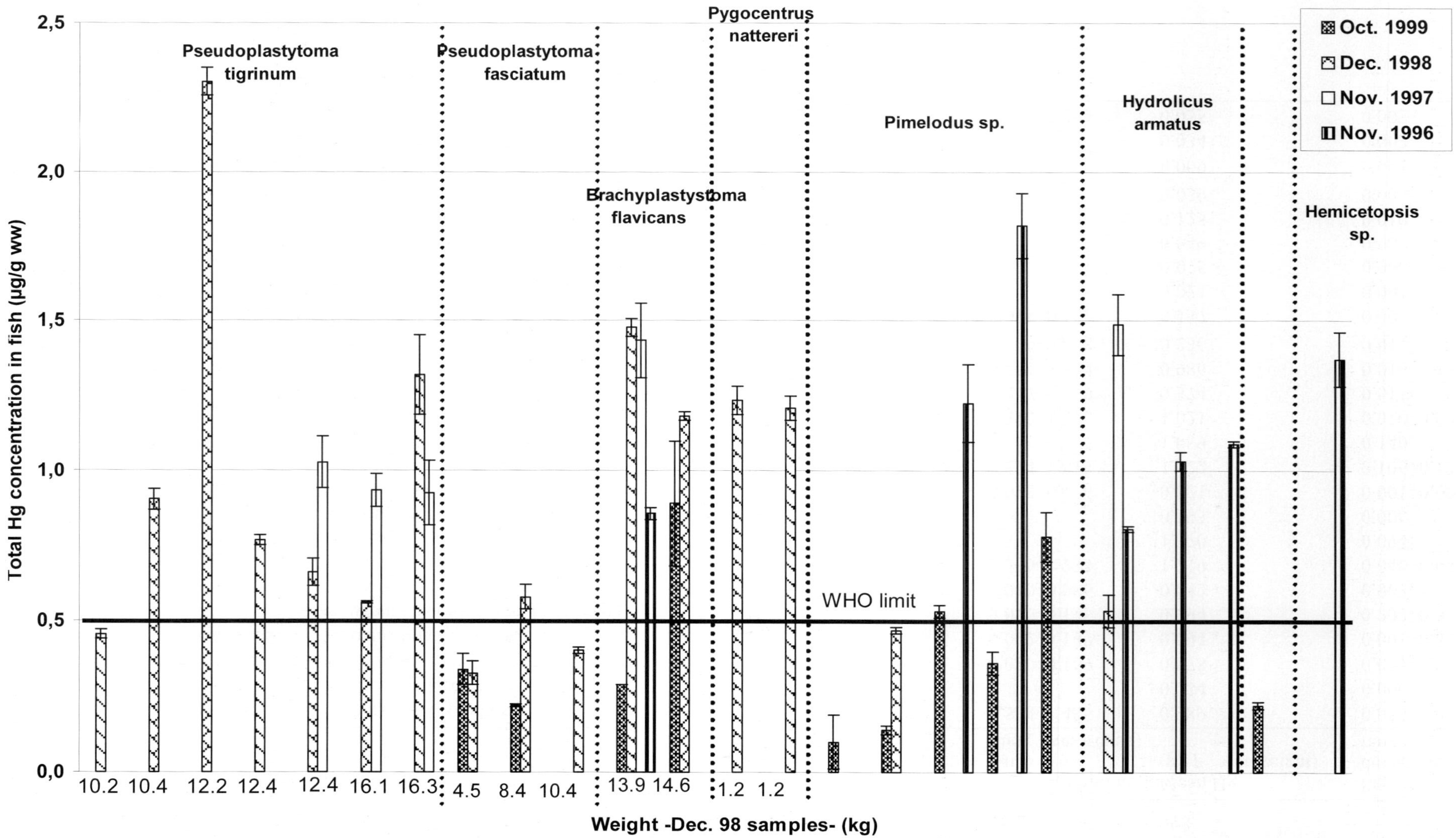

Fig. 4.4 Total mercury concentrations in piscivorous fishes (in µg g^{-1} wet weight) collected in the Beni River (Rurrenabaque area), Bolivian Amazon basin. For each species, results are presented by increasing individual total weight; error bars mean standard deviation among replicate analysis (six per sample).

in waters sampled at the end of the dry season of between $10^{4.19}$ and $10^{5.68}$. Data from the Tapajós River suggest a greater bioavailability of MeHg due to the seasonal increase in water level and the consequent inundation of the flood plain; inundation favours the development of large macrophyte floating mats which increase the bioavailability of epiphyton to herbivorous and detritivorous fish (Guimarães *et al.*, 1996, 1998; Roulet *et al.*, 2000).

BIOMAGNIFICATION IN THE AQUATIC FOOD CHAIN: BASIC CONCEPTS AND *in situ* EVALUATIONS. APPLICATION TO THE BOLIVIAN AMAZONIAN BASIN

Concentration range of Hg in aquatic organisms is very large: varying from 50 to 500 ng g^{-1} (dry weight) in zooplankton species and most invertebrates and from 100 to 20 000 ng g^{-1} (dry weight) (or from 20 to 4000 ng g^{-1} (wet weight), units usually employed for fish) for all fish species and all sizes of fish. The high variability of Hg content in fish is due to the variability in the ecological and physiological factors, their dietary habits which can vary with the season and the species, their age, their sex, their growth rate and sometimes their behaviour—migration for example (Meili, 1991).

The Hg accumulated in aquatic organisms is mainly present in its organic form. MeHg fixation in organisms is 10-fold faster than that of the inorganic Hg contained in food or waters (Meili, 1997). In contrast to inorganic Hg which is mainly accumulated in visceral tissue, MeHg is fairly evenly distributed within animals because of its high potential to cross biological membranes. The increase in the proportion of MeHg/THg, which reaches 0.5–10% in waters and sediments, 50–80% in invertebrates and 80–100% in fish, clearly shows the importance of the bioaccumulation process in the aquatic food chain (Meili, 1991).

Temporal changes in the Hg concentration in organisms for one species are slow relative to the generation or age. Biotic Hg concentrations can be regarded as steady-state quasi equilibria of uptake, depuration and growth dilution. Increases and decreases with age have been observed for different fish species in the Amazon basin. It should be noted that bioaccumulation can be consistent with a decrease in Hg concentration if a biological compartment grows more rapidly than its accumulated content of contaminants; this phenomenon is known as "growth dilution".

The total bioaccumulation can be described by the bioaccumulation factor (BAF), which is analogous to the BCF but includes the uptake from food. In natural ecosystems the BAF of MeHg in animals is much higher than the BCF for direct uptake alone, and can reach values on the order of 10^8 l (kg dry weight)$^{-1}$ in large predators.

As a result of physiological dynamics, MeHg is enriched during the transfer along natural food chains, which is a phenomenon known as biomagnification (Fig. 4.5). This enrichment can be described by the biomagnification factor (BMF) corresponding to the concentration increase for each trophic transfer, which is about 2 to 5-fold for total Hg. This applies to aquatic ecosystems from boreal lakes to tropical rivers and to all typical levels, including microorganisms, invertebrates, invertebrate feeders and piscivores. In freshwater fish, more than 80% of the Hg is MeHg whereas a substantial proportion of the Hg in zooplankton and benthic invertebrates is inorganic Hg, in particular invertebrates associated with the detrital food chain, even in humic waters where most of the inorganic Hg is associated with organic substances and unavailable for direct uptake (Meili, 1997). The BMF reflects a dynamic quasi equilibrium between where uptake is balanced by depuration and growth dilution. Owing to the enrichment of MeHg along food chains, the relative contribu-

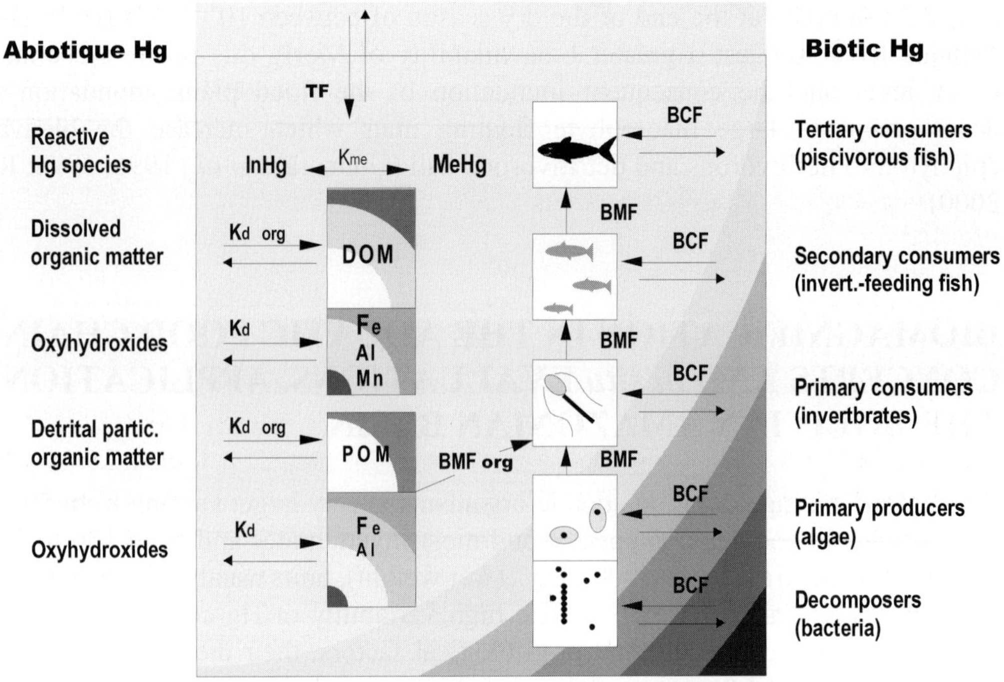

Fig. 4.5 Pathways and factors controlling the environmental partitioning of inorganic and methylated Hg in oxic freshwater ecosystems (from Meili, 1997).

tion from direct uptake from the water declines with trophic level and the uptake from food usually dominates even in primary consumers. A comparison of bioenergetics and MeHg turnover shows that the balance in natural fish populations is controlled by growth dilution rather than depuration of MeHg (Fig. 4.6). The BMF is therefore a function of growth efficiency (growth/ingestion), which in natural fish populations varies between 3 and 30%.

In Bolivia, at the Andean piedmont, Hg content in piscivorous and carnivorous fishes is 10–fold higher than in herbivorous species. The BMF between a piscivorous species which does not migrate, and an omnivorous fish species reaches 2.6.

In the Amazon basin, on the Tapajós River, only four fish species show a linear correlation between Hg concentration and length or weight of the fish (Roulet *et al.*, 1999b): *Cichla ocellaris*, *Hoplias malabaricus*, *Serrasalmus naterreri* and *Raphiodon vulpinus*. Other groups show a positive (*Pellona* sp. and *Hemiodus argenteus*) and negative (*Triportheus* sp., *Schizodon* sp., *Pseudoplatystoma fasciatum* and *Colossoma macropomum*) nonlinear correlation between the Hg content in muscle and the standard

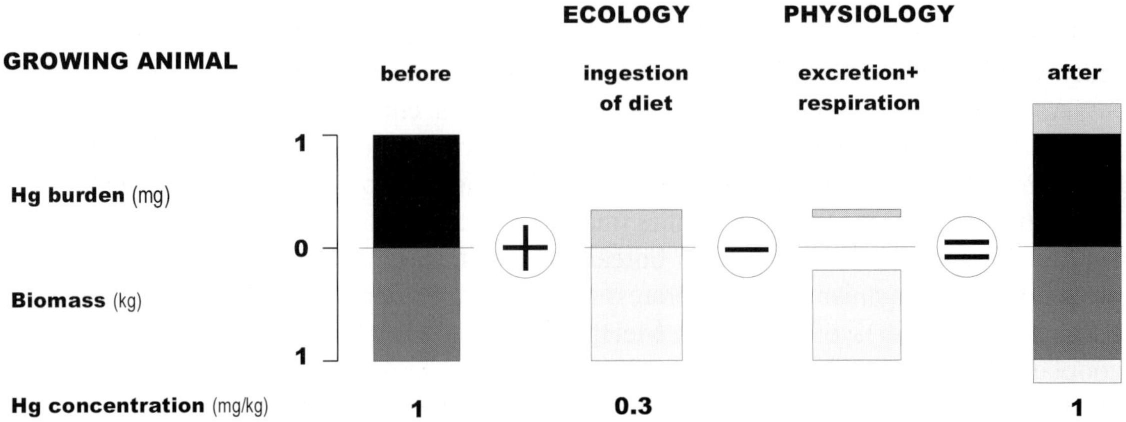

Fig. 4.6 Principal of bioaccumulation, biomagnification and growth dilution in animals illustrated with an example methylated Hg over a period of days and weeks (from Meili, 1997). The Hg concentrations in organisms reflects the balance between Hg and biomass turnover (assimilation, growth or retention, elimination or depuration). Hg concentrations in animals are 2 to 5 times higher than in their diet.

length of fish. A variation in Hg content for length appears to corresponds to the sexual maturity of numerous species, which occurs during migration and diet changes (Roulet *et al.*, 1999b). In Amazonian rivers, the fish growth rate seems to be highly correlated with the water level variations; the food availability tends to increase during the high water period. Sexual maturity and migration periods are the two main factors which highly influence the bioaccumulation of Hg in fish populations, during the year and the life cycle of a fish.

But the study of the bioaccumulation in the food chain in the Amazon basin is still unclear and complicated by the biodiversity and the evolution of the trophic chains in the ecosystem, specifically with the hydrological variations of the rivers and their flood plain. It is impossible to allocate a unique level in the food chain to one fish species.

The results obtained in the Bolivian Amazonian basin show an accumulation of mercury between omnivorous and piscivorous fish species within the food chain. Even if the particulate mercury concentration in the river water is within the range of the natural background level stated in the literature, 72% of carnivorous and piscivorous fish caught in the Beni River, in the Rurrenabaque area, exceed the maximum permissible level. But, we didn't observe a correlation between gold-mining sites and high Hg content in fish; the highest concentrations were not measured where gold mining takes place but 150 km downstream, at the edge of the Andean piedmont. This observation can be explained more by the geomorphology of the study area characterized by a high deposition rate of Andean sediments than by direct Hg discharges from gold-mining operations upstream.

This raises an important problem: as mining activities in this region are declining, many miners have turned to fishing as a livelihood. Fish constitutes an important part of the diet of the local population, especially during the wet season when roads are flooded, and these communities become isolated and economic activity ceases. The major health risk caused by mercury in this area, as in the Madeira River near the Brazilian border, does not just concern the miners by direct inhalation of mercury vapour during the burning process, but concerns all the local population, through the regular consumption of contaminated piscivorous fish.

CONSEQUENCES IN RIPARIAN POPULATIONS

Methodology

Human hair samples were collected from members of indigenous communities, fishermen, miners and from riparian people living near the Beni, Mamore and Madeira rivers. These samples were stored in plastic bags at 4°C.

Hair samples were thoroughly rinsed in 0.01% EDTA solution to remove dust particles, oily substances or other external contamination. Analyses have been undertaken on the total length of each hair sample.

Reproducibility (0.5–7%) of the mercury determination by CV–AAS were checked in triplicate and by intercalibration with standard samples supplied by the Swedish Food Administration. The detection limit for hair by the method used was 1 ng g^{-1}.

Mercury concentrations in 150 human hair samples collected downstream from gold-mining sites in the entire Bolivian Amazonian basin are presented in Fig. 4.7. People studied can be classified into three categories depending on their activities, their dietary habits and their residence areas: gold miners, indigenous communities and riparian people.

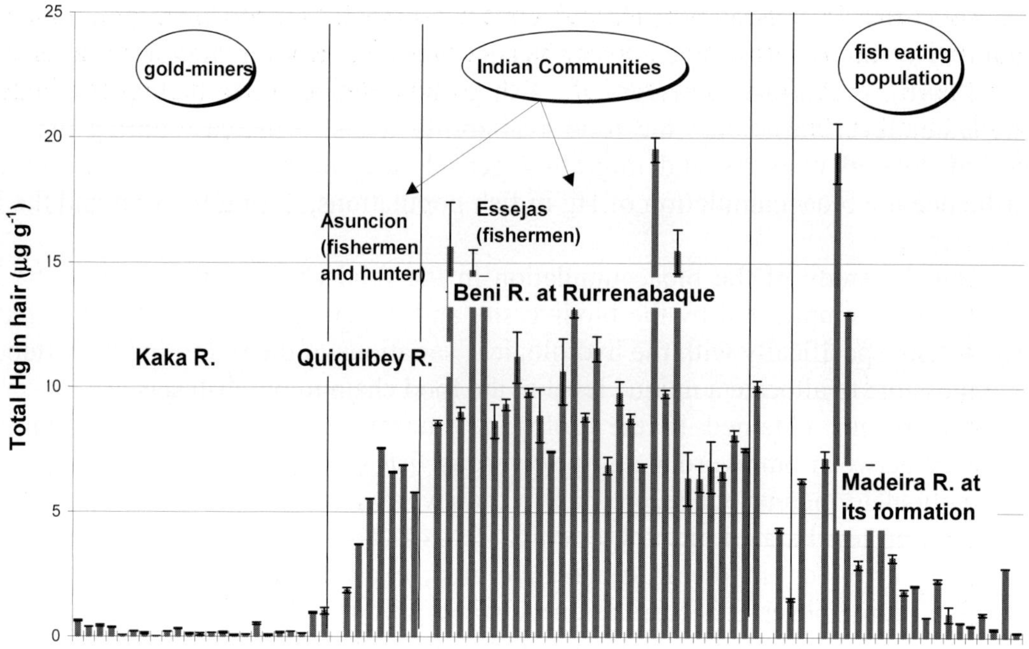

Fig. 4.7 Total mercury concentrations in the hair of people living by the Bolivian tributaries of the Madeira River. Error bars mean variance among nine replicate analysis for each hair sample (from Maurice-Bourgoin *et al.*, 2000a).

Mercury concentrations in miners

Gold miners from three different areas have been studied. The most contaminated people were encountered in the Illimani area, at 5000 m a.s.l. close to La Paz, the Bolivian capital. The total mercury concentrations in their hair ranged from 0.02 to 3.20 $\mu g\ g^{-1}$. In the Yani mining area and along the K'aka River, the maximum THg concentrations measured in gold-miners' hair was respectively 0.94 $\mu g\ g^{-1}$ and 1.87 $\mu g\ g^{-1}$ (with an average in this last population of 0.28 $\mu g\ g^{-1}$). The differences between the maximum values obtained for each study area can be explained by various factors, like (a) climate, which at high altitudes prevents miners from burning their amalgams in the open-air, and (b) their lack of knowledge on the Hg contamination effects and precautions to take.

During the 1998 rainy season, about 200 cases of malaria were registered each month in the K'aka River area, with as many as 500 cases reported for November alone. In many mining families, malaria has been diagnosed. The rainy season and the proximity of numerous shallow shafts dug to collect alluvial gold favour the development of insect vectors of tropical diseases.

Mercury concentrations in indigenous communities

From the data it appears that the Essejas indigenous communities living on the banks of the Beni River, a few kilometres downstream of Rurrenabaque village, present the highest mercury concentrations with 4.30 and 19.52 $\mu g\ g^{-1}$ as minimal and maximal values respectively (average: 9.81 $\mu g\ g^{-1}$). In women, the Hg concentration in hair tends to decrease with age, due to their gestation periods (Barbosa *et al.*, 1998) and less consumption of food with age. We note an increase in the mercury concentration in the last generation (Fig. 4.8), probably due to contamination during their gestation and less consumption of food with age. The mercury concentration in the hair of babies can be affected by maternal mercury contamination during pregnancy.

Especially in riparian people, high concentrations have been measured in members of a family living in Rurrenabaque, on the banks of the Beni River. Every son is a fisherman and fish being the common dish cooked in these households could explain the high mercury levels found in their hair.

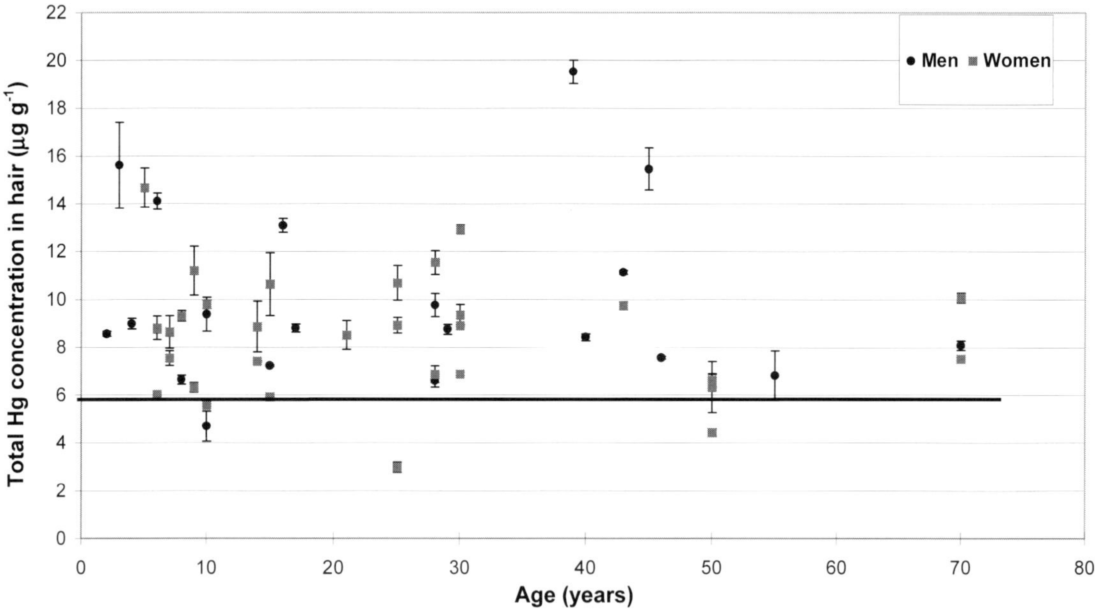

Fig. 4.8 Mean mercury concentrations in the hair of indigenous Essejas families living on the banks of the Beni River near Rurrenabaque (December 1998 and October 1999 campaigns).

Importance of the diet in the human exposure to mercury

More than 50% of the miners working in the Andean basin cooperatives report eating fish only twice a month (Villaviciencio, 1995), mainly in the dry season (Maurice-Bourgoin *et al.*, 1999). These results clearly show that the uptake of mercury by man, mainly in its methylated form, occurs as a result of consumption of carnivorous and piscivorous fish. It appears that mercury pollution affects an area 150 km downstream of the gold-mining activities and affects people who are not directly concerned with gold extraction.

All men of the Essejas community are fishermen. Fish constitutes their main source of protein and they told us they eat both omnivorous and carnivorous fishes, depending on their availability. Proportionally, more piscivorous fish are consumed during the rainy season but when the waters are too high to go fishing, people consume more fruit, eggs or chicken. This diversification and low Hg levels in omnivorous fishes can explain why the mean Hg concentration in hair observed in the Bolivian Amazonian basin is lower than the mean ranges published in the Brazilian Amazonian basin of from 8.7 to 75.5 $\mu g\ g^{-1}$ (Barbosa *et al.*, 1997). People who have a high and regular fish diet are at a greater risk of mercury poisoning because they are exposed to its organic form (Kehrig *et al.*, 1997), directly assimilated by the organism. The few results we have of levels of methylmercury in hair show that this organic form represents 73–98% of the total mercury.

CONCLUSIONS

The results presented here show that in Bolivia, the release of mercury into the Amazon basin contaminates both the hydrosystem and the ecosystem. The low quantity of mercury directly released into the streams, the local hydrological regimes and the dynamic of the sediment transport in the Bolivian Amazon basin explain the increase of mercury contamination in waters at the bottom of the Andean piedmont.

Mercury from soil erosion and recent agriculture practices, such as forest burning, and gold extraction represents a direct environmental and health threat, affecting both the ecosystem and the people living not only in the gold-mining area but more importantly downstream, at the Andean drainage basin outlet.

The mercury concentrations found in piscivorous fish from the Beni River are of great concern since they can exceed the WHO safety limit by a factor of 4. For a single species, the relation between Hg concentration and body weight depends on the food diversification with hydrological seasons of the species, their migrations (mainly for reproduction), and their movement between numerous oxbow lakes, especially in the low water season. Even if the total mercury concentration of the river water is within the range of the natural background level stated in the literature, in 86% of the piscivorous fish caught in the study area the maximum permissible level was exceeded. In contrast, the Hg content of all omnivorous fish never exceeded the WHO (1976) limit.

The regular consumption of mature adult fish which have accumulated high quantities of methylmercury through prolonged bioconcentration and biomagnification processes can be particularly dangerous. The indigenous communities living along the Beni River present quite elevated mercury levels in hair (average: 9.81 $\mu g\ g^{-1}$). Fish constitute an important part of their diet and consumption appears to be relatively constant from year to year, among the three generations analysed in two large families. The mercury levels in hair in this indigenous population reflect long-term exposure, as in other Amazonian populations in Brazil. Changes in the nervous system have been demonstrated in a riparian population of the Brazilian Amazon at mercury levels in hair below 50 $\mu g\ g^{-1}$ and Lebel *et al.* (1998) show a dose–effect relationship for certain motor and visual functions. Thus, the dose–effect relationship is most likely related to cumulative and long-term exposure but more studies are clearly needed in this type of population to determine the long-term significance of nervous system alterations. In the Essejas community, children under 3 years old present a Hg content in hair that is twice that of their grandfather's generation. So, the last generation presents substantial methylmercury exposure *in utero*. This finding is quite alarming because in this region, the proportion of these persons is increasing and more severe effects have been observed for *in utero* exposure, as compared to perinatal and adult exposure periods (Rice, 1989).

The major health impact caused by mercury affects people who are not working in gold cooperatives but who have a fish diet. This problem is well studied in Brazil (Malm *et al.*, 1990; Lacerda *et al.*, 1989; Nriagu *et al.*, 1992) and French Guyana (Cordier & Grasmick, 1994) but there were few data available for Bolivia.

The biogeochemistry of Hg in tropical freshwater ecosystems needs to be better understood given the severe threats from Hg pollution as well as the importance of freshwater fish as the main source of protein in tropical regions.

Acknowledgements In 1995, researchers from IRD (formerly ORSTOM), the French Scientific Research Institute for Development, in cooperation with Brazilian and Bolivian universities, initiated a basinwide study of the hydrogeochemistry of the Amazonian basin (HiBAm Programme) coordinated by Jean-Loup Guyot, I would like to thank Jean-Loup Guyot for the support he gave my research. Financial support for this study was provided by the FONAMA sponsored by the World Bank ("Cuentas iniciativas para las Americas") as a part of the Mercury Contamination of the Beni River Project. We thank Olaf Malm and

Helena Kehrig from the Universidade Federal do Rio de Janeiro, and Lucia Alanoca, Jaime Chincheros and Jorge Quintanilla from the Universidad Mayor de San Andres (La Paz) for their help in the laboratory and cooperation in our research. Last but not least, we express our gratitude to the riparian people of the Beni and Mamore rivers for their invaluable participation and collaboration in this study.

REFERENCES

Agemian, H. & Cheaner, V. (1978) Simultaneous extraction of mercury and arsenic from fish tissues and an automatic determination by atomic absorption spectrometry. *Analytical and Chemical Acta* **101**, 193–197.

Akagi, H., Malm, O., Branches, F. J. P., Kinjo, Y., Kashima, Y., Guimaraes, J. R. D., Oliveira, R. B., Haraguchi, K., Pfeiffer, W. C., Takizawa, Y. & Kato, H. (1995) Human exposure to mercury due to gold mining in the Tapajós River basin, Amazon, Brazil: speciation of mercury in human hair, blood and urine. *Water, Air and Soil Pollution* **80**, 85–94.

Alhers, W. W., Reid, M. R., Kim, J. P. & Hunter, K. A. (1990) Contamination-free sample collection and handling protocols for trace elements in natural freshwaters. *Australian Journal of Marine and Freshwater Research* **41**, 713–720.

Barbosa, A. C., Garcia, A. M. & de Souza, J. (1997) Mercury contamination in hair of riverine populations of Apiacas Reserve in the Brazilian Amazon. *Water, Air and Soil Pollution* **97**, 1–8.

Barbosa, A. C., Silva, S. R. L. & Dórea, J. G. (1998) Concentration of mercury in hair of indigenous mothers and infants from the Amazon basin. *Archives of. Environmental Contamination and Toxicology* **34**, 100–105.

Benes, P. & Havlik, B. (1979) Speciation of mercury in natural waters. In: *The Biogeochemistry of Mercury in the Environment* (ed. by J. O. Nriagu), 175–202. Elsevier/North-Holland Biomedical Press, Amsterdam.

Benoit, G. (1995) Evidence of the particle concentration effect for lead and other metals in freshwaters based on ultraclean technique analysis. *Geochimica et Cosmochimica Acta* **59**(13), 2677–2687.

Boudou, A., Durieu, G. & Maury-Brachet, R. (2000) Bioamplification du mercure. Niveaux de contamination des poissons. In: *Programme Mercure en Guyane.* Rapport CNRS-PEVS, CNRS, Meudon, France.

Cordier, S. & Grasmick, C. (1994) *Etude de l'imprégnation par le mercure dans la population guyanaise.* Réseau National de Santé Publique, France.

Gaudet, C., Lingard, S., Cureton, P., Keenleyside, K., Smith, S. & Raju, G. (1995) Canadian environmental quality guidelines for mercury. *Water, Air and Soil Pollution* **80**, 1149–1159.

Guimarães, J. R. D., Roulet, M. & Lucotte, M. (1996) Seasonal and spatial variations of Hg net methylation in a floodplain lake of the Tapajos River, Brazil. In book of abstracts: *Fourth International Conference on Mercury as a Global Pollutant* (Rio de Janeiro, Brazil), p. 418.

Guimarães, J. R. D., Meili, M., Malm, O. & Sousa Brito, E. M. (1998) Mercury methylation in sediments and floating meadows of a tropical lake in Pantanal, Brazil. *Science of the Total Environment* **213**, 165–176.

Guyot, J. L. (1993) *Hydrogeochemistry of the Bolivian Amazonian Rivers.* PhD Dissertation. ORSTOM, France.

Kehrig, H. A., Malm, O. & Akagi, H. (1997) Methylmercury in hair samples from different riverine groups, Amazon, Brazil. *Water, Air and Soil Pollution* **97**, 17–29.

Kehrig, H. A., Malm, O., Akagi, H., Guimarães, J. R. D. & Torres, J. P. M. (1998) Methylmercury in fish and hair samples from the Balbina reservoir, Brazilian Amazon. *Environmental Research,* Section A **77**, 84–90.

Lacerda, L. D., Pfeiffer, W. C., Ott, A. T. & Silveira, E. G. (1989) Mercury contamination in the Madeira River, Amazon—Hg inputs to the environment. *Biotropica* **21**(1), 91–93.

Langston, W. J. (1982) The distribution of mercury in British estuarine sediments and its availability to deposit-feeding bivalves. *Journal of the Marine Biological Association, UK* **62**, 667–684.

Lebel, J., Mergler, D., Branches, F., Lucotte, M., Amorim, M., Larribe, F. & Dolbec, J. (1998) Neurotoxic effects of low-level mercury contamination in the Amazonian basin. *Environmental Research* **A79**, 20–32.

Lowe McConnell, R. H. (1987) *Ecological Studies in Tropical Fish Communities.* Cambridge Tropical Biology Series, Cambridge, UK.

Malm, O., Pfeiffer, W. C., Souza, C. M. M. & Reuther, R. (1990) Mercury pollution due to gold mining in the Madeira river basin, Brazil. *AMBIO* **19**(1), 11–15.

Malm, O., Branches, F. J. P., Akagi, H., Castro, M. B., Pfeiffer, W. C., Harada, M., Bastos, W. R. & Kato, H. (1995) Mercury and methylmercury in fish and human hair from the Tapajós River basin, Brazil. *Science of the Total Environment* **175**, 141–150.

Maurice-Bourgoin, L., Guyot, J. L., Seyler, P., Quintanilla, J. & Courau, P. (1997) Mercury distribution in the Madeira and Amazon drainage basins (in French). In: *Freshwater Contamination* (ed. by B. Webb) (Proc. Rabat, Morocco, Symp., April–May 1997), 85–92. IAHS Publ. no. 243.

Maurice-Bourgoin, L., Quiroga, I., Guyot, J. L. & Malm, O. (1999) Mercury pollution in the upper Beni River basin, Bolivia. *AMBIO* **28**(4), 302–306.

Maurice-Bourgoin, L., Quiroga, I., Chincherros, J. & Courau, P. (2000a) Mercury distribution in waters and fishes of the Upper Madeira rivers and mercury exposure in riparian Amazonian populations. *Science of the Total Environment* **260**, 73–86.

Maurice-Bourgoin, L., Fraizy, P., Alanoca, L., Seyler, P. & Guyot, J. L. (2000b) Hydrological control on the temporal variability of mercury in surface waters of the Upper Madeira basin, Bolivia. In: *25th International Conference on Heavy Metals in the Environment* (ed. by J. Nriagu), contribution no. 1282. University of Michigan, School of Public Health, Ann Arbor, Michigan (CD-ROM).

Maurice-Bourgoin, L., Aalto, R., Fraizy, P., Barnaud, V. & Guyot, J. L. (2001) Modulation of mercury transport through fluvial systems by channel–floodplain interaction: a case study for the Beni River, Bolivian Amazonia. In: *6th International Conference on Mercury as a Global Pollutant* (Minamata, Japan) (ed. by H. Akagi & S. E. Lindberg).

Maurice-Bourgoin, L., Quémerais, B., Guyot, J. L. & Seyler, P. (2002) Transport and behaviour of total mercury in the Amazon River at the confluence of black and white waters. *Hydrological Processes* (in press).

Meili, M. (1991) Mercury in boreal forest lake ecosystems. *Acta Universatis Upsaliensis* **336**.

Meili, M. (1997) Mercury in lakes and rivers. In: *Metal Ions in Biological Systems*, vol. 34—*Mercury and its Effects on Environment and Biology* (ed. by A. Sigel & H. Sigel), 21–51. Marcel Dekker Inc., New York.

Mendes dos Santos, G., Jegu, M. & de Merona, B. (1984) *Catalogo de Peixes Comerciais do Baixo Rio Tocantins.* Projeto Tucurui, ElectroNorte SA, Manaus, Brazil.

Nolting, R. F. & De Jong, J. T. M. (1994) Sampling and analytical methods for the determination of trace metals in surface seawaters. *Journal. of Environmental and Analytical Chemistry* **57**, 189–196.

Nriagu, J. O., Pfeiffer, W. C., Malm, O., Souza, C. M. M. & Mierle, G. (1992) Mercury pollution in Brazil. *Nature, London*, 356–389.

Pfeiffer, W. C., Lacerda, L. D., Malm, O., Sousa, C. M. M., Silveira, E. G. & Bastos, W. R. (1989) Mercury concentrations in inland waters of Rondônia, Amazon, Brazil. *Science of the Total Environment* **87–88**, 233–240.

Pfeiffer, W. C., Lacerda, L. D., Malm, O., Sousa, C. M. M., Silveira, E. G. & Bastos, W. R. (1991) Mercury in the Madeira river ecosystem, Rondônia, Brazil. *Forest Ecology and Management* **38**, 239–245.

Pfeiffer, W. C., Lacerda, L. D., Salomons, W. & Malm, O. (1993) Environmental fate of mercury from gold-mining in the Brazilian Amazon. *Environmental Review* **1**, 26–37.

Porcella, D. B., Huckabee, J. W. & Wheatley, B. (1995) Mercury as a global pollutant. *Water, Air and Soil Pollution* (Special Issue) **80**.

Quémerais, B. & Cossa, D. (1995) Procedures for sampling and analysis of mercury in natural waters. *Rapport ST-31E, Environment Canada, Environment Conservation, Centre Saint Laurent, Montréal, Québec.*

Rae, J. E. & Aston, S. R. (1982) The role of suspended solids in the estuarine geochemistry of mercury. *Water Research* **16**, 649–654.

Rice, D. C. (1989) Delayed neurotoxicity in monkeys exposed developmentally to methylmercury. *Neurotoxicology* **10**, 645–650.

Roulet, M. & Lucotte, M. (1995) Geochemistry of mercury in pristine and flooded ferallitic soils of a dense tropical forest in French Guiana. *Water, Air and Soil Pollution* **80**, 1079–1088.

Roulet, M., Lucotte, M., Canuel, R, Rhéault, I., Tran, S., De Freitos Gogh, Y. G., Farella, N., Souza do Vale, R., Sousa Passos, C. J., De Jesus da Silva, E., Mergler, D. & Amorim, M. (1998) Distribution and partition of total Hg in waters of the Tapajós River basin, Brazilian Amazon. *Science of the Total Environment* **213**(1/3), 203–211.

Roulet, M., Lucotte, M., Farella, N., Serique, G., Coelho, H., Sousa Passos, C. J., De Jesus da Silva, E., Scavone de Andrade, P., Mergler, D., Guimaraes, J. R. D. & Armorim, M. (1999a) Effects of recent human colonization on the presence of mercury in Amazonian ecosystems. *Water, Air and Soil Pollution* **112**, 297–313.

Roulet, M., Lucotte, M.; Dolbec, J., Gogh, Y. F., Peleja, J. R. P., Silva, D. S., Lebel, J., Mergler, D & Guimarães, J. R. D. (1999b) Mercury bioaccumulation in two fish communities of the Tapajós River, Brazilian Amazon. In: *Vth International Conference on Mercury as a Global Pollutant* (Rio de Janeiro, Brazil, 23–28 May).

Roulet, M., Lucotte, M., Rheault, I & Guimarães, J. R. D. (2000) Methylmercury in water, seston and epiphyton of an Amazonian River and its floodplain, Tapajós River, Brazil. *Science of the Total Environment* **261**, 43–59.

Silva Forsberg, M. C., Forsberg, B. R. & Zeidemann, V. K. (1999) Mercury contamination in humans linked to river chemistry in the Amazon basin. *AMBIO* **28**(6), 519–521.

Telmer, K., Costa, M., Araujo, E. S., Angelica, R. S. & Maurice, Y. (1999) Mercury in the Tapajós River basin—the significance of suspended sediments from alluvial gold-mining, Pará, Brazilian Amazon. In: *Manaus 99—Hydrological and Geochemical Processes in Large-scale River Basins* (HiBAm Project, ed. by J. L. Guyot) (Proc. Manaus Symp., Brazil) (CD-ROM).

Villaviciencio, A. (1995) *Diagnostico Socio-economico de la Region del Rio K'aka.* International Report MEDMIN, Prog. de Manejo integrado del Medio Ambiente en la pequeña Minería, La Paz, Bolivia.

Vinson, D. S., Bonzongo, J. C., Lyons, W. B., Lechler, P., Miller J. & Lacerda, L. D. (1997) Gold-mining and mercury pollution in the Amazon river system, Brazil: Hg level in surface waters of the Madeira rivers from Porto Velho to the confluence with the Amazon river. 1997 Fall Meeting in *EOS, Transactions American Geophysical Union* **78**(46), F199.

WHO (1976) *Environmental Health Criteria*. I. *Mercury*, 1–131. World Health Organization, Geneva.

ABBREVIATIONS AND DEFINITIONS

BAF Bioaccumulation factor, defined by the concentration ratio between an organism and its abiotic environment from direct and food uptake (usually in units of volume per biomass)

BCF Bioconcentration factor, describing "biotic sorption", defined by the concentration ratio between an organism and its abiotic environment from direct uptake alone (usually in units of volume per biomass)

BMF Biomagnification factor, describing "biotic transfer" in the food chain, defined by the concentration ratio between an animal and its diet (dimensionless)

DHg Dissolved mercury concentration in waters (usually in ng l^{-1})

K_d Distribution coefficient (partition coefficient), describing "abiotic sorption", defined by the concentration ratio between particulate + colloidal and dissolved phase (usually in ng kg^{-1}/ng l^{-1})

MeHg Methylmercury = monomethylmercury + dimethylmercury (main Hg organic and bioavailable species)

PCE "particle concentration effect" which is caused by metals associated with colloidal particles but counted with the dissolved fraction (Benoit, 1995). The PCE could be caused by a decrease in surface area and surface complexation sites due to re-suspension of larger particles under high suspended solids concentrations.

PHg Particulate mercury concentration adsorbed on suspended solids or colloidal matter in water (usually in ng g^{-1})

THg Total mercury concentration in waters (= DHg + PHg) (usually in ng l^{-1})

CHAPTER 5

HYDRO-ECOREGIONS OF THE BOLIVIAN AMAZON: A GEOGRAPHICAL FRAMEWORK FOR THE FUNCTIONING OF RIVER ECOSYSTEMS

Jean-Gabriel Wasson, Soraya Barrera, Bénédicte Barrère, David Binet, David Collomb, Ismael Gonzales, Franck Gourdin, Jean-Loup Guyot & Giovanna Rocabado

The Bolivian Amazonian basin is not a small piece of the puzzle. With 724 000 km^2, it accounts for one half of the Rio Madeira basin and 10% of the whole Amazon basin. It further comprises one quarter of the Andean Amazon catchments whose influence upon the hydrological functioning of the larger basin is significant (Fig. 5.1). The geography of the Bolivian Amazon is extraordinarily diverse, from rainforest to semi-desert and from glacier peaks to wetlands, and it shelters some of the planet's most important biodiversity refuges. It still contains extensive wild and almost depopulated zones, but it also includes Bolivia's three largest cities and a fast growing agro-industrial area. Both development and inappropriate practices tied to under-development generate conflicts between the aquatic resource conservation and development. Being unevenly distributed in the basin, water acts also as a strong determinant of human settlements and constitutes a major social stake; recent bloody riots flared over a conflict about water distribution. While environmental legislation is being implemented with difficulty, there is no scientific support for a global policy towards aquatic ecosystem management and conservation, and the geography of water problems remains blurred. A global view of hydrosystems, including biological aspects, is necessary to define sustainable development policies at the basin scale.

Integrated research management requires a strong scientific grounding. At the beginning of our research programme, in 1996, running water ecosystems of the Bolivian Amazon were still poorly investigated. Scientific knowledge was largely limited to two fields: hydrology and fish. The hydrology was investigated by PHICAB (hydrological joint research programme (1982–1992) for the Bolivian Amazonian basin), the main results of which were summarized in a water balance atlas (Roche *et al.*, 1990) with a description of hydrological regimes given by Bourges *et al.* (1993) and a detailed study of the hydrochemistry and sediment transport reported in Guyot (1993). More limited investigations concerning dissolved organic carbon (DOC) (Guyot & Wasson, 1994) and sedimentology (Guyot *et al.*, 1999) have been documented. In ichthyology, most of the data concerned the lowland rivers investigated during the "Convenio Piscicola" (Fish Research Agreement) joint research programme (1982–1987) on fish distribution, taxonomy, biology and production, which provided an inventory of almost 400 fish species in the Beni plain, including halieutic aspects (Lauzanne & Loubens, 1985; Lauzanne *et al.*, 1990, 1991). Pioneering work in the Andean rivers was done by Pearson (1924), and Sarmiento & Barrera (1997) recently produced a simple inventory of fish. Little was known about the invertebrate fauna in the basin, with the exception of the benthic invertebrates in a few mountain streams (Wasson *et al.*, 1998). Moreover, hydrology and biota have traditionally been studied independently, and the functional links between hydrology-dependant factors and biological structures—which constitute the basis of ecohydrology—had not been investigated.

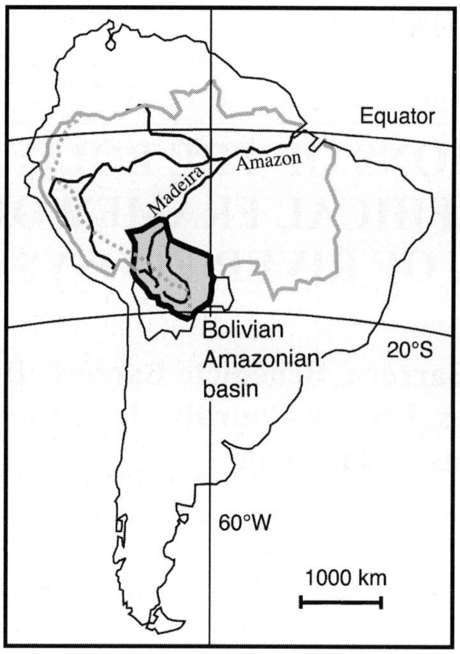

Fig. 5.1 Location of the Bolivian Amazonian basin (bold line) in South America; grey line: Amazon basin; dotted line: Andean border.

With its biological richness, its wilderness and geographical diversity, the Bolivian Amazon constitutes a fantastic natural laboratory for ecohydrological research. A topic of growing interest concerns the factors controlling the distribution and maintenance of aquatic biodiversity at both regional and local scales. This was a major aim of BIOBAB, a joint research programme (1996–2000) on aquatic biodiversity in the Bolivian Amazonian basin, which was initiated in 1996 to provide basic knowledge for management and conservation. The first step of this project was to develop a global view of running water ecosystems at the basin scale, which included the distribution of fish and invertebrates in the Andean zone; other topics concerned the hydrological dynamics of wetlands (Bourrel *et al.*, 1999) and hydrosystem functioning in the lowland flood plain (Pouilly *et al.*, 1999), bioindication (Fossati *et al.*, 2001), and mercury contamination (see chapter 4 of this book—Maurice-Bourgoin & Quiroga, 2002). Another limnological research programme initiated simultaneously was the PROLIMCO joint research project, which was more oriented towards aquatic resources management, with an emphasis on invertebrate fauna and bioindication in mountainous streams, and aquatic biodiversity distribution in lowland river systems. The first results of both projects were published in the proceedings of the Bolivian Congress on Limnology and Aquatic Resources (1999) held at Cochabamba.

We present in this chapter the regional approach of the BIOBAB project, the aim of which was to define homogeneous regions as regards the ecological functioning of running waters. The approach consisted of identifying at different hierarchical levels the ecohydrological links between physical control factors and ecosystem structures. Such an approach should provide a global view of the natural diversity of aquatic ecosystems in the basin, and by understanding the distribution of the main functional types, to allow some spatial prediction of their most important physical characteristics and associated biota. We shall define, map, and briefly describe the "hydro-ecoregions" of the Bolivian Amazonian basin and present some important aquatic ecosystem characteristics of the morphology, hydrochemistry, and fish and invertebrate communities in the Andean zone. This work is the first attempt of regionalizing running water ecosystems in South America, with the aim of providing a geographical framework for management and conservation.

CONCEPTUAL FRAMEWORK

Regionalization differs from a mere typology of known objects, as the purpose is to delimit geographical entities in which, at the present case, unknown running water ecosystems should *a priori* present similar characteristics. Usually, regionalization is a bottom-up process done by spatial aggregation of numerous sites that have only a few parameters. But in the case of aquatic ecosystems, we deal at best with a limited number of sites described by a complex set of variables. So the regionalization developed here is a top-down approach, linking ecosystem functioning to the geographical features of their catchments. The approach requires a strong theoretical background, based upon the hydrosystems hierarchical control concept.

The geographical classification of terrestrial ecosystems began in Europe and Russia at the beginning of the twentieth century and has been further developed in the USA by Bailey (1976) who popularized the term "ecoregion" (see Bailey, 1996). This approach was adapted to aquatic ecosystems by Omernick (1987), in order to define regional goals for water quality and management (Hughes & Larsen, 1988); a good agreement was observed between these ecoregions and hydrochemistry, fish communities, and to a certain extent with benthic invertebrates (Rohm *et al.*, 1987; Whittier *et al.*, 1988). This approach is obviously interesting for management (Omernick & Griffith, 1991; Warry & Hanau, 1993), but as the functional links between geographical characteristics and ecosystem structures are not explicitly described, an explanation of the observed patterns is somewhat lacking.

The conceptual framework for regionalization is substantially improved by the theories of hydrosystem hierarchical control, and particularly the nesting of physical structures from the basin to the microhabitat scale proposed by Frissell *et al.* (1986) and Naiman *et al.* (1992) and based on the ideas expressed by Hynes (1975) and Lotspeich (1980). All of these authors recognized geology, relief and climate as the primary determinants of running water ecosystem functioning at the basin scale. More recently the role of temporal variability and spatial heterogeneity in structuring aquatic communities has been emphasized (Resh *et al.*, 1988; Townsend, 1989; Ward & Stanford, 1983). The integration of these findings led to the hydrosystem concept, a four-dimensional dynamic structure ruled by hydrology and morphology (Amoros & Petts, 1993).

The following is a summary of the hierarchical nesting of factors that determine running water ecosystem functioning (Fig. 5.2). At the local scale, aquatic biodiversity and

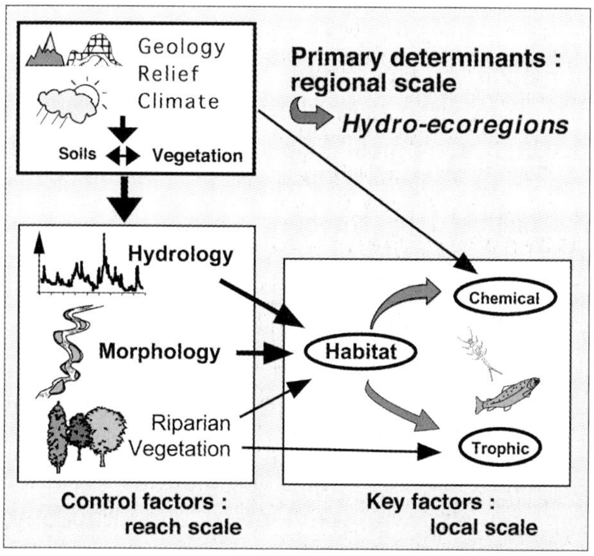

Fig. 5.2 Conceptual frame for the regionalization: schematic representation of the hierarchical control factors concept for running water ecosystems. See text for explanation.

productivity respond to "key factors" that can be grouped into three categories (Wasson, 1989): physical habitat (hydraulics, substrate), hydroclimate (light, temperature, dissolved gases, hydrochemistry etc.), and trophic resources (autochthonous production and allochthonous organic matter); however, the influence of physical conditions upon essential parameters such as oxygen or trophic resource availability leads on to assigning a central role to the habitat factor. At the reach scale, the distribution and dynamics of these key factors depend upon the hydro-sedimentological regime and valley geomorphology; these "control factors" rule the reach morphodynamics (river bed form and stability, substrate composition), the riparian vegetation structure, the lateral (flood plain) and vertical (hyporheic) connectivity, and the overall temporal variability of the system. Finally, at a regional scale, these control factors depend upon a limited set of "primary determinants": geology (lithology), relief (geomorphology) and climate (temperature and precipitation). Soils and vegetal cover obviously affect the hydro-sedimentological balance, but they are ultimately determined by the geophysical and climatic characteristics.

This conceptual framework justifies an attempt to regionalize aquatic ecosystems in a large basin by delineating "hydro-ecoregions" (HER) on the basis of the primary determinants identified above (geology, relief, climate) (see Wasson 1996, Wasson *et al.* 1993). The underlying hypotheses are:

(a) within each HER, running water ecosystems should present a limited range of variation (as compared to the whole basin) for physical, chemical and biological characteristics, together with a similar pattern of longitudinal change;

(b) ecosystems of various regions should differ in at least one important abiotic parameter, leading to a consistent and significant difference in their biological structure. This should open the way to the definition of a limited set of ecological types in each region.

METHODOLOGY

The methodology for regionalization is detailed in Wasson & Barrère (1999), and here we shall merely point out some relevant aspects.

The approach is clearly based on a top-down control concept, following the principle that a classification is "better done according to the causes of the class differences than according to the effects that differences produce" (Strahler, 1975 cited in Lotspeich, 1980). This method also allows the definition of several hierarchical regional levels, and the identification of the control factors.

Basic parameters for regionalization are natural characteristics: geology, relief and climate. However, as we deal with existing data, their validity and relevance have to be carefully evaluated. For instance, the geological map has been interpreted in the light of lithological characteristics, such as rock hardness and chemical properties. The geomorphological structure was analysed by comparing three different sources: a broad digital elevation model (DEM), a specific study of remote sensing scenes, and a physiographical map. The climatic maps are very imprecise in Bolivia, and we used the vegetation as a geographically exact indicator of the climate. As vegetation cover is also a regulating factor for water and sediment fluxes, it emerges as an important parameter for regionalization and was analysed by comparing three different sources: an ecological map, a forest and agriculture map and a NOAA scene.

Regionalization deals with some striking natural boundaries (geomorphological and climatic) obvious in the landscape but also with pluviometry and altitude gradients. For the latter, we have singled out regionalization by structural (geomorphological) massif, with the hypothesis of a predictable longitudinal evolution of running waters in these massifs.

Delineating regions along a climatic gradient is more arbitrary, and in this case we followed important changes in the vegetation cover.

Running waters integrate rapidly the microscale (*sensu* Bailey 1996, i.e. at a kilometric scale) heterogeneity of their basin, which constitutes a great advantage for their regionalization. Such heterogeneity, for instance in rock composition, can be considered statistically as a component of HER characteristics. Landscape heterogeneity (10–100 km^2), such as rain shadow effects in mountainous area, is more difficult to deal with, but this was considered in the same manner. Consequently we recognize a fairly high internal heterogeneity in some regions. In every case the variation range for a given parameter must be predictable, and lower within a HER than between HERs.

Finally, such a regionalization approach remains an hypothesis until it has been validated at the running water ecosystem level on the basis of independent physical, chemical and biological parameters. Obviously, the basic hypotheses described above are valid only for small- and medium-sized reaches occurring entirely within a single region, but these ecosystems, which are often poorly known, account for a very high percentage of the total hydrographic network. For high-order reaches or systems crossing various regions, the analysis—and thus the prediction of their characteristics—must evidently be done in considering the proportion of the various HERs encompassed by their catchment.

The following thematic layers were used and processed within a geographical information system (GIS) with ArcView® and IDRISI®: **Geology**: Geological map, 1/1 000 000, SERGEOMIN* 1998. **Relief**: DEM GTOPO30, 30″, US Geological Survey EROS Data Center (http://edcdaac.usgs.gov/gtopo30/gtopo30.html); physiographic map, 1/1 000 000, GEOBOL 1996; specific analysis of 1/450 000 ERTS scenes by Gourdin (1997) for the Andean zone and Binet (1998) for the lowlands. **Climate**: Temperature and precipitation map from the PHICAB Atlas, 1/5 000 000 (Roche *et al*., 1990). **Vegetation**: Vegetation and ecoregions map, 1/1 500 000 (Rivera *et al*., 1996); forest map including agriculture, 1/1 500 000, CUMAT 1996; NOAA scene (23 May 1998) and the normalized differential vegetation index (NDVI), ABTEMA. **Soils**: Preliminary soils map for the Beni lowlands, 1/1 000 000, MDS/OT 1999. **Hydrography**: Hydrographic map, 1/1 500 000, IGM 1986.

Some of these sources, such as the vegetation map, carry interesting typological information but are geographically very imprecise. Others, such as the physiographic map, present important distortions; the DEM is also really inaccurate in the humid regions. The hydrographic network and the forest map were used to correct some geographical distortions. All the maps were reduced to a 1/2 000 000 scale for the HER delineation. The most obvious geomorphological and climatic boundaries were first used to define a first level of hydro-ecoregions (HER-1). Then, we looked within the most heterogeneous HER-1 for identifiable regional structures to delineate a second level (HER-2) that could be considered at present to be the most precise significant regionalization attainable with existing sources. Finally, these HERs were grouped into a few large geoclimatic domains.

For validation at the hydrosystems level, three types of independent data were used:

(a) geomorphologic characteristics of valleys and river beds,
(b) hydrochemical parameters and
(c) a hydro-ecological study of 25 stream sites in the Andean zone.

The methods will be described in the corresponding sections.

* Acronyms used in this paragraph are as follows:
SERGEOMIN: Servicio de Geología Minería (Geology and Mining Service (formerly GEOBOL)
GEOBOL: Servicio de Geológico de Bolivia (Bolivian Geological Service)
CUMAT: acronym of a private environmental office
ABTEMA: Asociación Boliviana de Teledetección para el Medio Ambiente (Bolivian Association for Environmental Teledetection)
MDS/OT: Ministerio de Desarrollo Sostenible/Ordenamiento Territorial (Sustainable Development Ministry/Land Planning Board)
IGM: Instituto Geográfico Militar (Military Geographical Institute)

THE HYDRO-ECOREGIONS OF THE BOLIVIAN AMAZON BASIN

Geoclimatic domains

The main geomorphological and climatic structures delineate nine hydro-ecological domains (Fig. 5.3 and Table 5.1). The Andes cordillera is divided into two geomorphological zones, the first internal and the second external or "subandean". The mountainous internal zone is principally made up of primary rocks, more or less consolidated, interspersed with an intrusive granitic axis in the north, some volcanic deposits and localized calcareous formations in the south. The V-shaped valleys run perpendicular to the main cordillera axis, and the altitude extends from 500–1000 m up to more than 6000 m. The subandean zone comprises a succession of lower altitude ranges (2000–2500 m), parallel to the main axis, of secondary poorly consolidated rocks, often calcareous. These ranges, called "Serranías", delineate wide valleys and large depressions filled with Tertiary

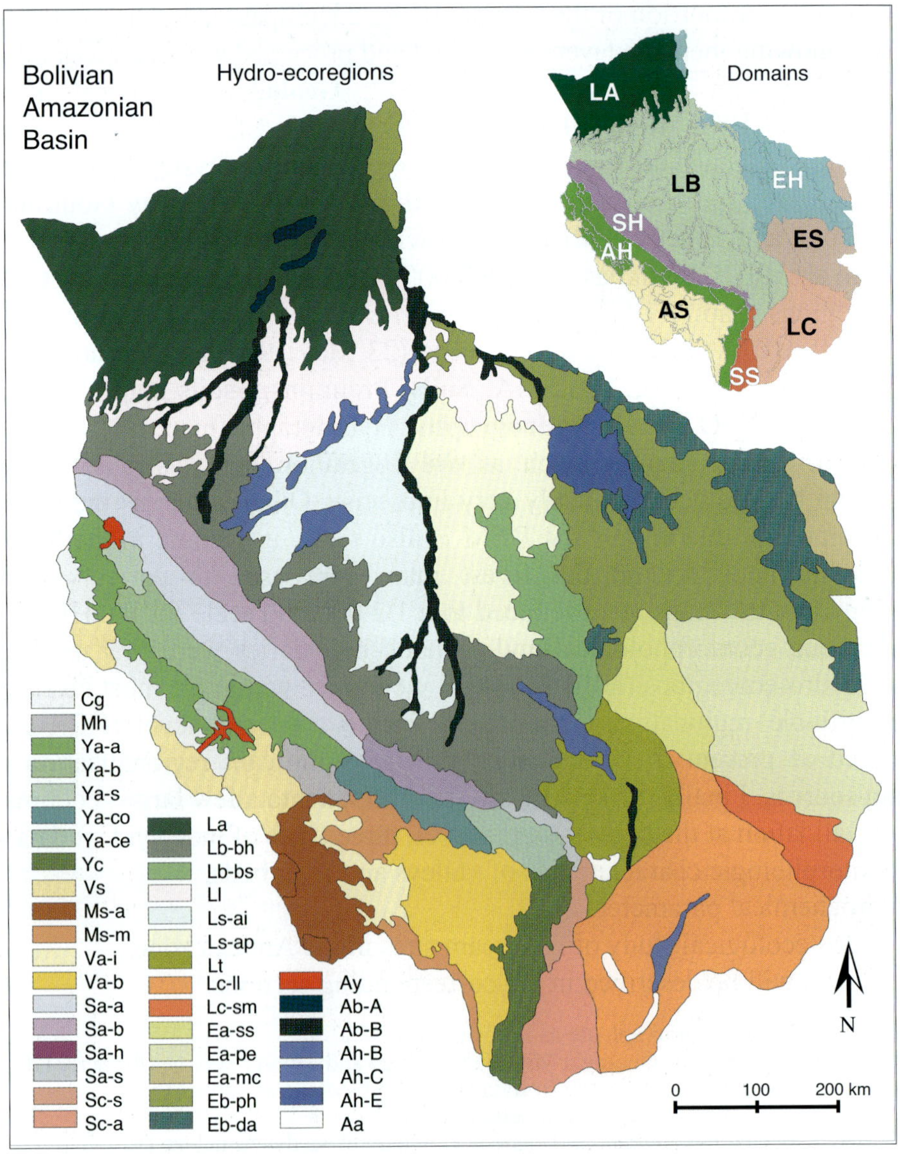

Fig. 5.3 Hydro-ecoregions of the Bolivian Amazonian basin, and geoclimatic domains (upper right). For the hydro-ecoregions map, the first two letters of the code correspond to the first level (HER-1) and the other letters to the second level (HER-2). See text for explanation and Table 5.1 for definition of codes.

sediments. On this longitudinal geomorphological structure is superimposed a well-defined transversal climatic boundary. The main axis of the eastern cordillera, oriented northwest–southeast, blocks the wet air mass coming from the central Amazon, thus separating a very humid domain northward from a semiarid one southward. This major divide corresponds to the limit between two (dry and tropical humid) of the four major climatic domains recognized at the global scale by Bailey (1996).

The lowlands comprise a vast central plain, called "Llanos", and the edge of the Brazilian shield in the east. The Llanos plain is filled with Quaternary sediments and is divided into three domains:

Table 5.1 Names and codes of hydro-ecoregions of the Bolivian Amazonian basin. Original Spanish terms have been conserved for local use and homogeneity with previous publications.

Domains	HER level 1	Code	HER level 2	Code
AH—Andino húmedo	Cordilleras glaciares	Cg	Cordilleras glaciares	Cg
	Montañas húmedas	Mh	Montañas húmedas	Mh
	Yungas peri-amazónicos	Ya	Ya—altos	Ya-a
			Ya—bajos	Ya-b
			Ya—subhúmedos	Ya-s
			Ya—"Chapare" hyper-húmedo	Ya-co
			Ya—"Chapare" subhúmedo	Ya-ce
	Yungas peri-chaqueños	Yc	Yungas perichaqueños	Yc
AS—Andino seco	Valles secos intra-andinos	Vs	Valles secos intra-andinos	Vs
	Montañas secas	Ms	Ms—altas	Ms-a
			Ms—medianas	Ms-m
	Valles semi-áridos	Va	Valles semi-áridos internos	Va-i
			Valles semi-áridos bajos	Va-b
SH—Subandino húmedo	Subandino peri-amazónico	Sa	Sa—húmedo alto	Sa-a
			Sa—húmedo bajo	Sa-b
			Sa—hyper-húmedo	Sa-h
			Sa—subhúmedo	Sa-s
SS—Subandino seco	Subandino peri-chaqueño	Sc	Sc—semi-seco	Sc-s
			Sc—seco	Sc-a
LA—Llanos amazónicos	Bosque amazónico del Pando	La	Bosque amazónico del Pando	La
LB—Llanos del Beni	Bosques inundables de llanos	Lb	Lb—Bosque húmedo peri-andino	Lb-bh
			Lb—Bosque subhúmedo peri-chiquitano	Lb-bs
	Sabanas lateríticas de llanos	Ll	Sabanas lateríticas de llanos	Ll
	Sabanas alcalinas de llanos	Ls	Ls—Sabanas alcalinas inundables	Ls-ai
			Ls—Sabanas alcalinas de palmeras	Ls-ap
	Zona de transición Beni Chaco	Lt	Zona de transición Beni Chaco	Lt
LC—Llanura del Chaco	Llanura seca del Chaco	Lc	Lc—Llanura xérica del Chaco	Lc-ll
			Lc—Serranías y mesas Chiquitanas	Lc-sm
ES—Escudo subhúmedo	Escudo subhúmedo alto	Ea	Ea—Serranías subhúmedas	Ea-ss
			Ea—Penillanura subhúmeda estacional	Ea-pe
			Ea—Meseta del Cerrado	Ea-mc
EH—Escudo húmedo	Escudo húmedo bajo	Eb	Eb—Penillanura húmeda boscosa	Eb-ph
			Eb—Depresiones aluviales	Eb-da
AH	Azonal—Valles secos de Yungas	Ay	Valles secos de Yungas	Ay
LA	Azonal—Bosque galería	Ab	Bosque galería Amazónicos	Ab-A
LB			Bosque galería del Beni	Ab-B
LB	Azonal—Humedales	Ah	Humedales del Beni	Ah-B
LC			Humedales del Chaco	Ah-C
EH			Humedales del Escudo	Ah-E
LC	Azonal—Arenales	Aa	Arenales	Aa

(a) the Amazonian Llanos in the north, a very humid region where tropical forest lies on undulating lateritic formations;
(b) the "Beni" Llanos in the centre, a flat plain where central savannas and wetlands are surrounded by humid forests; and
(c) a xeric domain in the south, the "Chaco" with its characteristic shrubby vegetation.

The Brazilian shield, made up of late Palaeozoic rocks, is inclined northward. It can be divided into a seasonally dry and more dissected plateau in the south and a humid peneplain surrounding wide sedimentary depressions and wetlands in the north.

Hydro-ecoregions level 1

Inside of the defined domains, the first level of regionalization results in 17 main hydro-ecoregions (nine HER-1s in the Andes and eight in the lowlands) and four types of "azonal" formations (Fig. 5.3).

Andes In the humid Andes we define a "glacial cordillera" region (Cg) and another "humid mountains" region (Mh) devoid of the glaciers that constitute an essential factor of hydrological regulation. In the intermediate zone of humid tropical mountains called "Yungas" we define the very humid "Amazonian Yungas" (Ya) in the north, and the somewhat dryer "Chaco Yungas" (Yc) in the south exposed to less abundant rainfall coming from Argentina. The same climatic dichotomy is encountered in the subandean zone (Sa and Sc). In the dry Andes, we must distinguish three HER-1s: northward, a "dry valleys" region (Vs) sheltered from Amazonian air masses, deeply incised in unconsolidated rocks with huge erosion rates; southward, in the Rio Grande basin, we define a peripheral "dry mountains" region (Ms) of rather high altitude, including calcareous and volcanic formations, that tightly surround a lower region of "arid valleys" (Va) often deeply incised and heavily eroded.

Llanos and Brazilian shield The "Amazonian Llanos" (La) and the "Chaco" (Lc) constitute homogeneous regions, but the central Beni plain must be divided into four HER-1s. A peripheral region of tropical humid "Llanos forest" (Lb), partially vulnerable to flooding, surrounds an extensive savanna landscape. The seasonally flooded savannas of the central zone can be separated according to very different soils and hydrogeology encountered on both sides of a geophysical fault, known as "the Bala-Rogagua line". This uprising fault divides the Beni plain into a northern region (Ll) characterized by lateritic acidic soils and a deeper (15 m) groundwater table, and a southern region (Ls) crossed by the Rio Mamore, lying upon recent sedimentary alkaline soils with more superficial phreatic waters (Hanagarth, 1993). Between the Beni plain and the Chaco Llanos is a transition region (Lt) characterized by fertile, unflooded soils favourable for intensive agriculture (the soybean belt, east of the city of Santa Cruz). The Brazilian shield is divided into a higher seasonally dry and a lower humid region (Ea and Eb), following the geoclimatic domains.

The "azonal" formations correspond to very particular mesoscale physical conditions, driving to such a point the hydrosystem properties that they override the ecoregional context. We have separated some very dry valleys interspersed in the Yungas region due to a rain shadow effect (Ay), and in the Llanos the gallery forests (Ab), the permanent wetlands (Ah), and some aeolian dune formations (Aa).

Hydro-ecoregions level 2

The majority of HER-1s have sufficient heterogeneity in geomorphological or climatic parameters to justify a second level of regionalization. For instance, the Amazonian Yungas (Ya) are climatically fairly heterogeneous, as rainfall varies from 1500 to 5000 mm year^{-1} due to the orientation of the massifs, and geomorphologically the landscapes are more or less incised due to variations in lithological composition. This region is divided into six HER-2s and one azonal formation. Similarly, the subandean zone can be separated into four HER-2s on the basis of rainfall and geomorphology. All the Andean regions are more precisely described in Wasson & Barrère (1999).

In the lowlands, the geographical structure is less heterogeneous. However, the savannas situated on both parts of the Rio Mamore differ sufficiently in terms of precipitation and vulnerability to flooding to be separated into the western region (Ls-ai) with significantly more flooding (Bourrel *et al.*, 1999) than the eastern region (Ls-ap) interspersed with palm-tree assemblages (Rivera *et al.*, 1996). The same phenomenon occurs for the eastern part of the Llanos forest (Lb-bs), which is dryer than the piedmont region (Lb-bh). In the Brazilian shield, the second level of hydro-ecoregions has been defined mainly on the basis of relief and geomorphology. In the lower region, the granitic peneplain (Eb-ph) is separated from the alluvial depressions (Eb-da), and in the higher region, we distinguished the peneplain (Ea-pe), the mountainous ranges (Ea-ss) and the very particular "Cerrado" plateau (Ea-mc). For the azonal formations, gallery forest and wetlands have been separated according to the geoclimatic domain to which they pertain.

On the whole, 33 HERs have been identified at the second level (19 in the Andes and 14 in the lowlands), and seven types of azonal formations. The global figures reflect some outstanding structures, such as the geological domains (Andes, Llanos, Brazilian shield), the wet/dry climatic boundary in the Andes or the savanna/forest limit in the Llanos. But other important boundaries are much less obvious in the landscape, and the transition between two regions is sometimes blurred. However, in every case the variations in the parameters used to delineate the regions are expected to involve significant differences at the river level in a combination of morphodynamic and chemical characteristics.

In the same way as a typology is best defined by the barycentre of the object classes rather than their limits, HERs are defined by their dominant characteristics rather than their boundaries. We have not derived an independent set of "predictions" about the aquatic ecosystem characteristics in each region, but we assume that such predictions are possible. Nevertheless, HERs could be used for predictive spatial extrapolation, i.e. the probability for an unknown system to match a given ecological type (and a limited range of variation for a set of important parameters) according to the HER it belongs to. In the following section we present the validation of the regional hypothesis at the aquatic ecosystem level.

VALIDATION: REGIONAL CHARACTERISTICS OF RUNNING WATER ECOSYSTEMS

Ideally, the validation should be made first on the hydrological and morphological control factors, but as hydrometric data are very poor in Bolivia for small and medium sized rivers, the characterization of hydrological regimes at the HER scale remains questionable. Conversely, the topographic map coverage, although incomplete, is sufficient to provide quantitative information on river and valley morphology for most HERs.

Geomorphologic characteristics

The methods and detailed results for the Andean and lowland zones are described in Gourdin (1997) and Binet (1998), respectively. Geomorphological parameters were measured from available topographic maps (1/50 000) for reaches about 5 km long, thus the data source is independent from those used for the regionalization. The database includes 520 reaches for the Andes and 306 for the lowlands where the map cover is more scattered. The parameters were selected to describe the valley morphology and as far as possible the structure and dynamics of the river beds. Structural parameters were the stream order (Strahler), the hill-slope gradient, the valley bottom slope and width, the fluvial form and sinuosity coefficient, and the transversal valley profile in the lowlands. The reach dynamics were evaluated differently in the two zones. In the Andes, the active strip was defined as a ratio between the total (active) and ordinary (wet) mapped river-bed width; in the lowlands, we used the nature and width of the river corridor, and the density of oxbow lakes. On the basis of these variables, a typical (statistical) longitudinal profile has been defined in all HERs with sufficient cartographic coverage (at least five reaches analysed by stream order classes). As a first conclusion, the geomorphological identity of the HER was validated on the basis of a dichotomous differentiation using four or five parameters. This result is quite clear in the Andes even at level 2, while in the lowlands the discrimination is good only at the first level, because of a much less heterogeneous landscape and a less complete database.

We present below the most significant results related to the structure and dynamics of medium-sized reaches. The parameters illustrated have been selected for their particular relevance in terms of ecological functioning: the slope determines hydraulic conditions and

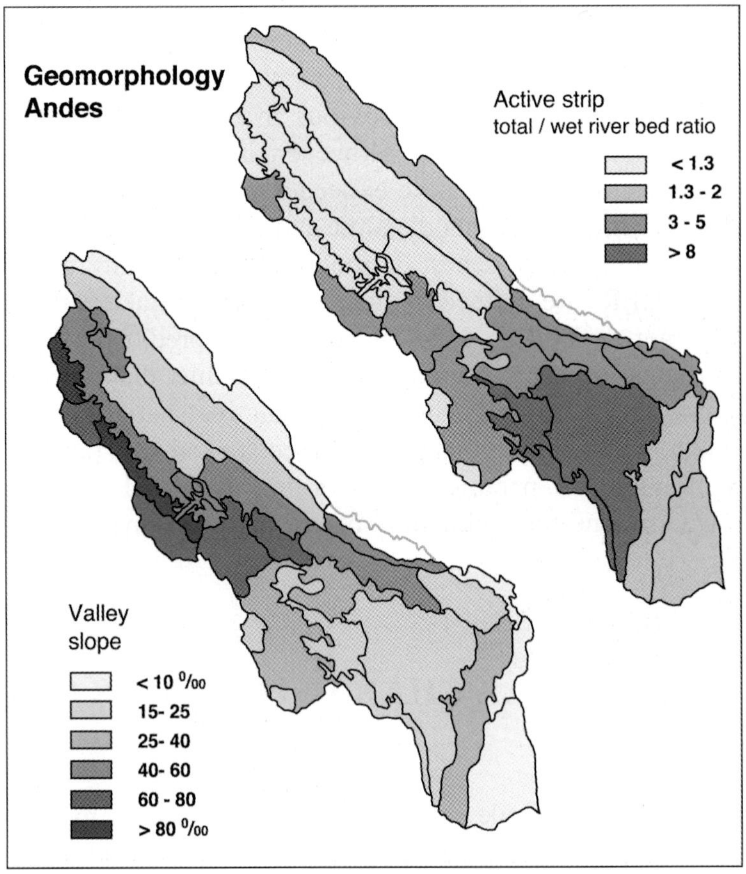

Fig. 5.4 Hydro-ecoregional pattern of the valleys' morphodynamics in the Andean zone: valley bottom slope (m km^{-1}) and active strip ratio (total bed width/wet width). Average values by hydro-ecoregion of the third and fourth order reaches characteristics, calculated from topographic maps; data from Gourdin (1997).

habitat structure; the active strip width reflects the sediment discharge and thus the fluvial dynamics and bed instability during high flow; for lowlands rivers, the sinuosity controls the physical habitat heterogeneity, and the corridor reflects both the dynamics and the connectivity with the terrestrial ecotone. Map layouts are justified by the number of measured (5–30 by HER) reaches and their good spatial distribution.

Andes The valley slopes define a very clear regional structure (Fig. 5.4). On the Amazon facing side of the Cordillera, a regular decrease is observed from the central axis to the subandean zone, and significant differences are noted at the HER-2 level within both Yungean and subandean regions (Ya and Sa). In the Rio Grande basin, facing the Chaco, the structure is different as the Yungean valleys (Yc) are steeper than the more internal semiarid ones (Va). The latter also have much more gentle slopes than the dry valleys of the Rio Beni basin (Vs). For the active strip, the main difference occurs between the humid and dry domains, as the river beds are much more unstable in the latter. But in the humid domain, the eastern Yungean regions (Ya-co, Ya-ce), and the lower subandean ones (Sa-b, Sa-h, Sa-s) also present wider active strips. Thus, if the slope is directly related to the geomorphological structure, the active strip mainly reflects the climatic dichotomy, and secondarily regional differences in rock hardness.

Lowlands The valley slopes (Fig. 5.5) distinguish first the Llanos from the Brazilian shield, and inside this latter the southern regions (Ea) from the northern peneplains and depressions (Eb-ph, Eb-da). This typology is completed using the width of the active corridor and the sinuosity that clearly set apart the Amazonian Llanos (La) from all the Beni

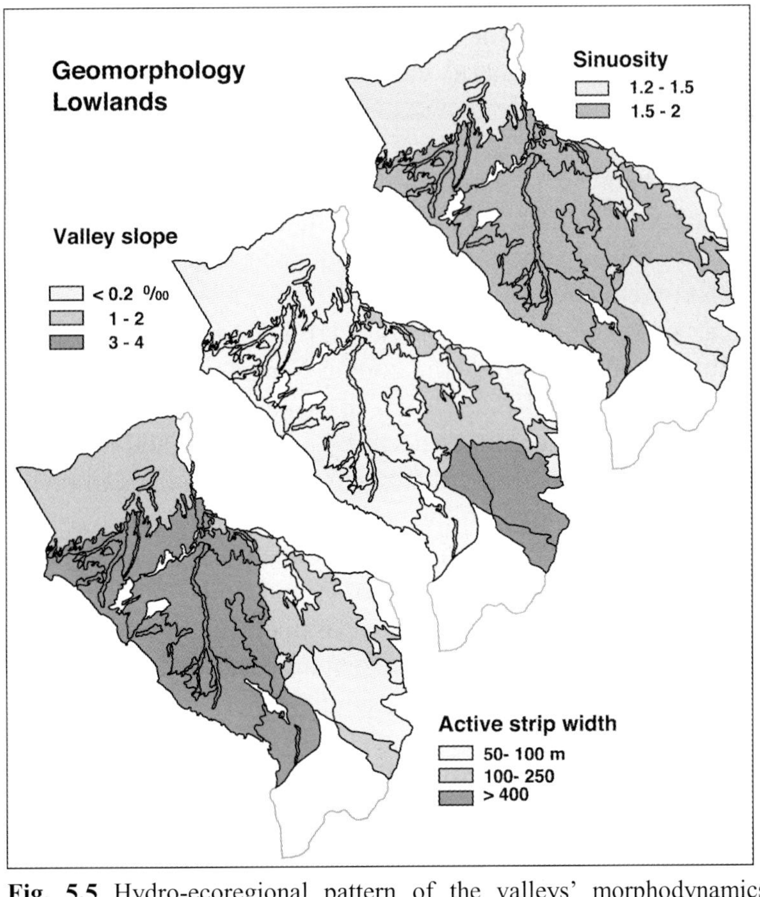

Fig. 5.5 Hydro-ecoregional pattern of the valleys' morphodynamics in the lowlands: sinuosity coefficient, valley bottom slope (m km^{-1}) and active strip width (m). Average values by hydro-ecoregion of the characteristics of third and fourth order reaches, calculated from topographic maps; data from Binet (1998).

Llanos (Lb, Ll, Ls) where dynamics and connectivity are higher, probably because of unconsolidated alluvial formations and intense flooding in these regions. On the Brazilian shield, the rivers of the northern peneplain (Eb-ph) exhibit wider corridors than in the surrounding regions, but the reason for that is unclear. The few rivers of the eastern serranías of the Chaco (Lc-sm) also have active mineral strips that can be related to the general instability of the river beds in the arid regions.

Hydrochemistry

We used the data collected and published by Guyot (1993) during the PHICAB programme, and all corresponding methods and results are available in Guyot's book. The database analysed here includes more than 260 sampling points, with as many as 2400 analyses for some parameters; we used only data from small- or medium-sized basins, generally occurring entirely within a single HER, or in some cases encompassing a maximum of two HERs. The Brazilian shield and the Chaco were excluded from the analysis due to insufficient data. Five parameters were selected for their high biological relevance: conductivity and calcium content for the global productivity, and pH and suspended solids (SS) which may act as limiting factors for aquatic communities; high SS contents are also indicative of river-bed instability affecting benthic assemblages (see Wasson *et al.*, 1998), and we added water colour as an indicator of the dissolved organic matter (DOM) content, another important parameter of the ecological functioning.

The regional differentiation on these five parameters is very clear from the scale of the geoclimatic domains down to the HER-2 level (Fig. 5.6). In the Andes, rivers have generally higher values for conductivity, calcium, pH and SS in the dry domain than in the humid one, and the most salient characteristic is the huge mean SS content of the rivers flowing through the dry (Vs) and lower semiarid valleys (Va-b), suggesting a strong limitation of the biological communities in these regions. These values are fairly reliable for the Vs region, as they are based upon 229 samples from 39 sites, with a mean value of 11.9 g l^{-1}. For the Va-b region, the data are less reliable for small basins with only 12 analyses for six sites, but our estimate for the region is based upon data collected at the output of the Andes from the main collector of these valleys (the Rio Grande at Abapo, 71 samples over 10 years), leading to a mean SS content of 5.9 g l^{-1}.

When looked at more closely, almost all HER-2s have their own hydrochemical identity. Some notable examples in the Amazonian Yungas are, the acidic pH of the lower region (Ya-b) due to humic soils (Rivera *et al.*, 1996) and the higher SS in the small semi-humid ones (Ya-s) due to greater erosion caused by aggressive land use for the coca culture. The "Chaco" Yungas (Yc) are also very distinct from the Amazonian ones, and noticeable differences appear at the HER-2 level in the semiarid valleys (Va-b, Va-i) or dry mountains (Ms-a, Ms-m). The factors governing regional hydrochemical characteristics seem to be, in order of importance: climatic conditions, geology, and soils. In the Llanos, the northern region (La) differs from the others for each parameter, thus confirming its particular identity. In the central zone, all the HER-2s present at least one important characteristic, the most obvious being the difference between the acidic waters of the northern savanna region (Ll) and the neutral waters of the central region (Ls-ai), as could be expected on the basis of the soil characteristics that led to their differentiation. Colour is an interesting parameter highlighting the higher DOM content of waters in the savanna regions, which could be related to the higher connectivity of the rivers with the surrounding wetlands and fluvial corridors, as previously assumed on the basis of the geomorphological study. On the humid side of the Andes, with tropical forest cover, the colour increases from the high Yungas to the subandean regions, suggesting a relationship with the mean hillslope gradient of the

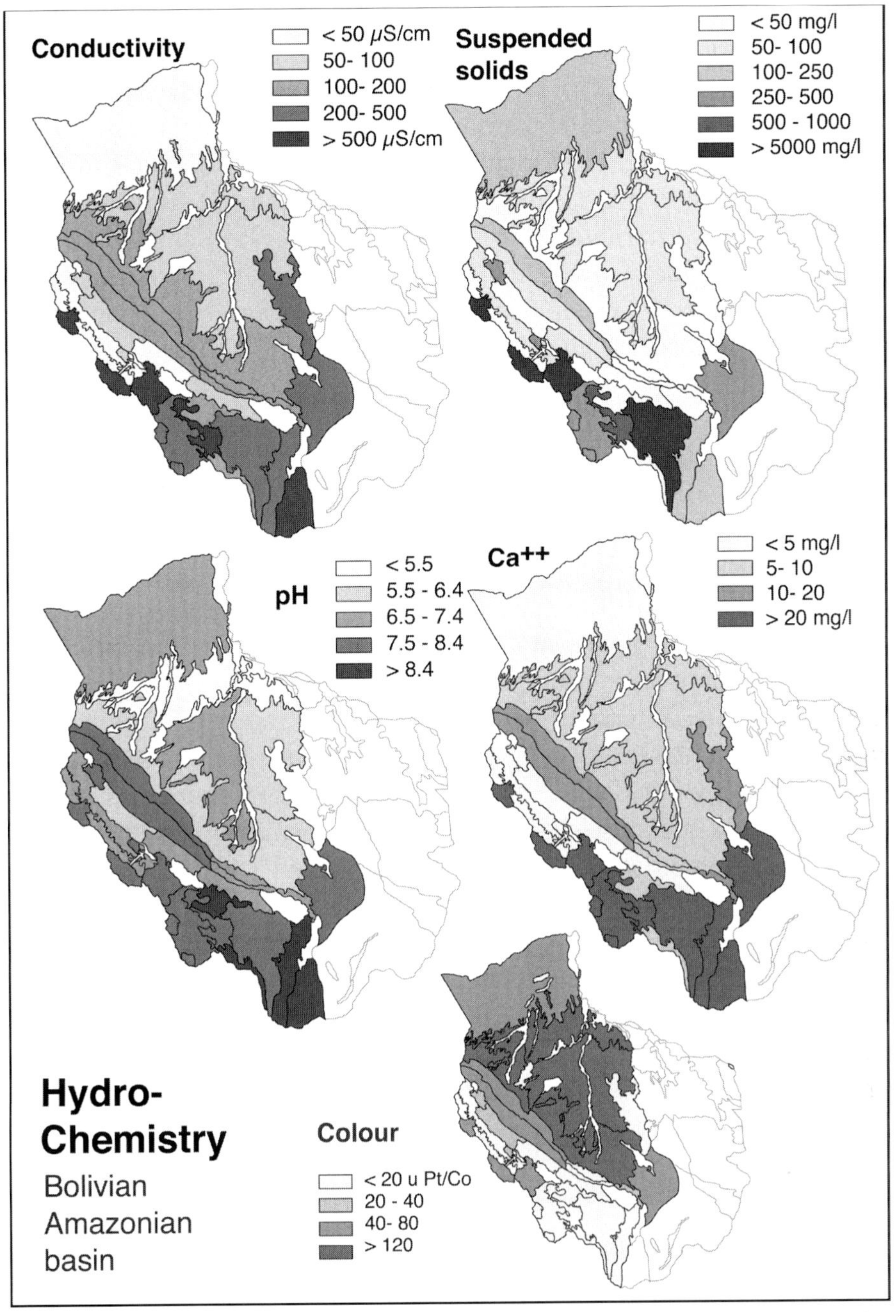

Fig. 5.6 Hydro-ecoregional pattern of river chemistry in the Bolivian Amazonian basin: conductivity (µS cm^{-1}), suspended solids (mg l^{-1}), pH, calcium (mg l^{-1}) and colour (Pt/Co units). Average values by hydro-ecoregion for endogenous basins (see text). Data from Guyot (1993).

different HERs; thus the interaction between the geomorphological structure and the vegetal cover could regulate the DOM content. In a former study of the regional DOC pattern in this basin, Guyot & Wasson (1994) hypothesized about the important role played by the connectivity of the lowlands rivers, but the possible importance of the geomorphological structure in the Andes has been overlooked.

Stream ecosystem structures

Ecological data were collected for 25 stream sites (24 sites for fish) representative of eight HER-2s of the Andean zone. As the purpose was to evaluate the relevance of the regional framework, independent sites (i.e. situated in different basins without any upstream–downstream relationship) were chosen having similar width (20–50 m at bankfull discharge) to avoid the effect of longitudinal variations. All the streams are undisturbed at the local scale (without direct contamination or physical alteration), but we cannot avoid completely the effects of traditional agriculture, which include water abstraction for irrigation in the driest region (Va-b). As sampling is almost impossible during high flow, the sites were sampled during the dry seasons (1997–1999) and data are representative of low flow conditions. The basins of the studied sites are generally limited to a single HER-2, or in some cases partially encompass an adjacent HER-2 belonging to the same HER-1; these sites were grouped for analysis according to the HER-2 in which they lie. But four stream sites of the Amazonian Yungas (Ya-a), whose basins come from the glacial Cordillera (Cg), were grouped separately for the analysis (Ya+Cg) to evaluate the possible differences between endogenous and heterogeneous streams. At each site data were taken on physical and chemical parameters, fish and invertebrates. Physical parameters are stream slope, proportion of lentic units, global stability evaluated by the Pfankuch index (Collier, 1992; Pfankuch, 1975), substrate characteristics (granulometry, porosity and mobility), and water temperature. Chemical parameters include conductivity, suspended solids (SS), pH, alkalinity, calcium and potassium. Fish communities were sampled at each site by electrofishing (15 sampling points per site), a battery of nets (from 10 to 110 mm mesh size), and ichtyotoxic (rotenone) when possible. For the benthic invertebrates, six Surber samples (0.1 m^2) representative of the main habitats were collected at each site. The detailed methodology for abiotic parameters and invertebrates, together with the results obtained at 13 sites are published in Rocabado & Wasson (1999) and Rocabado *et al.* (2001). All the multivariate analyses were performed with ADE-4 software (Chessel & Dolédec, 1996). We present below the first results of the regional differences of these 25 sites based upon physical and chemical parameters and some salient examples of the characteristics of the fish and invertebrates communities.

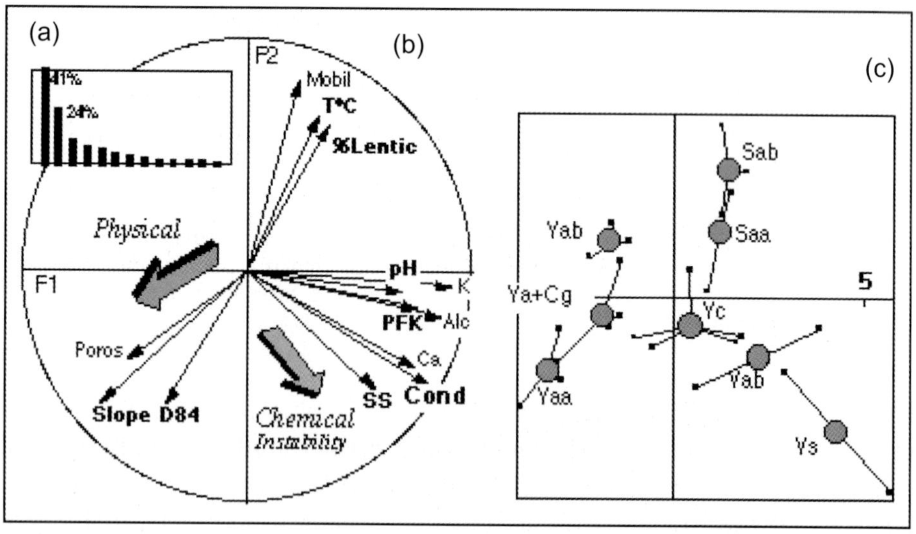

Fig. 5.7 Abiotic structure of the 25 stream sites sampled in the Andean zone: normalized principal component analysis (nPCA) of 13 physical and chemical parameters. (a) Relative contribution of the axis (eigenvalues). (b) Factorial map of the first two axis; Alc: alkalinity, Ca: calcium (mg l^{-1}), Cond: conductivity (µS cm^{-1}), D84: diameter (phi units) of the superficial granulometry 84th percentile, K: potassium (mg l^{-1}), %Lentic: proportion of lentic morphological units (pools), Mobil: ranking of substrate elements potential mobility, PFK: Pfankuch index, Poros: porosity of the substratum (%), Slope: stream slope (m km^{-1}), SS: suspended solids (mg l^{-1}), T°C: temperature in degrees Celsius. (c) site map of the first two axis; each star represents the sites (squares) grouped by hydro-ecoregions (circles projected at the barycentre of the related sites); see Table 5.1 and text for region codes.

Abiotic structure A normalized principal component analysis (nPCA) was performed for 13 physical and chemical parameters. The first two axes of the nPCA account for 65% of the total variance (Fig. 5.7(a)). The factorial map (Fig. 5.7(b)) indicates that the streams are grouped according to two independent (orthogonal) axes; the first opposing slope, granulometry and porosity with temperature, substrate instability and percentage of lentic units, is clearly related to a gradient of physical conditions; the second corresponds to the chemical characteristics and the Pfankuch index. The pattern of the sites grouped by HER-2 (Fig. 5.7(c)) reveals a satisfactory discrimination of the different regions on the basis of the abiotic characteristics of the streams. Regions are spread out over the physical axis, from the high Amazonian Yungas (Ya-a) to the subandean regions (Sa-a, Sa-b), according to a gradient of physical conditions mainly governed by the stream slope. The chemical axis clearly separates the dry and arid valleys (Vs, Va-b) with high conductivity and SS, from all the humid regions, among which the low Amazonian Yungas (Ya-b) lies apart due to a significantly lower pH. The pattern of stream instability is also interesting: the substrate mobility, estimated on the basis of size and form of the elements, is inversely correlated to the slope and thus higher in the subandean regions. But the Pfankuch index, which measures a global morphological instability, is correlated to the chemical parameters and not to the physical ones, thus indicating highly unstable river beds in the dry regions due to high sediment transport. The main conclusion here is that the abiotic conditions observed at the local scale are clearly consistent with the hydro-ecoregional structure: the physical and chemical axis of the nPCA correspond respectively to the geomorphological and climatic determinants used for the HER construction.

Biological communities: quantitative metrics The fish and invertebrate data set is still under analysis, but some consistent regional trends in diversity and density are already recognizable (Fig. 5.8). There is evidence of a very distinctive pattern between the humid and dry domains.

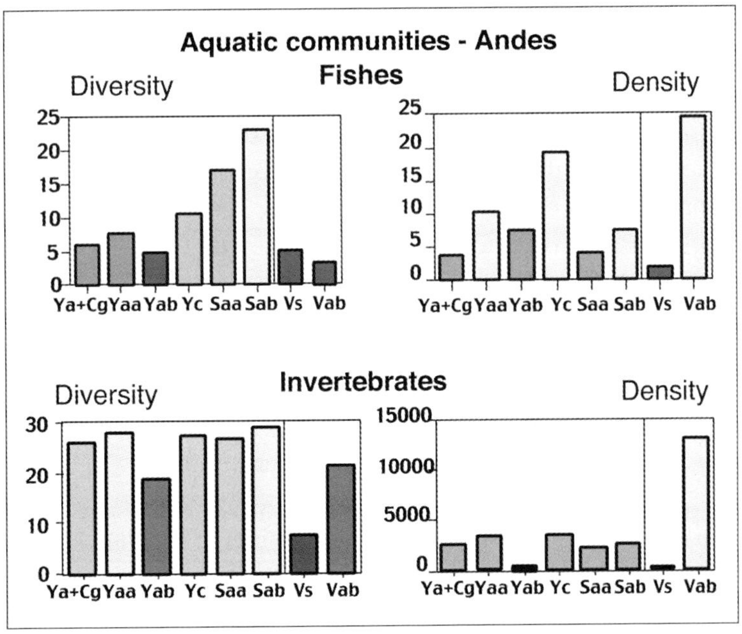

Fig. 5.8 Aquatic communities quantitative structure in the Andean hydro-ecoregions: mean values by hydro-ecoregion for four metrics measured at the stream sites. Bars with different fill colours are significantly different ($P < 0.05$) on the basis of nonparametric tests (Mann-Whitney) performed at the sample level. *Fishes*: 24 stream sites, quantitative data from 15 electrofishing points by site; diversity: number of genera; density: number of individuals 10 m^{-2} min^{-1}. *Invertebrates*: 25 stream sites, six Surber samples (0.1 m^2) by site; diversity: number of families; density: number of individuals m^{-2}. See Table 5.1 and text for region codes.

In the humid domain, the most salient feature is the progressive diversification of the fish community (number of genera) from the high Yungas (Ya-a) to the low subandean regions (Sa-b), which can be directly related to the decrease in the geomorphological gradient (lower slope, higher proportion of lentic units, see Fig. 5.7). Conversely, the invertebrate diversity pattern (number of families) is fairly homogeneous, with only a slight increase in both high Yungas and low subandean. A noticeable deviation to this pattern appears in the low Yungas region (Ya-b) that presents for both communities a significantly lower taxonomic richness related to the limiting effect of the acidic pH. When looking at fish density, the regional pattern is unclear, also reflecting high variability at the local scale; the low densities observed in different HERs cannot be related to the same limiting factor, and this is not explained. The pattern of invertebrates' density is homogeneous, with the same exception of the low Yungas region where densities are lowest because of acidic pH. However, for Ephemeroptera, a positive relationship has been observed between the geomorphological gradient and the densities, confirmed by a significant correlation with the stream slope (Rocabado *et al.*, 2001).

In the dry domain, the consistent figure for both fish and invertebrates is a low biodiversity and a strong contrast between the lowest densities observed in the dry valleys (Vs) and the highest densities in the arid valleys (Va-b). These biotic structures illuminate a common limiting factor for both regions but also an important difference in terms of temporal variability of aquatic ecosystem functioning. In both regions, during high flow, the river bed instability and huge SS contents act as strong limiting factors, but during low flow the hydrological conditions are very distinct. In the dry valleys (Vs), the main water courses fed by the high cordillera maintain a sustained baseflow during the dry season, and because of their high slope, the waters are permanently turbid and river beds unstable. Conversely, rivers in the arid valleys (Va-b) have lower slopes, and during the dry season the discharge drops to a few litres per second, which is just sufficient to feed extensive pools where some species can proliferate. Thus the biological data reflect the important difference in stream ecosystem functioning, with a marked seasonality in the one case (Va-b), and permanently adverse conditions in the other (Vs).

Biological communities: qualitative structure The composition of the communities was first examined by means of correspondence analysis, and the strength of the partition assessed by discriminant analysis with a Monte-Carlo simulation test. For fish, the regional structures are very well discriminated at the HER-1 level, but the HER-2 does not show a much clearer partition. For invertebrates, the figure is more blurred. However, for both communities, any regional classification (domain, HER-1 or HER-2) explains significantly

Table 5.2 Discrimination strength of various classification factors for fish and invertebrate communities in the Andean sites: results of discriminant analysis performed over a correspondence analysis (see text). Fish—semi-quantitative data from electrofishing and net samples; invertebrates—log transformed densities from six Surber samples. % Inertia—total inertia explained by each classification. *P*—significance level (Monte-Carlo simulation) fixed at 0.05 (*), 0.01 (**), 0.001 (***), or not significant (n.s.).

Factors	No. of classes for each factor	Fish: % Inertia	*P*	Invertebrates: % Inertia	*P*
Catchment	2	11.3	***	6.6	*
Altitude	3	17.2	***	12.3	*
Altitude	6	-	n.s.	-	n.s.
Slope	3	20.5	***	12.2	*
Slope	6	32.7	**	26.1	*
Domains	3	21.2	***	15.9	**
HER-1	6	39.5	***	29.4	**
HER-2	8	46.4	***	38.3	**

more of the total inertia than any classification based on a single factor such as slope, altitude or basin, with the same number of classes (Table 5.2). Among these factors, the clear basin effect (Rio Beni *vs* Rio Grande) observed for fish is mainly a biogeographical effect, as at present these two basins are connected near the Brazilian border, at a distance of about 1000 km from the Andean piedmont. Sarmiento & Barrera (1997) have reported noticeable dissimilarities in fish fauna between the Andean basins of the Rio Beni and Rio Mamore, to which pertains the Rio Grande, and such biogeographical dissimilarities interfere with the regional effect. Conversely, to explain the regional differences in the community composition, slope is a factor much more powerful and significant than altitude which allows only the lowest sites to be distinguished from the others, as partition into six classes does not give a significant result. Thus, we can infer that the geomorphological structure used as a primary determinant for HER delineation also plays a major role in explaining faunistic patterns.

Correspondences between hydro-ecoregions and biological structures Fish and invertebrate communities present clear, and sometimes strong regional differences when taking into account a combination of density, diversity, and taxonomic composition. The distinction is good at the HER-1 level and still relevant for HER-2, at least with the quantitative metrics (density and diversity). Regional faunistic patterns are stronger for fish than for invertebrates, a result consistent with similar studies in the USA (Rabeni & Doisy, 2000; Whittier *et al.*, 1988). However, the distinction based upon the endogenous *vs* heterogeneous basins (Ya-a *vs* Ya+Cg) did not lead to a decisive differentiation at the site scale, perhaps because in the studied case the particular characteristics of the upstream HER (glacial cordillera), mainly related to stream slope, temperature and hydrological regime, is not likely to have a strong effect on the downstream characteristics. The figure would probably be more contrasted in the case of a drier upstream region, expected to strongly influence the water chemistry and sediment transport downstream.

A rather surprising result here is the low level of differentiation in the composition of the invertebrate community when compared to the strong differences observed in the abiotic structures: why does the invertebrate fauna not respond more clearly? The answer should be sought in three directions: (a) the influence of local *vs* regional factors upon invertebrate communities, (b) the taxonomic resolution and (c) the seasonal variability. For (a), a rapid survey of existing papers reveals some conflicting conclusions. Some authors found a good or fair agreement between various ecoregional classifications and invertebrate communities (Feminella, 2000; Gerritsen & Barbour, 2000; Ivol, 1998; Rabeni & Doisy, 2000; Sandin & Johnson, 2000), but in many cases the fit is poor (Hawkins & Vinson, 2000; Quinn & Hickey, 1990; Waite *et al.*, 2000; Whittier *et al.*, 1988; and see review in Rabeni & Doisy 2000). All authors emphasize the importance of local factors in structuring benthic communities. Ecoregional discrimination is significantly higher when streams of similar size are compared between ecoregions (Waite *et al.*, 2000), and obvious differences are generally noted between mountains and plains; in our study, the first condition is observed, but not the second. Nevertheless, when comparing the range of geographical variations encompassed in the Andes, the regional discrimination of the benthic communities is somewhat weaker than expected, and we suspect that other factors could have reduced the fit. For taxonomy, all the studies comparing ecoregional discrimination at different taxonomic levels concluded that the genus/species level generally improves the discrimination, although not strongly, when compared to the family level (Feminella, 2000; Hawkins & Vinson, 2000; Ivol, 1998; Waite *et al.*, 2000). But these works deal with northern countries where the fauna is well known, and it remains questionable whether the present knowledge of neotropical fauna, where many genera and species are currently awaiting detailed description, leads to the same degree of comparability of different taxonomic levels (family/genus/species) as in temperate countries. In other words, does the family level give the same degree of discrimination for fauna in North and South America? The identification at genus level, presently in progress, will help answer this question. Finally, another important bias possibly specific to the humid tropics is the high inter-seasonal variability of the stream fauna, following dry/wet climate alternations. For the few specialists of the Yungas regions, the wet and dry seasons are two distinct "faunistic worlds" (E. Dominguez, personal communication). Because sampling is impossible at most sites during high flow, we might have captured only a limited part of the invertebrate fauna, perhaps the most resilient and widespread, thus reducing its ability to discriminate hydro-ecoregional differences.

DISCUSSION AND PERSPECTIVES

We shall briefly broach four questions.
(a) Is the hydro-ecoregion approach an ecohydrological question?
(b) Does it work?
(c) What are the scientific perspectives?
(d) What is the practical interest?

Ecohydrological questions

This point refers to the specificity of running water ecosystems: water flowing in a bed, i.e. physically driven, linear, oriented, dynamic. When Hynes stated in 1975 "the valley rules the stream", the concept was rapidly raised to the basin control level by "applied" scientists of the Corvallis (Oregon) US Environment Protection Agency team (Frissell *et al.*, 1986; Lotspeich, 1980). These authors focused their approach upon the geomorphological control of running water ecosystems by nesting the physical factors in a hierarchical scheme from the basin to the instream habitat level. From here arose the idea of using an ecoregional framework to classify running waters, first developed by Omernick (1987). This approach was also advocated and developed in France in an "applied" research institute, CEMAGREF (Wasson, 1989, 1996; Wasson *et al.*, 1993). But the undiscriminating use of the term "ecoregion" (without the qualifier "aquatic" initially used by Omernick) might have blurred the concept, as in its original meaning this word referred to terrestrial ecological units principally characterized by their vegetation. In this sense, climate is seen as the main controlling factor, and physiography only as a secondary one modifying the climate (Bailey, 1996). Even if mountains and landforms are finally taken into account respectively at the third ("provinces") and fourth level ("sections") of Bailey's classification, i.e. the levels generally used to compare aquatic ecosystems, Bailey's and Omernick's ecoregions do not correspond to one another because they do not refer to the same hierarchy of controlling factors. For the same reasons, our regionalization does not coincide with the geographical classification of the "aquatic and swampy systems of Bolivia" proposed by Navarro (1999), which was mainly based upon bioclimatic factors and vegetation features.

Following the concepts of the Corvallis team, we advocate that for running waters the geophysical factors (geomorphology, geology) must be raised *at least* to the same hierarchical level as the climatic ones. For instance, we expect more similarity between mountainous running water systems over a broad climatic range than between mountain and plain systems under the same climate. This conceptual difference is strong enough to justify a specific approach exclusively dedicated to the regionalization of running waters, for which we prefer the term "hydro-ecoregion". In this perspective, the hydro-ecoregion approach is not a mere (or one more) classification system, but relies on real ecohydrological questions dealing with the hierarchy of ecosystem controlling factors. This comes to investigating the degree of dependence between a running water ecosystem and its basin. The central questions refer less to the accuracy of the classification than to "how much" and "by means of what factors" regional features explain ecological functioning. Another important difference is that in the terrestrial approach, the vegetation itself serves as a basis to delineate ecoregions: i.e. we classify objects we can see. The figure is completely different for running waters. The regions are not based on aquatic biota but upon their potential determinants. Thus hydro-ecoregions are an hypothesis, whose validation is made *a posteriori* and independently with the challenge to relate assumed causes to observed effects. In this view, the regional framework cannot be seen as definitively fixed, and like any scientific hypothesis it might be improved by taking into account the results of the validation process.

Another misleading point is to consider the ecoregional approach as an alternative to the classical longitudinal "zonation" or "continuum" concepts to classify ecosystems and predict their biological features. In reality, the longitudinal position of a site in the hydrographic network, i.e. its size, has been recognized for half a century as the first factor governing the physical structure and biotic pattern of running waters (see review by Wasson 1989). No regional characteristic is expected to erase or even override this structure. Thus, as pointed out by many authors, a regionalization must be seen as complementary to any longitudinal classification and not as an alternative as it is sometimes depicted in the literature (Van Sickle & Hughes, 2000). The right way to assess the validity of a regionalization is to compare either sites of similar size, or the longitudinal evolution patterns between different regions. And its usefulness must be evaluated on the basis of "how much" the prediction of ecological features at a site of a given size is improved.

Does it work?

In summarizing the preceding results and comparing them to the literature referenced above, three important conclusions can be drawn. First, HERs provide a valid framework to regionalize ecologically important abiotic characteristics of running water ecosystems. In the Bolivian Amazon basin, morphodynamic and hydrochemical parameters measured at the reach scale are consistent with the geomorphological and climatic regional features. In the Andes, regions are well discriminated by the abiotic structures observed at the site scale. The predictive capacity of the HER approach is not yet evaluated but is expected to be good for a limited set of abiotic parameters. Secondly, for aquatic biota, the regional pattern is stronger for fishes than for invertebrates. This can be related to the spatio-temporal scale of the physical habitat required by each community. Fish have a greater life span, a larger body size and a greater mobility than invertebrates, so they are more dependent on the bi-dimensional (longitudinal and transversal) morphological structures and on the seasonal dynamics of habitat availability. We can infer that fish react to the inter-annual morphodynamic characteristics at the reach scale, while invertebrates react seasonally to micro-habitat structure and stability. We thus formulate the hypothesis that the broader the spatio-temporal habitat exploited by a community, the higher its dependence on large-scale physical factors. This could explain the weaker regional patterns observed for invertebrates, and we expect on this basis even weaker structures for communities of micro-organisms, as observed in the Loire basin (see Ivol, 1998). Third, we can validate the hypothesis that among the primary determinants, geophysical factors are *at least* as important as climatic factors. In the study of Andean sites, the main discrimination is given by the physical structure governed by the site gradient, even though the study area comprises two major global-scale climatic domains (Bailey, 1996), but does not encompass the major physical border between mountainous and lowland zones. In fact, we could hypothesize a higher importance of geophysical factors, but this must be validated with more biological data. However, we can note that river-bed instability, acting as a major physical control factor at the reach scale, is mainly related to the aridity, i.e. a climatic determinant.

Scientific perspectives

The first improvement of the approach would be to shift from a mere classification based upon structural parameters to a regional typology expressing the ecological functioning in relation to the controlling factors. As an example, we could combine only six abiotic and biotic functional traits to characterize the seven HERs studied in the Andes. These traits refer to physical harshness (due to slope, current velocity), chemical harshness (due to conductivity, pH, SS), instability (mainly morphological), global biological productivity, and fish and invertebrate biodiversity. Each trait can be ranked and seasonally differentiated (Table 5.3), highlighting the opposition between favourable milieus with high biological potential (Sa-a, Sa-b) and harsh systems with poor aquatic communities (Ya-b, Vs, Va-b). We emphasize the *tentative* nature of this typology, due to the limitation and incomplete treatment of the data set, but it gives an idea of what could be achieved.

A second perspective is the extension of the regionalization to surrounding basins. By its geographical position, the Bolivian Amazonian basin comprises samples of most ecoregions encountered in the inter-tropical range of the eastern side of the Andes, from northern Argentina to Venezuela; so a spatial extrapolation would be relatively easy without increasing excessively the number of HERs, at least at the first level. The methodology is solid and the only limitation concerns the geographical data availability. Such a regional framework should greatly favour the comparison of hydro-ecological studies in this zone.

Further developments concern the possibility of linking the basin and reach scales by the construction of hydro-ecoregional models. The basis for this approach derived from the conceptual scheme (Fig. 5.1) is the following: regionalization simplifies the geographical space by dividing it into homogeneous areas where some dominant physical processes are expected to govern ecosystem structure and dynamics. This stabilizes many fundamental "input parameters" for modelling, including geology and geomorphology (and thus sediment input rates), precipitation regime (and hydrological variability), hydrochemistry (controlling biological productivity), etc. Thus, in each region a limited set of parameters mainly related to longitudinal variation (slope, drainage area) can emerge as powerful control variables, leading to robust statistical models allowing the prediction of the physical

Table 5.3 Hydro-ecoregional typology of running water ecosystems (*tentative*). Abiotic and biotic functional traits are ranked from very low (+) to very high (+++++); S indicates a high seasonal variation of the trait.

HER-2	Ya-a	Ya-b	Yc	Sa-a	Sa-b	Vs	Va-b
Physical harshness	++++	+++	+++	++	+	+++	+++
Chemical harshness	++	++++	+	+	+	+++++	S +++++
Morphological instability	+	+	++	++	++	+++++	S +++++
Biological productivity	++	+	+++	++	++	+	S ++
Biodiversity of invertebrates	++++	++	+++	+++	++++	+	S ++
Biodiversity of fish	++	+	+++	++++	+++++	+	+

and chemical characteristics at the reach scale (see Cohen *et al.*, 1998). Then, the development of functional models, coupling the biological responses to the physical and chemical habitat characteristics, might enable the prediction of important biotic community features at the reach scale (Lamouroux *et al.*, 1999). This approach is still under development but could lead to integrated modelling across spatial scales.

Practical interest

The "ecoregional" approach of running waters was first developed in the USA with the aim of establishing realistic quality standards based upon regional reference conditions, and to define regionally coherent management policies (Hughes *et al.*, 1990). These objectives apply now to the European Community with the recent "Water Framework Directive" in which a similar approach is recommended. This remains the main practical application of the hydro-ecoregions. However, with our experience in Bolivia, we can point out two major outcomes for developing countries. First, the top-down approach may save time and money in achieving a preliminary description of the characteristics and potential of poorly known ecosystems. A limited panel of sampling sites representative of the main HERs might give a fairly good preview of the existing conditions, allowing further spatial extrapolation by means of hydro-ecoregional modelling. Secondly, cartographic representation of the results is a powerful communication tool for national decision makers. An attendant regionalization of anthropogenic activities and colonization dynamics, presently in progress, will reveal rising "water problems" related to both resource management and ecosystem conservation. We hope that this integrated approach will help Bolivian leaders in establishing sustainable development policies taking into account the regional potentialities and vulnerabilities of aquatic ecosystems.

Acknowledgements This research was made possible thanks to IRD funding, and the fruitful collaboration of La Paz University—Ecological Research Institute (UMSA—Instituto de Ecologia). A special acknowledgement to Ruben Marín who warmly accommodated us in his laboratory and provided constant support for the BIOBAB project. We are extremely grateful to the numerous people, Bolivian and French technicians and students, who helped us in field sampling and laboratory processing, and particularly to Fabrice Cathelin, Jean-Louis Menou and Gina Zurita. Many thanks also to Odile Fossati and François Gibon for their precious help in invertebrate identification and data analysis, and to Marc Pouilly who gave us the first impulse for sampling and analysing fish communities in the Andes. We send our cordial regards to the now retired initiator of this project, Laurent Lauzanne, to whom Bolivian hydrobiology is greatly indebted.

REFERENCES

Amoros, C. & Petts, G. E. (1993) *Hydrosystèmes Fluviaux*. Masson, Paris.

Bailey, R. G. (1976) *Ecoregions of the United States*, 1/7 500 000, in colour. US Dept Agriculture, Forest Service, Intermountain Region, Ogden, Utah.

Bailey, R. G. (1996) *Ecosystem Geography*. Springer-Verlag, New York.

Binet, D. (1998) *Régionalisation et Caractérisation Géomorphologiques des Cours d'eau dans le Bassin Amazonien de Bolivie. Mémoire de Fin d'études*. ENGEES, Strasbourg, and ORSTOM, La Paz.

Bourges, J., Hoorelbecke, J., Cortes, J. & Carrasco, L. M. (1993) Los regimenes hidrológicos de la cuenca Amazonica de Bolivia. In: *Actas del Seminario PHICAB*, 125–134. ORSTOM, La Paz.

Bourrel, L., Moreau, S. & Philips, L. (1999) Dinámica de las inundaciones en la Cuenca Amazónica Boliviana. *Revista Boliviana de Ecología y Conservación Ambiental* **6**, 5–17.

Chessel, D. & Dolédec, S. (1996) Programmathèque ADE: analyses multivariées et représentations graphiques de données écologiques, v. 4.0. Université Lyon I, France. Free download at: http://biomserv.univ-lyon1.fr/ADE-4F.html.

Cohen, P., Andriamahefa, H. & Wasson, J. G. (1998) Towards a regionalization of aquatic habitat: distribution of mesohabitats at the scale of a large basin. *Regulated Rivers: Research and Management* **14,** 391–404.

Collier, K. (1992) *Assessing River Stability: Use of the Pfankuch Method.* Department of Conservation, Science and Research Internal Report no. 131, Wellington, New Zealand.

Feminella, J. W. (2000) Correspondence between stream macroinvertebrate assemblages and four ecoregions of the southeastern USA. *Journal of the North American Benthological Society* **19**(3), 442–461.

Fossati, O., Wasson, J. G., Héry, C., Salinas, G. & Marín, R. (2001) Impact of sediment releases on water chemistry and macroinvertebrate communities in clear water Andean streams (Bolivia). *Archiv für Hydrobiologie* **151**(1), 33–50.

Frissell, C. A.,. Liss, W. J, Warren, C. E. & Hurley, M. D. (1986) A hierarchical framework for stream habitat classification: viewing streams in a watershed context. *Environmental Management* **10**, 199–214.

Gerritsen, J. & Barbour, M. T. (2000) Apples, oranges, and ecoregions: on determining pattern in aquatic assemblages. *Journal of the North American Benthological Society* **19**(3), 487–496.

Gourdin, F. (1997) *Régionalisation des Déterminants Morphologiques des Hydrosystèmes dans le Bassin Amazonien Bolivien—Cas de la Zone Andine. Mémoire de Fin d'études.* ENGEES, Strasbourg, & ORSTOM, La Paz.

Guyot, J. L. (1993) *Hydrogéochimie des Fleuves de l'Amazonie Bolivienne.* ORSTOM, Coll. Etudes et Thèses, Paris.

Guyot, J. L. & Wasson, J. G. (1994) Regional pattern of riverine dissolved organic carbon in the Amazon drainage basin of Bolivia. *Limnology and Oceanography* **39**, 452–458.

Guyot, J. L., Jouanneau, J. M. & Wasson, J. G. (1999) Characterization of river bed and suspended sediments in the Rio Madeira drainage basin (Bolivian Amazonia). *Journal of South American Earth Sciences* **12**, 401–410.

Hanagarth, W. (1993) *Acerca de la Geoecología de las Sabanas del Beni en el Noreste de Bolivia.* Instituto de Ecología, La Paz.

Hawkins, C. P. & Vinson, M. R. (2000) Weak correspondence between landscape classifications and stream invertebrate assemblages: implications for bioassessment. *Journal of the North American Benthological Society* **19**(3), 501–517.

Hughes, R. M. & Larsen, D. P. (1988) Ecoregions: an approach to surface water protection. *Journal of Water Pollution Control Federation* **60**(4), 486–493.

Hughes, R. M., Whittier, T. R., Rohm, C. M. & Larsen, D. P. (1990) A regional framework for establishing recovery criteria. *Environmental Management* **14**, 673–683.

Hynes, H. B. N. (1975) The stream and its valley. *Verhandlungen Internationale Vereinigung fur Theoretische und Angewandte Limnologie* **19**, 1–15.

Ivol, J. M. (1998) Hydro-écorégions du bassin de la Loire et variabilité des communautés macrobenthiques: essai de typologie régionale et référentiel faunistique. Doctoral Thesis. Université Lyon I & CEMAGREF, Lyon.

Lamouroux, N., Olivier, J. M., Persat, H., Pouilly, M., Souchon, Y. & Statzner, B. (1999) Predicting community characteristics from habitat conditions: fluvial fish and hydraulics. *Freshwater Biology* **42**, 275–299.

Lauzanne, L. & Loubens, G. (1985) *Peces del Río Mamoré.* ORSTOM, Travaux et Documents no. 192, Paris.

Lauzanne, L., Loubens, G. & Le Guennec, B. (1990) Pesca y biología pesquera en el Mamoré medio (región de Trinidad, Bolivia). *Interciencia* **15**(6), 452–460.

Lauzanne, L., Loubens, G. & Le Guennec, B. (1991) Liste commentée des poissons de l'Amazonie bolivienne. *Revue d'Hydrobiologie Tropicale* **24**(1), 61–76.

Lotspeich, F. B. (1980) Watersheds as the basic ecosystem: this conceptual framework provides a basis for a natural classification system. *Water Resources Bulletin* **16**(4), 581–586.

Maurice-Bourgoin, L. & Quiroga, I. (2002) Total mercury distribution and importance of the biomagnification process in rivers of the Bolivian Amazon. Chapter 4 in: *The Ecohydrology of South American Rivers and Wetlands* (ed. by M. McLain). IAHS Special Publ. no. 6 (this book).

Naiman, R. J., Lonzarich, D. G., Beechie, T. J. & Ralph, S. C. (1992) General principles of classification and the assessment of conservation potential in rivers. In: *Rivers Conservation and Management* (ed. by P. Boon, P. Calow & G. Petts), 93–123. Wiley & Sons, Chichester, UK.

Navarro, G. (1999) Aproximación de la tipificación biogeográfico–ecológica de los sistemas acuáticos y palustres de Bolivia. *Revista Boliviana de Ecología y Conservación Ambiental* **6**, 95–110.

Omernik, J. M. (1987) Ecoregions of the conterminous United States. *Annals of the Association of American Geographers* **77**(1), 118–125.

Omernick, J. M. & Griffith, G. E. (1991) Ecological regions versus hydrologic units: frameworks for managing water quality. *Journal of Soil and Water Conservation* **46**(5), 334–340.

Pearson, N. E. (1924) The fishes of the eastern slope of the Andes. I. The fishes of the Rio Beni basin, Bolivia, collected by the Mulford expedition. *Indiana University Studies* **11**(64), 1–58.

Pfankuch, J. (1975) *Stream Reach Inventory and Channel Stability Evaluation.* US Dept Agriculture, Forest Service, Region 1, Missoula, Montana, USA.

Pouilly, M., Ibañez, C., Guttierrez, M. & Yunoki, T. (1999) Funcionamiento ecológico de las lagunas de la zona de inundación del Rio Mamoré (Beni—Bolivia). *Revista Boliviana de Ecología y Conservación Ambiental* **6**, 41–54.

Proceedings of the Bolivian Congress on Limnology and Aquatic Resources (in Spanish) (1999) (Cochabamba, Bolivia, 12–14 June 1999). *Revista Boliviana de Ecologia y Conservación Ambiental* no. 6. Centro de Ecologia Simon I Patiño, Cochabamba, Bolivia.

Quinn, J. M. & Hickey, C. W. (1990) Characterisation and classification of benthic communities in 88 New Zealand rivers in relation to environmental factors. *New Zealand Journal of Marine and Freshwater Research* **24**, 387–409.

Rabeni, C. F. & Doisy, K. E. (2000) Correspondence of stream benthic assemblages to regional classification schemes in Missouri. *Journal of the North American Benthological Society* **19**(3), 419–428.

Resh, V. H., Brown, A. V., Covich, A. P., Gurtz, M. E., Li, H. W., Minshall, G. W., Reice, S. R., Sheldon, A. L., Wallace, J. B. & Wissmar, R. C. (1988) The role of disturbance in stream ecology. *Journal of the North American Benthological Society* **7**, 433–455.

Rivera, M. O., Libermann, M., Beck, S. & Moraes, M. (1996) Vegetación de Bolivia. In: *Comunidades, Territorios Indígenas y Biodiversidad en Bolivia* (ed. by K. B. Mihotek), 169–222. Univ. Gabriel R. Moreno & CIMAR, Santa Cruz de la Sierra, Bolivia.

Rocabado, G. & Wasson, J. G. (1999) Regionalización de la fauna bentónica en la cuenca Andina del río Beni (Bolivia). *Revista Boliviana de Ecología y Conservación Ambiental* **6**, 121–132.

Rocabado, G., Wasson, J. G. & Lino, F. (2001) Distribution of Ephemeroptera in the Andean part of the Rio Beni drainage basin (Bolivia): regional pattern or control at the local scale? In: *Trends in Research in Ephemeroptera and Plecoptera* (ed. by E. Dominguez), 161–172. Kluwer Academic–Plenum Press, New York.

Roche, M. A., Fernandez-Jauregui, C., Aliaga Rivera, A., Peña Mendez, J., Salas Rada, E. & Montaño Vargas, J. L. (1990) *Balance Hídrico Superficial de Bolivia—Atlas PHICAB*. ORSTOM–UNESCO, La Paz.

Rohm, C., Giese, J. W. & Bennett, C. C. (1987) Evaluation of an aquatic ecoregion classification of streams in Arkansas. *Journal of Freshwater Ecology* **4**, 127–140.

Sandin, L. & Johnson, R. K. (2000) Ecoregions and benthic macroinvertebrate assemblages of Swedish streams. *Journal of the North American Benthological Society* **19**(3), 462–474.

Sarmiento, J. & Barrera, S. (1997) Observaciones preliminares sobre la ictiofauna de la vertiente oriental andina de Bolivia. *Revista Boliviana de Ecología y Conservación Ambiental* **2**, 77–99.

Townsend, C. R. (1989) The patch dynamics concept of stream community ecology. *Journal of the North American Benthological Society* **8**(1), 36–50.

Van Sickle, J. & Hughes, R. M. (2000) Classification strengths of ecoregions, catchments, and geographic clusters for aquatic vertebrates in Oregon. *Journal of the North American Benthological Society* **19**(3), 370–384.

Waite, I. R., Herlihy, A. T., Larsen, D. P. & Klemm, D. J. (2000) Comparing strengths of geographic and nongeographic classifications of stream benthic macroinvertebrates in the Mid-Atlantic Highlands, USA. *Journal of the North American Benthological Society* **19**(3), 429–441.

Ward, J. V. & Stanford, J. A. (1983) The intermediate-disturbance hypothesis: an explanation for biotic diversity patterns in lotic ecosystems. In: *Dynamics of Lotic Ecosystems* (ed. by T. D. Fontaine & S. M. Bartell), 347–356. Ann Arbor Science, Ann Arbor, Michigan.

Warry, N. D. & Hanau, M. (1993) The use of terrestrial ecoregions as a regional-scale screen for selecting representative reference sites for water quality monitoring. *Environmental Management* **17**(2), 267–276.

Wasson, J. G. (1989) Eléments pour une typologie fonctionnelle des eaux courantes. 1. Revue critique de quelques approches existantes. *Bulletin d'Ecologie* **20**(2), 109–127.

Wasson, J. G. (1996) Structures régionales du bassin de la Loire. *La Houille Blanche* **6/7**, 25–31.

Wasson, J. G. & Barrère, B. (1999) Regionalización de la Cuenca Amazónica Boliviana: las hidro-écoregiones de la zona andina. *Revista Boliviana de Ecología y Conservación Ambiental* **6**, 111–120.

Wasson, J. G., Bethemont, J., Degorce, J. N., Dupuis, B. & Joliveau, T. (1993) Approche écosystémique du bassin de la Loire: éléments pour l'élaboration des orientations fondamentales de gestion. Phase I: Etat initial—problématique. Rapport d'étude. CEMAGREF, Lyon, & Université de St Etienne (CNRS 260).

Wasson, J. G., Marin, R., Guyot, J. L. & Maridet, L. (1998) Hydro-morphological variability and benthic community structure in five high altitude Andean streams (Bolivia). *Verhandlungen Internationale Vereinigung für Theoretische und Angewandte Limnologie* **26**(3), 1169–1173.

Whittier, T. R., Hughes, R. M. & Larsen, D. P. (1988) Correspondence between ecoregions and spatial patterns in stream ecosystems in Oregon. *Canadian Journal of Fisheries and Aquatic Sciences* **45**, 1264–1278.

CHAPTER 6

ECOHYDROLOGY OF RIPARIAN FORESTS IN THE ORINOCO RIVER BASIN

Judith Rosales, Conrad Vispo, Nelda Dezzeo, Ligia Blanco-Belmonte, Claudia Knab-Vispo, Nirson Gonzalez, Félix Daza, Chris Bradley, David Gilvear, Gladys Escalante, Noemí Chacón & Geoffrey Petts

The Orinoco is the world's third largest river according to discharge (36 000 m^3 s^{-1}) with a basin area of 1 110 000 km^2 (Fig. 6.1). Its physical and biological characteristics have been widely described in the literature (Colonello, 1990a; Hamilton & Lewis, 1990; Nordin & Perez-Hernandez, 1989; Rosales *et al.*, 1999a; Vásquez & Wilbert, 1992; Weibezahn *et al.*, 1990). The Orinoco basin can be divided into three major drainage regions: the Andes, the Llanos, and the Guiana Shield, each with different geochemical and hydrological characteristics (Johnson *et al.*, 1991; Meade, 1994; Stallard *et al.*, 1990). Rivers draining from the Andes are called white-water rivers due to their high suspended-solids concentrations, electrical conductivities and pH. This differs from the clear- and black-water rivers draining the Llanos and the Guiana Shield with low suspended-solids concentrations and conductivities, but variable pH (Fig. 6.2, unpublished data from Rosales, Bradley, Gilvear & Petts). As the main stem of the Orinoco travels downstream from the triple confluence with the black-water Atabapo and the white-water Guaviare it is increasingly characterized by high-nutrient, high-sediment white waters. This becomes more pronounced below the junctions with the main sources of sediments draining from the Andes, the Meta and the Apure.

The Orinoco's fringing flood plain is dominated by riparian forest vegetation, with flooded savannas frequent in the middle and lower reaches. The strongly unimodal seasonal variation in river level becomes more pronounced as one goes downstream, from a 14 m difference between high and low water levels at Puerto Ayacucho to a 16 m difference at Ciudad Bolívar. This results in a highly predictable flood pulse to which flooded riparian forest and savanna vegetation are adapted. Differences in river geochemistry also affect plant communities. Flooded riparian forests have been differentiated into *igapó* and *várzea* forest types, due to differences in floristic composition in forests flooded by nutrient-poor and nutrient-rich waters, respectively (*sensu* Prance, 1979). Orinoco flood plain forests have many tree species in common with the Amazonian flood plain forests, apparently because of historical fluvial connections that date back to the Quaternary and the Tertiary (Rosales *et al.*, 1999a; Rosales, 2000). Comparatively few studies of riparian forests have been conducted in the Orinoco basin, and only one study, that of the *várzea* forest of Laguna de Mamo (Colonello, 1990a,b, 1991), can be considered as representative of a typical lower-Orinoco fringing flood plain away from direct effects of tributaries. The majority of riparian forest studies focusing on ecohydrological interactions have been conducted within the confluence zones of the Caura and the Mapire, two tributaries in the lower Orinoco.

Tributary confluence zones along the Orinoco's mainstem are largely influenced by flood pulses. High flows in the Orinoco impede drainage from its tributaries producing a backwater effect that creates strong hydrological and geochemical gradients. These gradients vary with flood pulse intensity but exert a strong influence on the composition and

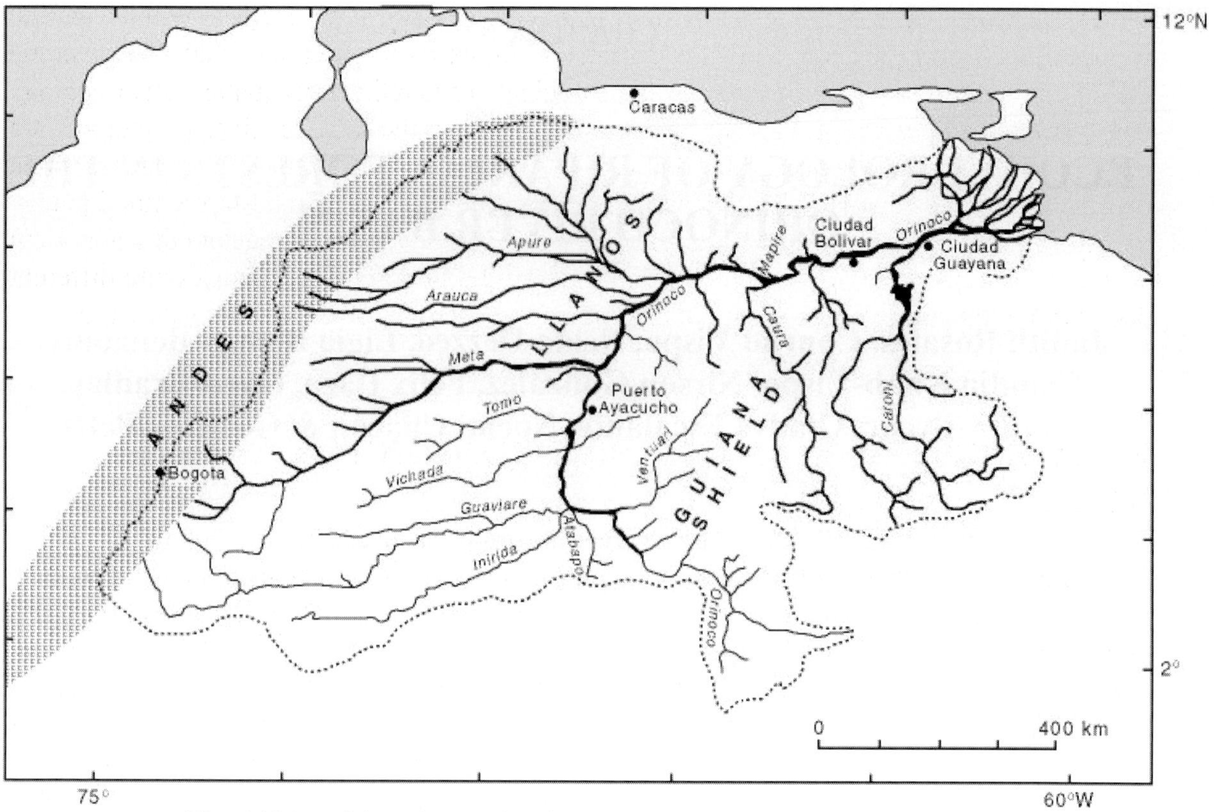

Fig. 6.1 Map of the Orinoco River basin and its main tributaries.

diversity of the flood plain vegetation (Rosales, 1990; Rosales *et al.*, 1999b, 2001). An example is in the Caura/Orinoco confluence zone (Hamilton & Lewis, 1990) where intermixing of the Orinoco's nutrient-rich waters with the Caura's nutrient-poor waters is thought to cause an atypical assemblage of species from both *várzea*- and *igapó*-type forests (Rosales *et al.*, 2001). A similar pattern was found in the smaller Mapire River confluence (Rosales, 1990).

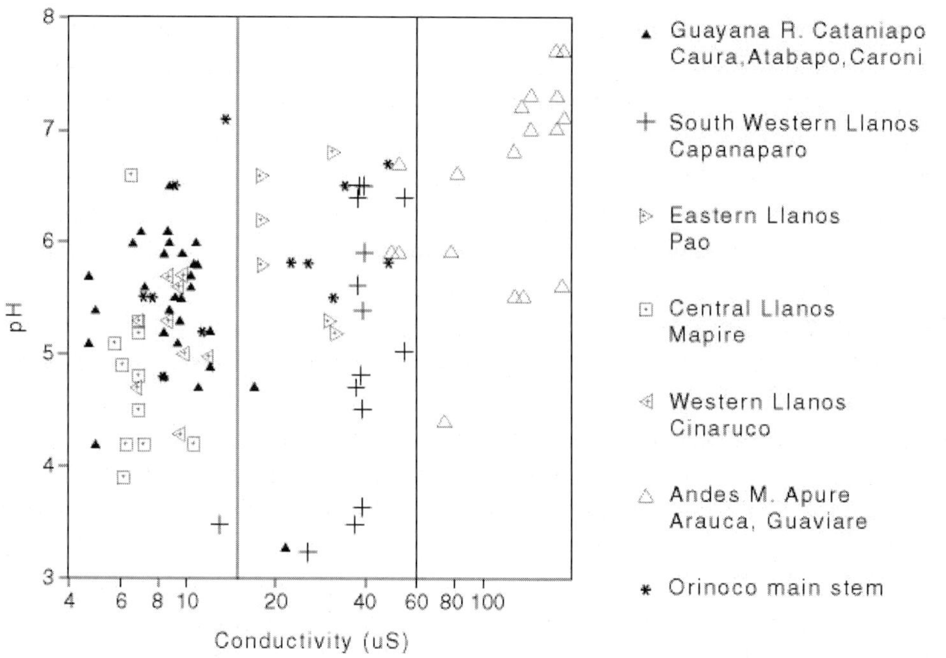

Fig. 6.2 Distribution of plots at confluence zones along the Orinoco mainstem according to geochemical variation in waters.

In this chapter, we describe some important ecohydrological interactions within riparian forests in the lower Orinoco using the Caura and Mapire rivers as case studies. We examine the importance of riparian forests and hydrological seasonality for aquatic faunal ecology and human use patterns in the Caura River. We then use the Mapire River to demonstrate how ecohydrology influences the composition, structure and diversity of riparian forest communities. By comparing decomposition processes of the Caura and Mapire flood plains, we show how biogeochemical gradients along river confluence zones regulate this important function of flooded forests. We hope to underscore the importance of considering different ecohydrological interactions when developing a strategy for riparian forest conservation.

ECOHYDROLOGICAL INTERACTIONS IN THE CAURA RIVER FLOOD PLAIN

The Caura River is a north flowing tributary of the Orinoco draining an area of 45 336 km^2. As is typical of Guiana Shield rivers, the Caura is generally considered a nutrient-poor river. Nonetheless, it has relatively high pH (averaging slightly above 6) and relatively low dissolved carbon values in comparison with classic "black-water" rivers. Its riparian corridor, which includes the river channel and adjacent flood plain, can be differentiated into distinct sectors according to their physical and hydrological characteristics (Rosales, 2000). The southern basin, upstream of La Mura Rapids, is dominated by bedrock-confined channels. Downstream, the flood plain is more developed and the river becomes increasingly influenced by contemporary and historical Orinoco River water-level fluctuations (Rosales *et al.*, 1999b). Superimposed on these broad differences in physical and hydrological characteristics are a range of riparian habitats associated with the different fluvial landforms present; these include overbank and in-channel geomorphic forms such as confluence swamps, back-swamps, abandoned channels, levees, point-bar systems, alluvial islands, lateral sand bars, rocky margins, bank benches, slack-waters and oxbow lakes (for descriptions see Rosales *et al.*, 2001). The riparian vegetation biota varies according to this diverse range of habitats.

The Caura's riparian forests experience pronounced wet and dry seasons that are reflected in the unimodal flood hydrograph. Differences in hydrological seasonality increase northwards to the confluence zone where more than half of the annual rain falls during the four wettest months, and maximum mean monthly rainfall is almost 15 times that of the driest month. Figure 6.3 shows the variation in annual flows for the San Luis gauging station, representative of the upstream portion of the confluence zones. A flood plain cross-section at this point is reproduced in Fig. 6.3, illustrating the extent to which flooding depths and duration are likely to vary. This strongly influences the types of vegetation found along the cross-section (unpublished data from Bradley, Rosales & Knab-Vispo). For comparison, data from the Musinacio gauging station on the Orinoco mainstem shows the variation in river stage just upstream of the Caura River confluence (Fig. 6.4). The differences in vegetation types occurring along the flood plain, in environments varying from open water (<10 m a.s.l.) to infrequently flooded savanna (>23 m a.s.l.), are shown in Fig. 6.4. The vegetation type and diversity vary in relation to flood regime, rate of sedimentation, nutrient availability and the stress caused by high water flows along the elevational gradient. Along the lower Caura, Rosales *et al.* (2001) found that the number of species (S) is negatively correlated with inundation depth (IN) and Brady's soil phosphorous availability (P) ($S = 29.45 - 2.52IN - 0.16P$, $p = 0.05$, Brady, 1990). The strength of both variables increases downstream. At the Caura's mouth, human-induced disturbance associated with fire and cattle ranching, is an important factor affecting vegetation composition and diversity (Knab-Vispo, unpublished data).

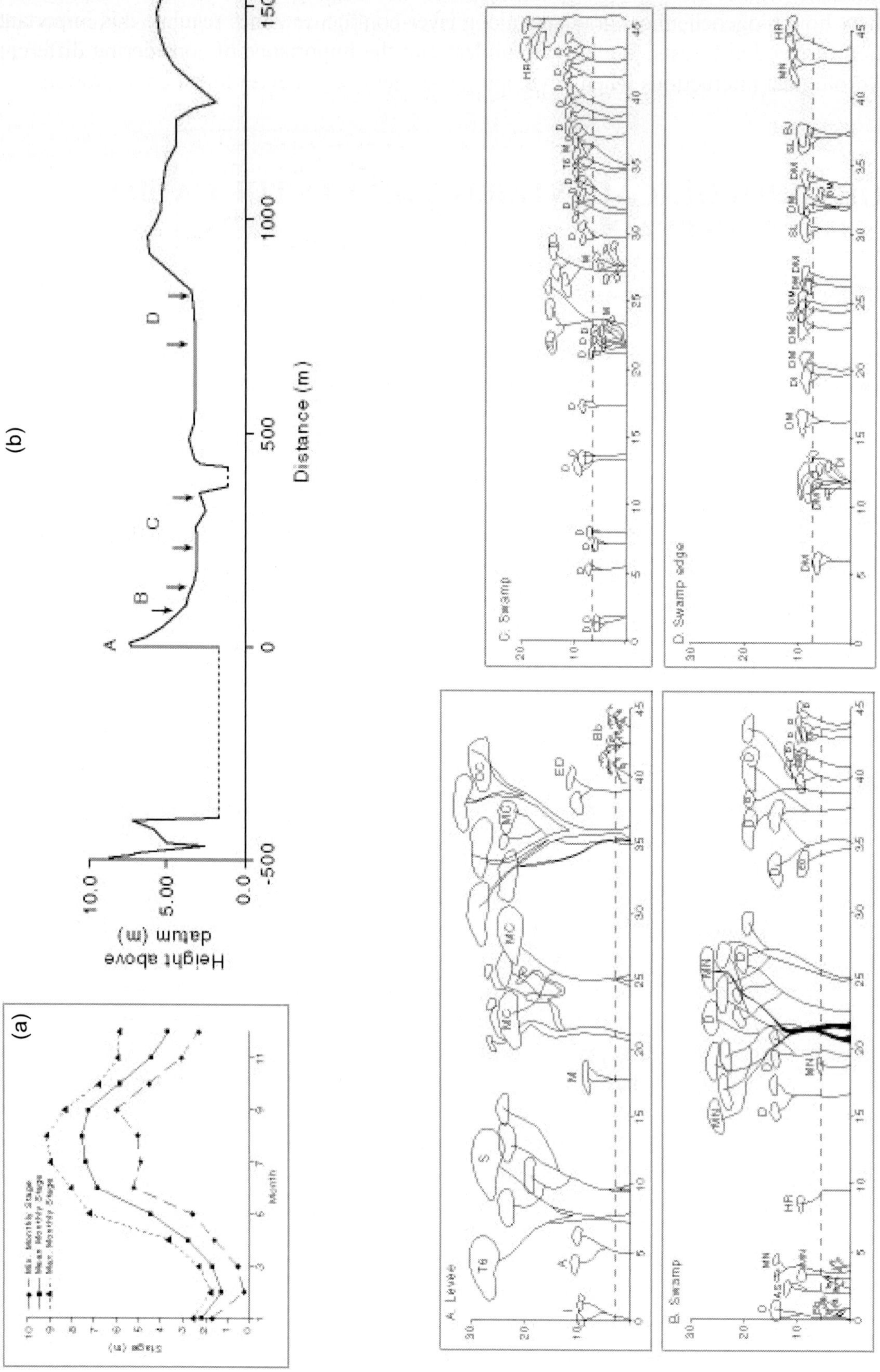

Fig. 6.3 (a) Hydrograph of San Luis gauging station on the Caura River. (b) Perpendicular topographical profile along the flood plain at the left margin of the Caura at San Luis gauging station showing forest plots.

The seasonal flood regime emphasizes the linkage between the river and flood plain. During high flows in July and August, extensive tracts of flood plain are inundated by up to 10 m of water. Such flooding provides for extensive nutrient interchange between the river and its flood plain and gives aquatic organisms access to temporary, food-rich habitats that are often important for key life cycle stages. As these flood waters recede, portions of the aquatic biota once again become confined to the fringing lake systems comprised of backswamps, oxbow lakes and abandoned channels. High rates of evapotranspiration and

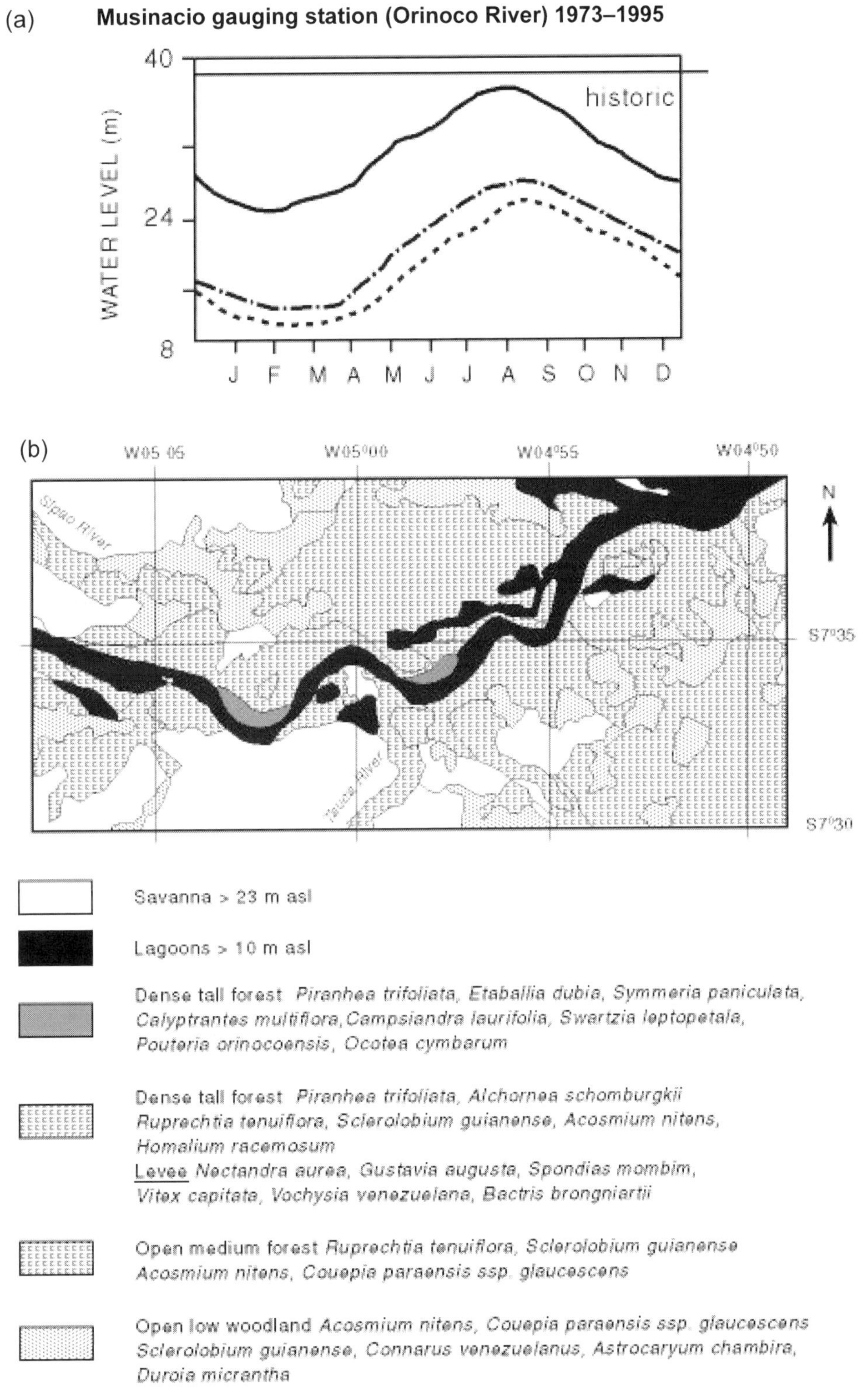

Fig. 6.4. (a) Hydrograph of Musinacio gauging station on the Orinoco River. (b) Vegetation at the lower part of the Caura's confluence zone.

plant productivity are likely in flood plain areas exposed by falling waters. In this way, the annual flood pulse of sediments and nutrients is critical to large flood plain river ecosystems such as the Caura (Junk *et al.*, 1989; Petts & Amoros, 1996).

AQUATIC INVERTEBRATES AND FISH COMMUNITIES OF THE LOWER CAURA RIVER

Caura River aquatic invertebrate communities have been shown to be strongly associated with substrate type (Fig. 6.5; Blanco-Belmonte, unpublished data). It is important to note that at least three of these substrates (woody debris, leaf litter and green, living leaves) are directly provided by the riparian forest. Substrate specificity appears to vary among invertebrate taxa (see Table 6.1). For example, while Diptera and Ephemeroptera were the main groups on all substrates, one Ephemeroptera (denoted as E4 in Table 6.1) was abundantly found in woody debris, but was scarcer on aquatic vegetation and green, living leaves of the flooded forest. Meanwhile another Ephemeroptera (denoted as B1) was a frequent dweller in the litter substrate, although it was found in much lower numbers in the benthos and on aquatic vegetation. Likewise, the Ephemeroptera CAM1 was almost exclusively a benthos inhabitant. In the rocky pools, another unique and diverse Ephemeroptera community was found. The value of woody debris as habitat and food source is quite important, especially given the relative lack of attention that this substrate has received in previous tropical studies. In the Caura flood plain, woody debris was found to house one of the highest invertebrate densities and biomasses. Submerged woody biomass has not yet been estimated within the Caura flood plain although ample woody debris has been observed. Eighty-eight aquatic macroinvertebrate species were encountered in woody debris, a diversity high in comparison with reports from the traditionally better studied benthos.

Invertebrates are important food sources for some fish. The average fish's diet was composed of about 15% invertebrates by volume. Diet analysis coupled with data describing invertebrate distribution by substrate were used to deduce where invertebrate-eating fish were feeding at different times of the year (Table 6.1; Gonzalez & Blanco-Belmonte unpublished

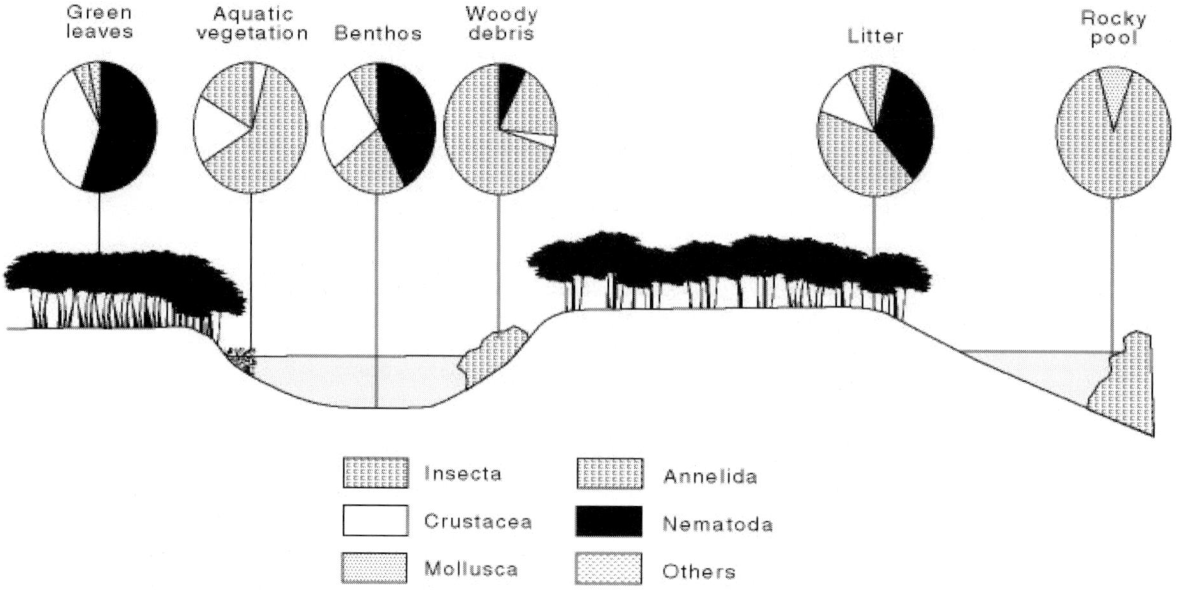

Fig. 6.5 The distribution of aquatic invertebrates in the different habitats of the flood plain in the lower Caura.

Table 6.1 Fish species that fed, at least partially, on invertebrates. This table was produced by comparing stomach contents with density of macroinvertebrates on various substrates found in flood plain lakes of the lower Caura River. *Note*: not all fish species were caught during all seasons, and not all of the listed fish consumed invertebrates during all seasons. Phase abbreviations: HW, high water; RW, rising water; LW, low water; FW, falling waters.

Fish species	Phase	Habitat	Main food items (invertebrates codes)
Triportheus angulatus	HW	Flooded forest	Terrestrial insects, flowers, fruits
	FW	Remaining flooded forest	Terrestrial insects
		Littoral benthos	Ephemeroptera (CAM1)
		Woody shores	Ephemeroptera (E4)
	LW	Littoral benthos	Ephemeroptera (CAM1)
		Woody shores	Ephemeroptera (E4)
		Lake shores	Terrestrial insects
T. elongatus	HW	Flooded forest	Terrestrial insects, flowers, fruits
	FW	Remaining flooded forest	Terrestrial insects, flowers, fruits
		Woody shores	Ephemeroptera (E4)
Pellona flavipinnis	HW	Woody flooded forest	Peces: Doradidos, microcharacids, terrestrial insects
	FW	Woody shores	Odonata (O1), Ephemeroptera (E4), Trichoptera (BRA1), Coleoptera (GYR1)
P. castelneana	HW	Flooded forest	Fish, Odonata (O5), (O6), (O10), Diptera (CHI1), Lepidoptera (LEP2)
Brycon bicolor	FW	Remaining flooded forest	Flowers, fruits, leaves, terrestrial insects
	LW	Vegetated shores	Flowers, leaves
		Woody shores	Ephemeroptera (E4), Diptera (CHI 2)
Brycon sp.	FW	Remaining flooded forest	Leaves, flowers, Hemiptera (TRI1), Aracnidae
Pimelodus blochii	HW	Flooded forest	Seeds, Gasteropods (PLA1), Coleoptera (ELM4), (CUR1),
		Woody shores	Lepidoptera (LEP2), Trichoptera (BRA1)
Raphiodon vulpinus	HW	Flooded forest	Fish, terrestrial insects, Diptera (CHI1), (CHI2), Odonata (O3?)
	FW	Remaining flooded forest	Fish, terrestrial insects
Leporinus friderici	FW	Woody shores	Ephemeroptera (E4), leaves and twigs
		Littoral benthos	Ephemeroptera (CAM1)
	LW	Vegetated shores	Diptera (CHI1) and leaves
Leporinus fasciatus	HW	Woody flooded forest	Terrestrial insects, Ephemeroptera (E4), Diptera (CHI1), (CHI3)
	DW	Woody shores	Ephemeroptera (E4), Diptera (CHI1)?
Hemiodus gr. inmaculata	RW	Woody shores	Coleoptera (D5), Branchiopoda (CYC1), Ephemeroptera (E4), (B1), (CAL1), leaves, roots
		Littoral benthos	Ephemeroptera(CAM1), Ostracoda (OS3), (OS8), (OS9)
Hydrolicus sp.	HW	Woody flooded forest	Fish, terrestrial insects, Ephemeroptera (E4), Hemiptera (TRI1), (NEO1), (NEP1), Aracnida, Nematoda (ME2), Odonata (O1)
	FW	Remaining flooded forest	Fish, terrestrial insects
		Littoral benthos	Ephemeroptera (CAM1), Diptera (CH1)
	LW	Woody shores	Fish, Ephemeroptera (E4)
Pygocentrus cariba	HW	Flooded forest	Fish, Terrestrial insects, Lepidoptera, Hemiptera (BEL1)
	FW	Woody shores	Fish, Ephemeroptera (E4), Coleoptera (GYR1), (TR)
		Littoral benthos	Ostracoda (OS9)
	LW	Woody shores	Fish, Ephemeroptera (E4), Diptera (CHI1)
Serrasalmus rhombeus	LW	Woody shores	Fish fins, Ephemeroptera (E4), Diptera (CHI1)
Platydora costatus	LW	Woody shores	Ephemeroptera (E4), Trichoptera (BRA1) and pieces of wood
Chaetobranchus flavescens	LW	Vegetated shores	Fish, Hemiptera (TRI2), Hydracarina, Diptera (CHI1),
		Littoral benthos	(CHI2), Ostracoda (OS9), Copepoda (COP2)
Piaractus brachypomus	FW	Lake shores	Fruits, seeds, leaves
		Littoral benthos	Ephemeroptera (CAM1)
		Woody shores	Ephemeroptera (E4)
	LW	Lake shores	Fruits, seeds, leaves
		Woody shores	Ephemeroptera (E4)
Plagiosum squamosisimus	FW	Littoral benthos	Fishes, Ephemeroptera (CAM1),
		Woody shores	Ephemeroptera (E4) (CHI2), (CER), Odonata (O11), Coleoptera (HUC3), Mollusca (PIS1), (DRE1)
Hassar iheringii	HW	Benthos	Ephemeroptera (CAM1), Diptera (CHI), (CER), Mollusca (PIS1)
Argonectes scapularis	RW	Litoral benthos	Diptera (CHI1), (CHI2), (CER4), Ephemeroptera (CAM1),
		Lake shores	terrestrial insect

data). These data show that during high water, almost all the fish species studied took advantage of flooded forest invertebrates (i.e. terrestrial invertebrates or invertebrates found on flooded debris). Furthermore even during low water, terrestrial insects probably derived from the lake littoral environment and invertebrates found on woody debris were important in the diets of several species. Thus, insects and/or invertebrate-substrate produced by terrestrial environments played important roles in the feeding ecology of invertebrativorous fish species.

Interesting ecohydrological interactions among the flooded forest, the aquatic fauna and human populations, are evidenced in Vispo's study of *Piaractus brachypomus* (unpublished data), This large member of the piranha family is the most important species by weight for local consumption and the lower Caura commercial fishery. *Piaractus brachypomus* is characterized by strong jaws and crushing teeth that are well adapted to a diet of fruits and seeds. In fact, its diet in the Caura alone includes fruits and seeds from more than 100 species (Knab-Vispo, unpublished data). The annual life cycle of this species appears to be timed to take advantage of the rich food stores provided during part of the year by the flooded forest (Fig. 6.6). During high waters when access to the flooded forest is greatest, fish condition (measured by gutted weight relative to length) is at a maximum as is mean stomach fullness. The increase in mean stomach volume between dry season (25.3 ml; standard error of the mean (SEM) = 7.8 ml) and wet season (60.7 ml; SEM = 9.9 ml) is due almost entirely to augmented fruit and seed consumption. Not only does volume increase but so too does the diversity of consumed fruits (Table 6.2). This is most likely due to: (a) the increased diversity of fruits available during rising and high water when most flood-plain plant species fruit, and (b) the concurrent access by fish to flood plain habitats. Fish appear to store up reserves (i.e. increase in biomass relative to length) during this period which is then invested in reproduction at the end of the dry season. After the food-limited dry season when stomach fullness is at a minimum and the reproductive period, relative fish weight (here equated with condition) is at a minimum but improves again when fish can enter the flooded forest. In tandem with this, human commercial fishing of *P. brachypomus* appears to intensify during high water; monthly catch (in kilogrammes) is

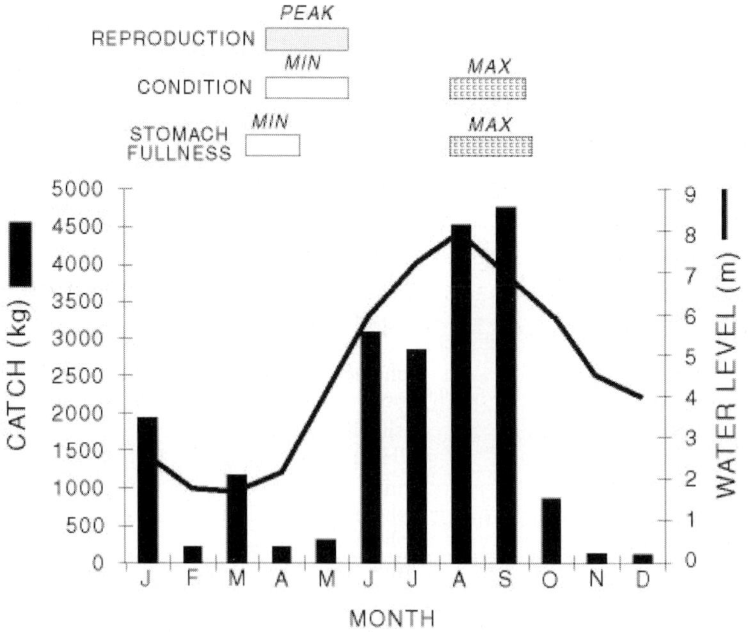

Fig. 6.6 Select aspects of *Piaractus brachypomus* biology in relation to the hydrological cycle (derived from Daza and Vispo unpublished data). Condition was indexed by comparing observed, gutted weight to weight predicted by fish length according to a length:weight relationship derived from the complete data set. "Catch" refers to total commercial capture estimated for 1998. Hydrological information was derived from Vargas & Rangel (1996).

Table 6.2 The diversity of fruits (or seeds) found in the stomach of *Piaractus brachypomus* during four hydrological periods. Hydrological periods were defined according to published hydrographs for the Caura (Vargas & Rangel, 1996). The total number of fruit species encountered is uncorrected for variation in the number of stomachs inspected. The index of fruit species availability is derived from the percentage of fruit species in a 34 species subsample which, during the given period, were observed to have mature fruits and to be standing over water (Knab-Vispo, unpublished data).

Hydrological period	Number of stomachs inspected	Mean number of fruit species per stomach	Total number of fruit species encountered	Index of fruit species availability
Low water (February–April)	51	1.26	21	12%
Rising water (May–June)	33	3.34	61	44%
High water (July–September)	98	4.32	94	53%
Falling water (October–January)	7	3.86	18	9%

greatest during the high water period (Fig. 6.7). During high waters, the most intense *P. brachypomus* fishing appears to occur along those portions of the river with the largest expanses of wooded flood plain (sectors II and III in Fig. 6.7), while both the river mouth and sector IV (which contain the rockier, clearer tributaries of the lower Caura) appear to be the prime capture sites during the dry season. This is corroborated by fishermen reports that during rising water *P. brachypomus* moves towards the wooded flood plains, while during low water they move into the headwaters or back towards the Orinoco. We are currently exploring the movements of *P. brachypomus* in greater detail using radio telemetry.

ECOHYDROLOGICAL INTERACTIONS IN THE CONFLUENCE ZONE OF THE MAPIRE RIVER

Riparian forest communities at the Mapire River confluence zone are related to longitudinal and perpendicular gradients of flooding depth and duration (Table 6.3). Like on the Caura these factors are associated with local topography. Vegetation was found to respond to topographic variation largely due to its influences on soil physical and chemical characteristics caused by differences in sedimentation dynamics and mineralization processes. Sedimentation rates are highest in in-channel geomorphic units such as point-bar systems exposed regularly to flow (Dezzeo *et al.*, 2000). All geomorphic units, from the in-channel units up through those located at the lower and higher topographical limits of the confluence zone (lateral limits) experience the highest interannual variability in flooding duration and depth. Higher altitude lateral edges can experience years without flooding and be highly affected by water shortages in the dry season, whereas lower altitude edges close to the river can experience several years of continuous flooding. The variability described is stronger in the longitudinal edges of the confluence zone. Vegetation communities are associated with different flooding regimes as shown in Fig. 6.8. For example, in point-bar interiors flooding depth can reach 8–10 m for a duration of 8–12 months, a lower diversity is observed and adventitious roots in the trees are profuse and close to the level of flooding. Levees experience maximum flooding depths of 3–6 m, but a more constant duration of about 4–5 months, having higher diversity values. The edges of the flood plain present maximum flooding depths of 1.5 m, one to two months of flooding duration or even years without flooding. One important factor affecting vegetation mainly in the flood plain edges is fire and cattle-raising, both of which are associated with topographic position.

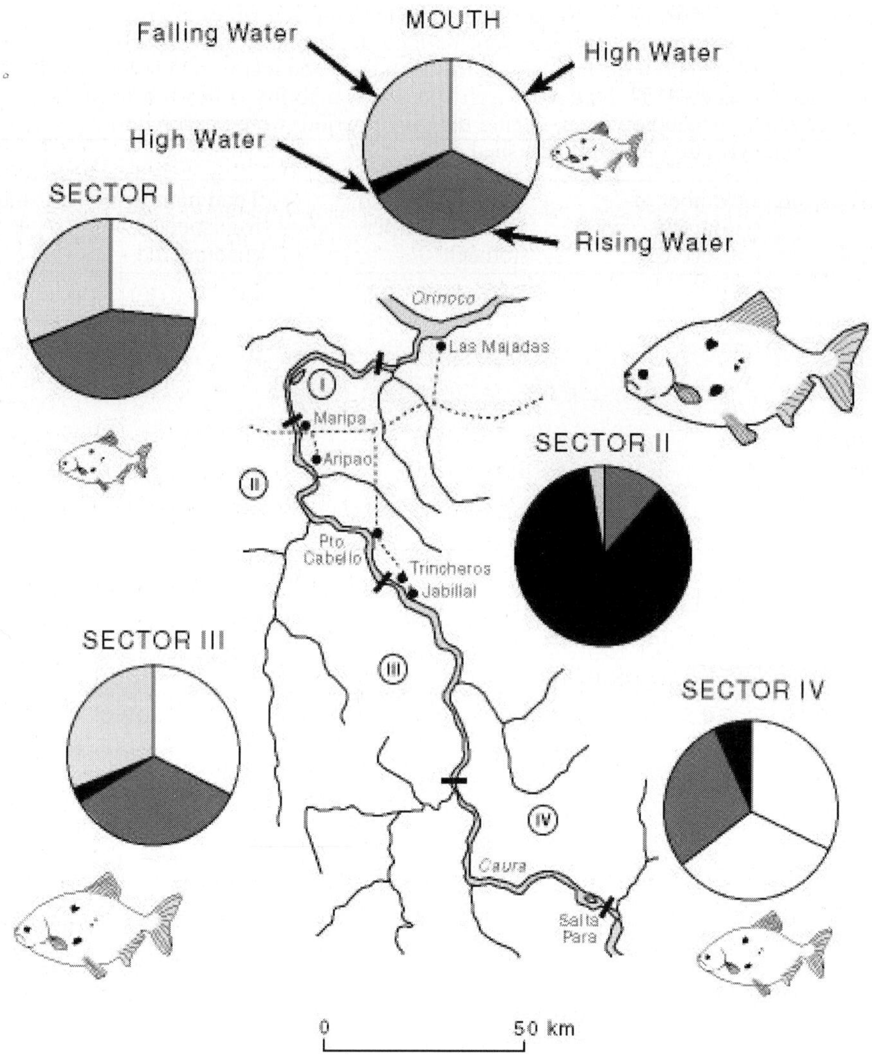

Fig. 6.7 Distribution of *Piaractus brachypomus*, the commercial catch along the lower Caura. The pie charts indicate the percentage of individuals caught in each sector during each hydrographic period. The size of the *P. brachypomus* illustration indicates the approximate relative number of individuals in the total capture for the given sector. Unpublished data from Vispo and Daza.

For the flooded forest communities present in the highly oligotrophic central area of the Mapire's confluence zone, mycorrhizal symbiosis and nitrogen fixation have been shown to be very important functional mechanisms for improving nutrient uptake (De Andrade, 1991; De La Rosa, 1988; Barrios & Herrera, 1994). Mycorrhizal activity is positively correlated with percent soil moisture. During the dry season, maximum activity is associated with

Table 6.3 Characteristics of communities along a perpendicular gradient in the central area of the confluence and at the upstream limit.

	Central area of the confluence zone ◄——— Flooding depth and duration			Upstream limit
Community	*Psidium-Campsiandra*	*Pouteria-Couepia*	*Astrocaryum-Caraipa*	*Quiina rhytidopus*
H′ Shannon-Wienner species diversity index	1.38	2.87	2.16	3.00
Evenness	0.63	0.84	0.64	0.84
Interannual variability in flooding	high	low	high	medium
Sedimentation rate	high	low	low	medium
Fire and cattle-raising effect	low	low	high	low

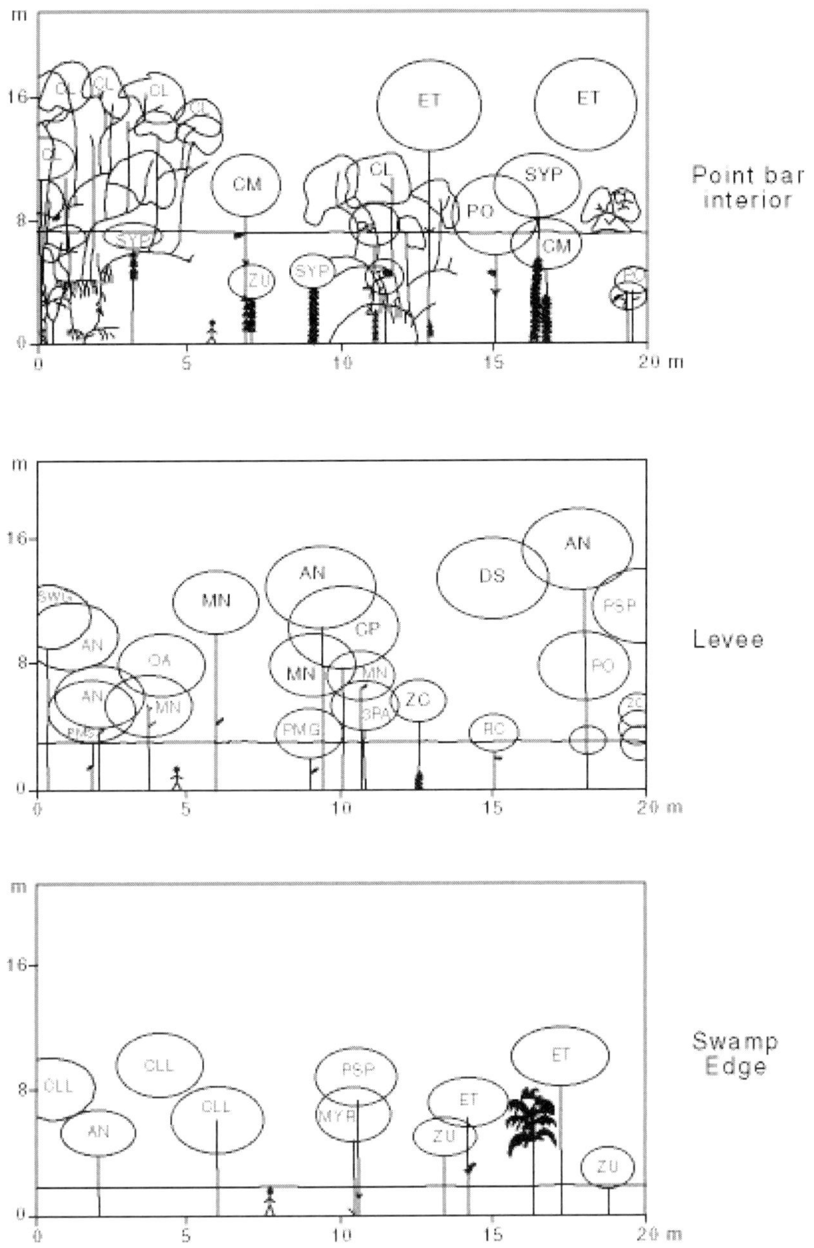

Fig. 6.8 Plots of forest profiles along a perpendicular gradient in the central area of the Mapier's confluence zone.

maximum soil moisture that occurs at the beginning and end of the flooding period (De La Rosa, 1988). As these forests are under semiarid conditions, both flooding during the wet season and water deficit during the dry season occur, creating better conditions for mineralization processes and nutrient uptake. There is no direct relationship between the local wet season and the flooding provided the Orinoco's backwater controls the hydrology of the Mapire's flood plain. For individual tree species such as *Acosmium nitens*, mycorrhizal symbiosis is very important for seedling survival during flooding. Nitrogen fixation also plays an important role in plant survivorship (De Andrade, 1991). Resident plants like *Psidium ovatifolium* and *Campsiandra laurifolia* have adapted physiological and morphological strategies to resist extended flooding and flow-induced mechanical damage (Fernandez *et al.*, 1999). Xylem water potential increased with depth of flooding in these species indicating that flooding did not cause water stress.

When the soil is waterlogged (flooded to the soil surface), gas exchange between the soil and atmosphere becomes negligible. Initially, the flood water contains oxygen, but this is depleted within hours, depending on temperature and respiration rates. Therefore, the plant experiences hypoxia prior to anoxia, and this gradual depletion has two major effects. The first is a stimulation of ethylene synthesis in response to depression of internal oxygen concentration (Jackson, 1982). Ethylene initiates and regulates many adaptive responses that allow the plant to avoid anaerobiosis by increasing oxygen availability to the roots in a flooded or waterlogged soil. In addition, ethylene triggers other symptoms that allow the plant to succeed with low amounts of gas exchange in the roots, such as epinasty, chlorosis and leaf senescence.

Although Fernandez *et al.* (1999) reported decreasing photosynthetic rate and leaf conductance with increasing flooding in *Campsiandra laurifolia* (characteristics that could be indicative of a physiological reduced flood tolerance), the presence of adventitious roots, lenticel hypertrophy plus vegetative sprouting capability, are probably related to the success of this species in areas of higher flooding intensity. These are all vegetative strategies that have been indicated by Gill (1970) to increase the avoidance of flooding stress in woody species.

The possibility of physiological tolerance to or avoidance of flooding stress in different tree species is related to the number of species observed in plots having different flooding regimes. In the flooded forest of the Mapire confluence zone, data from 26 plots of 0.01 ha each showed that species richness for individuals with DGH (diameter at ground height) above 2 cm (data from Rosales, 1990) was significantly correlated with inundation depth (*IN*), the soil carbon/nitrogen ratio (C/N) and the sum of soil cations (SUMCAT). $S = 11.05 - 0.13\text{C/N} - 0.98IN + 14.96\text{SUMCAT}$ ($R^2 = 0.78$, all coefficients significant, $\alpha = 0.05$). The inundation depth is related to the topographical position whereas the C/N ratio is inversely related to decomposition processes. The sum of cations is related to the quality of the inorganic and organic material. As was shown by Rosales *et al.* (1999b), the backwater effect created by the Orinoco's damming of the Mapire results in higher nutrient concentrations in the upstream and downstream portions of the confluence zone. In the lower reaches, close to the junction with the Orinoco, sedimentary deposits are characteristic of the Orinoco River. They are finer and richer in nutrients (Rosales *et al.*, 1999b). Furthermore, in the upper reaches of the impounded area, the coarser sediments, characteristic of the Mapire basin, settle out of suspension, but finer silt and colloidal-sized particles are transported downstream toward the central area where they are deposited due to the very slow-flowing, impounded waters. Two areas of nutrient enrichment thus result: one at the longitudinal limits of the confluence zone upstream and another downstream towards the river mouth. An area of nutrient depletion occurs towards the central area (Rosales *et al.*, 1999b). These variations in nutrient status have implications on the biogeochemistry of flooded forest communities occurring in the flood plain.

ORGANIC LITTER DECOMPOSITION IN THE MAPIRE *vs* THE CAURA

We have shown how the woody debris and leaf litter in the Caura forests host important invertebrate communities and how important these invertebrates are to fish diets. These materials also play an important role in the biogeochemistry of the flooded forest. Invertebrates are important early in the decomposition process as they set the stage for the work of the soil microbial community. Organic litter plays an important role in the cycling of carbon and other nutrients in tropical forest ecosystems (Vitousek & Sanford, 1986). The importance of this organic layer for both the nutrient pool and the nutrient turnover in seasonally flooded forests of the Amazon flood plain has been documented (Furch & Junk, 1997; Furch & Klinge, 1989).

The soil surface of the flooded forests accumulates variable amounts of organic material from woody debris and leaf litter that is both produced *in situ* and transported fluvially. The nutrients in this organic litter are then released during decomposition. Little information exists about the decomposition of such material under seasonally flooded conditions. In the Amazon flood plain, inundation decomposition occurs in the forests in an acid, anaerobic environment which after a time may stagnate further decay (Stark & Holley, 1975). During the emersion phase of these forests the fresh leaf litter is degraded more rapidly in the white-water flood plains (*várzea*) than in the black-water flood plains (*igapó*) (Irmler & Furch, 1980). Leaf litter has been shown to be of generally higher quality, defined by C/N, than woody debris. In other similar ecosystems leaves represent about 80% of fine litterfall flux while woody debris (twigs) represents about 15% (Franken *et al.*, 1979).

In order to evaluate decomposition rates of leaf litter in flooded forests, Dezzeo *et al.* (unpublished data) determined mass and nutrient loss from leaf litter along perpendicular topographical gradients in the flooded forests of the Mapire and Caura rivers. The flood plains studied included *igapó* forests in the central area of the Mapire River confluence zone and mixed *várzea–igapó* forests in the lower Caura confluence zone downstream of the Sipao tributary. The topographical gradient in the Mapire flood plain included the following sites: (i) Mapire channel bar (CB), inundated 8 months a year, (ii) Mapire lower flood plain (LF), inundated 6 months a year and (iii) Mapire upper flood plain (UF), inundated 3 months a year. The topographical gradient in the lower Caura included two sites: (i) Caura channel bar (CB), inundated 5 months a year and Caura lower flood plain (LF), inundated 4 months a year.

Dezzeo *et al.* (2000) studied leaf litter decomposition using litterbags (Crossley & Hoglund, 1962). Litterbags were filled with 30 g of relatively freshly fallen leaf litter of different species, which was collected from the forest floor of each study site during low waters in the dry season of 1995. Each study site had a determined species assemblage, and therefore the litter samples were different for each site. Collected litter was analysed to determine weight, moisture and initial chemical constituents (including N, P, Ca, K and Mg). These litterbags were then placed randomly on the forest floor late in the same dry season ("initial condition"; IC). Samples were collected during falling waters, just after the end of the flooding period of 1995 (AF), and then another sample collection occurred during the 1996 dry season (DS). For each site and sampling period, oven-dried material (at 70°C to constant mass) free of sediments was milled and digested. Total nitrogen was determined by the micro-Kjeldahl method (Jackson, 1976); total phosphorus was measured colorimetrically (Murphy & Riley, 1962); and magnesium, potassium and calcium were determined by atomic-absorption spectrophotometry. Statistical differences in dry weight and nutrient loss were tested using three-way ANOVAs followed by Tukey's HSD (α = 0.05).

Results indicated that the pattern of mass loss over time was similar in the Mapire and Caura forests, with the highest losses at the end of the flooding phase (first sample period) followed by relatively small changes during the dry season (Fig. 6.9). However, a comparison of leaf litter mass loss rates showed highly significant differences between the two tributaries; the leaf litter in Caura forests lost mass almost twice as rapidly as that of the Mapire forests. At the end of the 11-month decomposition period the mass loss fluctuated between 59–61% for the Caura forests and 16–33% for the Mapire forests. The marked loss of dry matter observed in both study sites at the end of the first sampling period is similar to the general pattern observed in other decomposition studies and is explained by the leaching of soluble materials coupled with microbial decomposition. In the first few days to weeks of decay, the detritus rapidly begins to lose soluble organic and inorganic materials (Webster & Benfield, 1986) mobilized either by passive physical processes or by decomposer activity (Brinson *et al.*, 1981; Singh & Gupta, 1977).

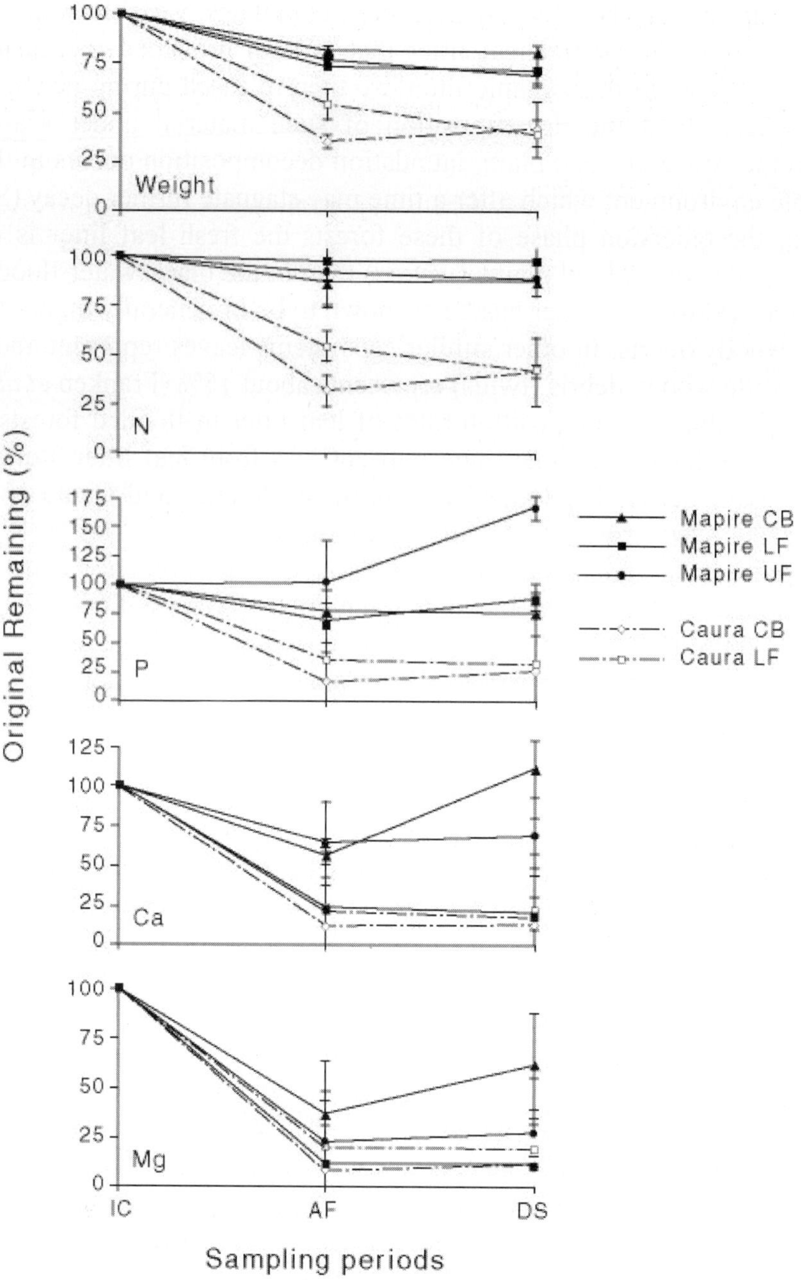

Fig. 6.9 Percent of mass (dry weight) and some nutrients concentrations (N = nitrogen, P = phosphorous, Ca = calcium, Mg = magnesium) remaining of leaf litter expressed as percentage of the initial value (n = 4). IC = initial condition; AF = after flooding ; DS = dry season; CB= channel bar; LF = lower flood plain; UF = upper flood plain.

With few exceptions, there were marked differences between Mapire and Caura forests in terms of initial nutrient concentration of the leaf litter (expressed in dry weight), this was particularly true for K and Ca (Table 6.4). Fresh leaf litter from the Caura was 30–76% richer in P, 61–75% richer in K, 67–90% richer in Ca, 4–50% richer in Mg and 4–28% richer in N than fresh leaf litter from the Mapire. The patterns of nutrient change in the decomposing leaf litter during the 11-month period of the field experiment also showed differences between the Mapire and Caura sites. At the end of the first sampling period 4–15% of the N was lost from leaf litter at Mapire, and 52–67% from leaf litter of Caura. The loss of P in this period was 64–83% in leaf litter of the Caura, and 0–30% in leaf litter of the Mapire. Between the first and the last sampling period, the amounts of N and P remained almost constant, except in Mapire UF, where the amount of P was 66% higher than the original amount. In the Caura the loss of Ca in the leaf litter at the end of the

Table 6.4 Initial element concentration (mg g^{-1}) (mean ± SD; $n = 4$) in dry and fresh leaf litter. For single periods, values between the columns of the Mapire forests or of the Caura forests that share a common letter are not significantly different; comparison of the values between Caura and Mapire sharing a common number are not significantly different (3-ways ANOVA followed by Tukey's HSD ($P \geq 0.95$). Dezzeo, Escalante and Chacon, unpublished data.

	Mapire flooded forests: Depositional bar	Lower terrace	Upper terrace	Caura flooded forests: Depositional bar	Lower terrace
Dry litter					
Weight	$26.05 \pm 1.2^{a}_{2}$	$29.08 \pm 1.0^{a}_{1}$	$28.25 \pm 0.1^{a}_{2}$	$25.62 \pm 1.3^{a}_{2}$	$27.23 \pm 0.8^{a}_{2}$
N	$12.69 \pm 0.6^{a}_{1}$	$8.87 \pm 0.3^{b}_{2}$	$8.78 \pm 0.2^{b}_{2}$	$13.44 \pm 0.6^{a}_{1}$	$11.29 \pm 0.7^{b}_{1}$
P	$0.32 \pm 0.1^{a}_{2}$	$0.20 \pm 0.1^{a}_{3}$	$0.13 \pm 0.0^{b}_{3}$	$0.59 \pm 0.1^{a}_{1}$	$0.45 \pm 0.2^{a}_{12}$
K	$1.41 \pm 0.1^{a}_{2}$	$1.74 \pm 0.3^{a}_{2}$	$1.67 \pm 0.2^{a}_{2}$	$6.08 \pm 0.8^{a}_{1}$	$4.75 \pm 0.7^{a}_{1}$
Mg	$0.80 \pm 0.1^{b}_{2}$	$1.48 \pm 0.1^{a}_{2}$	$1.37 \pm 0.2^{a}_{2}$	$2.19 \pm 0.2^{a}_{1}$	$1.52 \pm 0.3^{a}_{2}$
Ca	$0.50 \pm 0.1^{c}_{2}$	$0.72 \pm 0.1^{b}_{2}$	$1.35 \pm 0.1^{a}_{2}$	$4.56 \pm 0.2^{a}_{1}$	$5.02 \pm 0.3^{a}_{1}$
Fresh litter					
N	$330.4 \pm 28.7^{a}_{1}$	$257.9 \pm 8.8^{b}_{2}$	$247.8 \pm 7.9^{b}_{2}$	$344.3 \pm 15.9^{a}_{1}$	$307.5 \pm 19.2^{a}_{1}$
P	$8.5 \pm 3.0^{a}_{2}$	$5.7 \pm 1.9^{a}_{2}$	$3.7 \pm 0.8^{b}_{2}$	$15.1 \pm 2.9^{a}_{1}$	$12.2 \pm 5.3^{a}_{1}$
K	$38.9 \pm 4.4^{a}_{2}$	$50.6 \pm 8.9^{a}_{2}$	$47.1 \pm 6.6^{a}_{2}$	$155.9 \pm 30.4^{a}_{1}$	$129.3 \pm 17.4^{a}_{1}$
Mg	$20.7 \pm 1.9^{b}_{3}$	$42.9 \pm 1.5^{a}_{2}$	$38.8 \pm 5.0^{a}_{2}$	$56.1 \pm 5.3^{a}_{1}$	$41.3 \pm 15.6^{a}_{2}$
Ca	$12.9 \pm 1.8^{c}_{2}$	$21.0 \pm 2.2^{b}_{2}$	$38.2 \pm 2.2^{a}_{2}$	$117.0 \pm 9.9^{a}_{1}$	$136.7 \pm 17.2^{a}_{1}$

flooding phase was very similar in both sites: 76% in Caura CB and 88% in Caura LF. These percentages remained almost unchanged in the subsequent sampling period. In the Mapire sites, however, the pattern for Ca in leaf litter showed marked differences. After the flooding period the loss of Ca was 79% in Mapire LF, 44% in Mapire CB and 36% in Mapire UF; between the first and the last sampling period the amounts of Ca showed only slight changes, except in Mapire CB, where the amount of Ca increased by 53%. Potassium was the most rapidly lost element, followed by Mg. The loss pattern of these elements was very similar in the forests of both Mapire and Caura.

Differences in decomposition and related nutrient flux along the flooding gradients of Caura and Mapire may be attributable to differences in the flooding duration and to differences in substrate quality. As Fig. 6.8 shows, a slower rate of decay occurred in the depositional bar where prolonged flooding was typical whereas higher decay rates occurred on the terraces where flooding duration was shorter. The substrate quality, as defined by the chemical composition of the decomposing material, is a critical factor in determining the decomposition rates of plant detritus (Day, 1982; Tietema *et al.*, 1985; Webster & Benfield, 1986; Furch *et al.*, 1989). The freshly fallen leaf litter of Mapire and Caura sites is in general nutrient-poor, and their elemental content is comparable with those reported for leaf litter in the black-water flood plains of Amazonas (Furch & Junk, 1997; Furch & Klinge, 1989). The leaf litter of Caura forests, however, shows higher concentrations of nutrients compared to the litter of Mapire forests (Table 6.3).

Differences between the Mapire and Caura samples reflect differences in the nutrient cycles of both tributaries. Flooded forests contain many complex nutrient pathways, each with different magnitudes and qualities (Brinson *et al.*, 1981). Sediment is one of the major storage compartments of such ecosystems, and its role as a pathway in nutrient cycling covers essentially the non-biologically mediated exchanges associated with through-flowing water. Deposition of suspended sediments and their associated nutrients were higher at the Caura sites than at the Mapire sites (Dezzeo *et al.*, 2000). These differences derived from the above-mentioned backwater effect, given that the Caura sites were relatively closer to the junction with the Orinoco and were therefore more *várzea*-related, whereas the Mapire sites were more *igapó*-related given their position in the central area of the confluence zone.

CONCLUSIONS

Here, we have briefly described some ecohydrological interactions between the river and the flooded forest. We have shown how resident biota in aquatic and riparian communities reflect variable flood regimes and geographically unique channel processes. Although present knowledge of the Orinoco flood plain's ecohydrological interactions is far from complete, our current level of understanding can help assess the sustainability of development in the Orinoco basin. Evaluations of the likely consequences of development should consider the alteration of ecological processes related to the physical template, e.g. hydrological dynamics, and specifically flow characteristics such as velocity, flooding depth, duration and frequency, that varies in any cross-section of the river, both in low and high flows. This physical template strongly influences different riparian and river-bed biota communities, as well as quality and quantity of water and suspended sediments. Furthermore, hydraulic conditions modulate and at the same time are modulated by riparian vegetation density and structure in the flood plains. Finally, soil chemistry and nutrient cycling are strongly related to microbiological dynamics in a system that seasonally changes from anaerobic to aerobic conditions, thus influencing riparian biogeochemistry and primary and secondary productivity. Managers need to consider (a) the location in time and space of immediately relevant ecological processes (e.g. the importance to fisheries of the high water period and of regions of flooded forests; the importance of woody debris for the aquatic invertebrate community; the importance of flooding for rates of leaf litter decay), and (b) the sometimes-overlooked extensive temporal and spatial scales at which important functional aspects of flood plain ecosystems occur (e.g. the long-term, geohydrological developments that help determine the variation of soil nutrients in the riparian corridor; the spatially extended biogeochemical gradients influenced by backwater effects in the confluence zones of oligotrophic rivers).

Management practices to maintain biodiversity not only at compositional but also at functional levels need to be tailored to the conditions of each particular area. In areas with low degradation, integrated management designed to conserve the existing flooded forests is needed, especially if one wishes to preserve the fisheries. In more degraded areas where biodiversity is greatly decreased, the priority should be to enhance the value of the system through restoration and moderation of degrading influences. Primary needs in these areas are data-banks of suitable riparian plant species to enable their re-establishment on the riparian flood plains. Plant suitability for restoration purposes should be measured not only in terms of capability of reestablishment but also in terms of providing food for fish and other flood plain forest animals. The incorporation of human populations into activities of management and restoration should be encouraged through education and the promotion of ecologically sustainable socioeconomic alternatives.

Acknowledgements We want to acknowledge the institutions we represent (Universidad Nacional Experimental de Guayana, Fundación La Salle de Ciencias Naturales, Instituto Venezolano de Investigaciones Científicas, University of Birmingham and University of Stirling) for their support in conducting our research. Also acknowledged is the thorough review and important comments from Anne J. Glauber of the University of Washington, and the help of Kevin Burkhill with the drawings.

REFERENCES

Barrios, E. & Herrera, R. (1994) Nitrogen cycling in a Venezuelan tropical seasonally flooded forest: soil nitrogen mineralization and nitrification. *Journal of Tropical Ecology* **10**, 399–416.

Brady, N. C. (1990) Soil erosion and its control. *The Nature and Properties of Soils*, 10th edn, 431 470. Macmillan, New York.

Brinson, M., Lugo, A. & Brown, S. (1981) Primary productivity, decomposition and consumer activity in freshwater wetlands. *Annual Review of Ecology and Systematics* **12**, 123–161.

Colonello, G. (1990a) A Venezuelan floodplain study on the Orinoco River. *Forest Ecology and Management* **33**, 103–124.

Colonello, G. (1990b) Elementos fisiográficos y ecológicos de la cuenca del Río Orinoco y sus rebalses. *Interciencia* **15**, 476–485.

Colonello, G. (1991) Observaciones fenológicas y producción de hojarasca en un bosque inundable (Várzea) del Río Orinoco, Venezuela. *Interciencia* **16**, 202–208.

Crossley, D. & Hoglund, M. (1962) A litter bag method for the study of microarthropods inhabiting leaf litter. *Ecology* **43**, 571–573.

Day, F. P. (1982) Litter decomposition rates in the seasonally flooded Great Dismal Swamp. *Ecology* **63**, 670–678.

De Andrade, Z. (1991) Efectos de la inundación sobre la asociación simbiótica micorrizas-VA y rhizobium-leguminosa durante el primer año de vida de *Acosmium nitens* (Vog.) Yakoul, especie arbórea de un bosque estacionalmente inundable del Río Mapire, bajo Orinoco. MSc Thesis, Instituto Venezolano de Investigaciones Científicas, Caracas, Venezuela.

De La Rosa, T. (1988) Asociación micorrícica vesículo-arbuscular en un bosque estacionalmente inundable en las riberas del Río Mapire (Edo. Anzoategui). Licenciate Thesis, Universidad Simón Bolívar, Caracas, Venezuela.

Dezzeo, N., Herrera, R., Escalante, G. & Chacón, N. (2000) Deposition of sediments during a flood event on seasonally flooded forests of the lower Orinoco River and two of its blackwater tributaries, Venezuela. *Biogeochemistry* **49**, 241–257.

Fernandez, M., Pieters, A., Donoso, C., Herrera, C., Tezara, W., Rengifo, E. & Herrera, A. (1999) Seasonal changes in photosynthesis of trees in the flooded forest of the Mapire River. *Tree Physiology* **19**, 79–85.

Franken, M., Irmler, U. & Klinge, H. (1979) Litterfall in inundation, riverine and terra firme forest of central Amazonia. *Tropical Ecology* **20**, 225–235.

Furch, K. & Junk, W. (1997) The chemical composition, food value, and decomposition of herbaceous plants, leaves, and leaf litter of floodplain forests. In: *The Central Amazon Floodplain. Ecology of a Pulsing System* (ed. by W. Junk), 187–205. Ecological Studies 126, Springer-Verlag, Berlin, Germany.

Furch, K. & Klinge, H. (1989) Chemical relationships between vegetation, soil and water in contrasting inundations areas of Amazonia. In: *Mineral Nutrients in Tropical Forests and Savanna Ecosystems* (ed. by J. Proctor), 189–204. Blackwell Scientific Publications, Oxford, UK.

Furch, K., Junk, W. J. & Campos, Z. (1989) Nutrient dynamics of decomposing leaves from Amazonian floodplain forest species in water. *Amazoniana* **XI**, 91–116.

Gill, C. (1970) The flooding tolerance of woody species—a review. *Forestry Abstracts* **31**, 671–688.

Hamilton, S. & Lewis, W. (1990) Physical characteristics of the fringing floodplain of the Orinoco River, Venezuela. *Interciencia* **15**, 491–500.

Irmler, U. & Furch, K. (1980) Weight, energy, and nutrient changes during the decomposition of leaves in the emersion phase of central Amazonian inundation forests. *Pedobiologia* **20**, 118–130.

Jackson, M. (1976) *Análisis Químico de Suelos*, 3era edn. Editorial Omega, Barcelona.

Jackson, M. (1982) Ethylene as a growth promoting hormone under flooded conditions. In: *Plant Growth Substances* (ed. by P. Wareing), 291–301. Academic Press, London, UK.

Johnson, M., Stallard, R. & Lundberg, N. (1991) Controls on the composition of fluvial sands from a tropical weathering environment: sands of the Orinoco River drainage basin, Venezuela and Colombia. *Geological Society of American Bulletin* **103**, 1622–1647.

Junk, W., Bayley, P. & Sparks, R. (1989) The flood pulse concept in river–floodplain systems. In: *Proceedings of the International Large River Symposium* (ed. by D. Dodge). *Canadian Special Publication in Fisheries and Aquatic Sciences* **106**, 110–127. NRC Research Press, Ottawa.

Meade, R. (1994) Suspended sediments of the modern Amazon and Orinoco Rivers. *Quaternary International* **21**, 29–39.

Murphy, J. & Riley, J. (1962) A modified single solution method for the determination of phosphate in a natural waters. *Analytical Chemistry Acta* **27**, 31–36.

Nordin, C. & Perez-Hernandez, D. (1989) Sand waves, bars and wind-blown sands of the Rio Orinoco, Venezuela and Colombia. *US Geological Survey Water Supply Paper 2326-A*, 1–74.

Petts, G. & Amoros, C. (eds) (1996) *Fluvial Hydrosystems*. Chapman & Hall, London, UK.

Prance, G. (1979) Notes on the vegetation of Amazonia III. The terminology of Amazonian forest types subject to inundation. *Brittonia* **31**, 26–38.

Rosales, J. (1990) Análisis florístico-estructural y algunas relaciones ecológicas en un bosque estacionalmente inundable en la boca del Río Mapire, Edo. Anzoategui. MSc Thesis, Instituto Venezolano de Investigaciones Científicas, Caracas, Venezuela.

Rosales, J. (2000) Ecohydrological basis for riparian biodiversity conservation in large rivers. PhD Thesis, University of Birmingham, UK.

Rosales, J., Petts, G. & Knab-Vispo, C. (2001) Ecological gradients in riparian forests of the lower Caura River, Venezuela. *Plant Ecology* **152**(1), 101–118.

Rosales, J., Petts, G. & Salo, J. (1999a) Riparian flooded forests of the Orinoco and Amazon river basins: a comparative review. *Biodiversity and Conservation* **8**, 551–586.

Rosales, J., Bradley, C., Gilvear, D. & Petts, G. (1999b) A summary model of hydrological and ecological interactions within the confluence zone of major tropical rivers: tributaries of the Orinoco River, Venezuela. In: *Proceedings of the Symposium of Hydrological and Geochemical Processes in Large Scale River Basins: with Special Emphasis on the Amazon and other Tropical Basins* (Manaus, Brazil, 15–19 November 1999). CD-ROM extended abstracts.

Singh, J. & Gupta, S. (1977) Plant decomposition and soil respiration in terrestrial ecosystems. *The Botanical Review* **43**, 449–528.

Stallard, R., Koehnken, L. & Johnsson, M. (1990) Weathering processes and the composition of inorganic material transported through the Orinoco River system, Venezuela and Colombia. In: *The Orinoco River as an Ecosystem* (ed. by F. Weibezahn, H. Alvarez & W. Lewis Jr), 81–119. Universidad Simón Bolívar, Caracas, Venezuela.

Stark, N. & Holley, C. (1975) Final report on studies of nutrient cycling on white and black waters areas in Amazonia. *Acta Amazonica* **5**, 51–76.

Tietema, A., Kuikman, P. & Berendse, F. (1985) Mass loss, nutrient dynamics and influence of Diptera larvae in decomposing litter of *Erika tetralix* and *Molinia caerulea*. *Pedobiologia* **28**, 389–397.

Vargas, H. & Rangel, J. (1996) Hidrología y sedimentos. In: *Ecología de la Cuenca del Río Caura. I. Caracterización General* (ed. by J. Rosales & O. Huber), 48–53. Scientia Guianae Series 6, Ediciones Tamandua, Caracas, Venezuela.

Vásquez, E. & Wilbert, W. (1992) The Orinoco: physical, biological and cultural diversity of a major tropical alluvial river. In: *The Rivers Handbook* (ed. by P. Calow & G. Petts), vol. 1, 448–469. Blackwell Scientific Publishers, Oxford, UK.

Vitousek, P. & Sanford, R., Jr (1986) Nutrient cycling in moist tropical forests. *Annual Review of Ecology and Systematics* **17**, 137–167.

Webster, J. & Benfield, E. (1986) Vascular plant breakdown in freshwater ecosystems. *Annual Review of Ecology and Systematics* **17**, 567–594.

Weibezahn, F., Alvarez, H. & Lewis, W., Jr (eds) (1990) *The Orinoco River as an Ecosystem.* Universidad Simón Bolívar, Caracas, Venezuela.

CHAPTER 7

BENTHIC MACROINVERTEBRATES AND PHYSICAL HABITAT RELATIONSHIPS IN THE PARANA RIVER FLOOD-PLAIN SYSTEM

Mercedes Marchese, Inés Ezcurra de Drago & Edmundo C. Drago

A review of world literature shows that progress in stream ecology has been substantial over the past three decades, with studies focused mainly on the holistic character of these hydrosystems. The importance of physical processes in fluvial systems has drawn increasing attention, as the interrelationships between the physical habitat and the organisms became better understood. New paradigms emerged from the increasing comprehension of stream ecology, such as the river continuum concept (Vannote *et al.*, 1980), the nutrient spiralling concept (Newbold *et al.*, 1981, 1982; Elwood *et al.*, 1983), the serial discontinuity concept (Ward & Stanford, 1983), the potential applications of hydraulics in stream ecology (Statzner *et al.*, 1988), the role of disturbance in rivers (Resh *et al.*, 1988), the four-dimensional nature of lotic ecosystems (Ward, 1989), and the patch dynamics concept (Pringle *et al.*, 1988; Townsend, 1989).

It should be emphasized that the aforementioned theories have been formulated and applied almost exclusively to low-order streams. In relation to large river–flood-plain systems, an important concept is that of the flood pulse, which describes the periodic, two-way exchange of water and nutrients between the main channel and riparian ecosystems (Junk *et al.*, 1989). In this concept the factor of overriding importance that controls the interactions between the channel and the flood plain is river level fluctuation. Later, additional concepts regarding the potamophase and limnophase of these pulsing regimes were developed (Neiff, 1990,1999).

From studies on the benthos of the Paraná River and their flood-plain environments, we have defined fundamental controlling forces such as the high and low water phases (degree of connectivity) and the hydraulic complexity of the river–flood-plain system. The paradigms developed for small streams, however, also facilitate the comprehension of some processes in large rivers. The benthos of large flood-plain rivers has received little attention in the international literature, especially for those rivers that remain in a "natural" state. The majority of studies are on a particular component of the fluvial system such as the main channel of the Volga River (Mordukhai-Boltovskoi, 1979; Dudgeon, 1995) or the flood-plain lakes of the Amazon and Orinoco rivers (Fittkau *et al.*, 1975a,b; Reiss, 1977; Vázquez *et al.*, 1990; Lewis *et al.*, 1995; Junk & Robertson, 1997; Nessimian *et al.*, 1998). Lewis *et al.* (1995) observe that there is "the perception that these rivers lack benthic fauna because of their rapidly shifting, fine-grained substrates".

The hydro-sedimentological dynamics of large rivers create a large amount of habitat heterogeneity, and this is reflected by the spatial and temporal structure of the benthic community. This heterogeneity is clearly demonstrated by benthic studies in the main channel of the Paraná River and in the lotic and lentic components of its flood plain. We have analysed the longitudinal, lateral and temporal dimensions of benthic communities, as well as their relationships with the physical, chemical and biological characteristics of the fluvial habitats (Marchese & Ezcurra de Drago, 1992). These studies demonstrated how the topographic structure of the river bottom strongly influences many ecological interactions. In this chapter, we synthesize research on the response of the benthos community structure to physical characteristics of fluvial environments of the Paraná River, including the main channel, secondary channels of various hierarchical levels, and flood-plain lakes with direct or indirect connections to the river.

GEOGRAPHICAL SETTING

The Paraná is the second largest river in South America, in terms of catchment area (2.8×10^6 km^2), the second longest (4400 km from the headwaters of the Grande River in Brazil to the Río de la Plata estuary), and third largest, in terms of mean discharge to the ocean (21 000 m^3 s^{-1}). It drains nearly 32% of Argentina, 100% of Paraguay, and 8% of Bolivia, and 50% of its drainage area is within southern Brazil (Fig. 7.1). The different geological

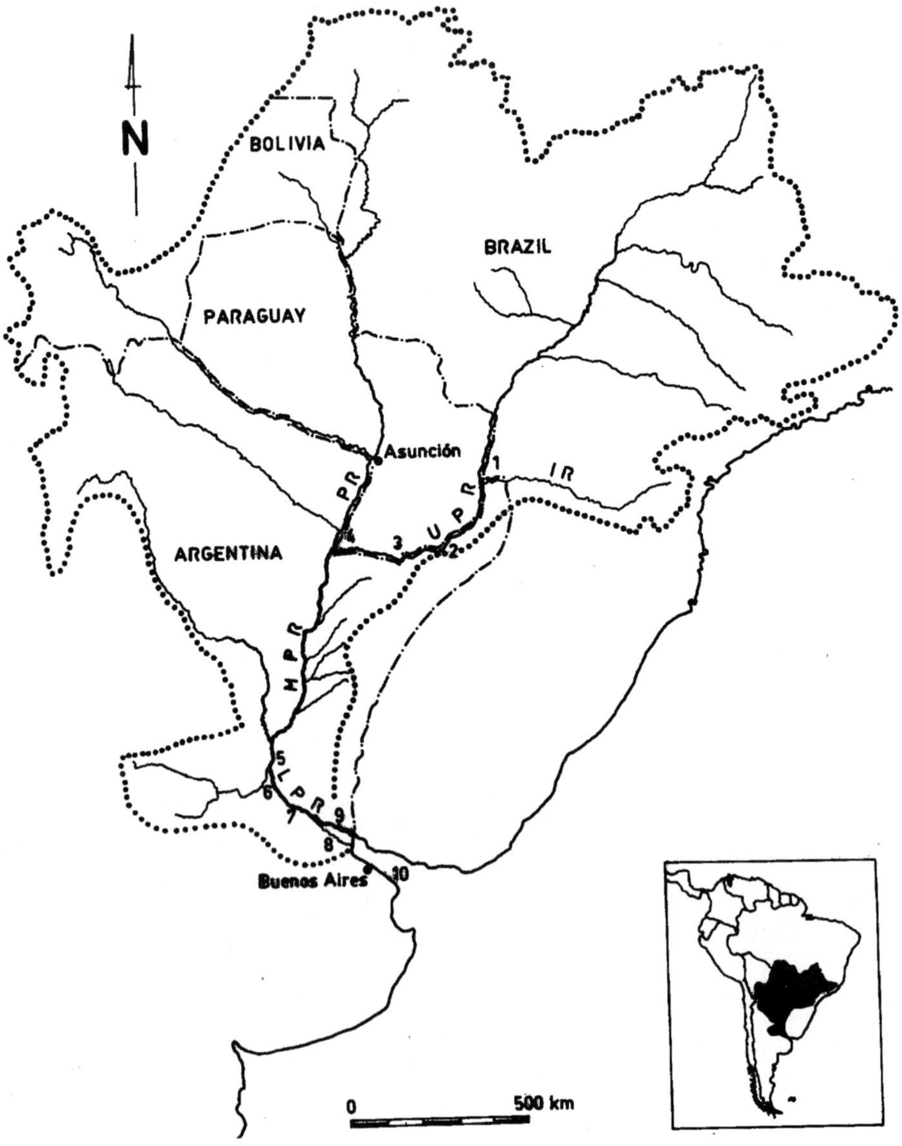

Fig. 7.1 Paraná River basin. IR: Iguazú River; UPR: upper Paraná River; PR: Paraguay River; MPR: middle Paraná River; LPR: lower Paraná River. 1: Iguazú River mouth; 2: Posadas city; 3: Yacyretá Dam; 4: Paraguay River mouth; 5: Diamante city; 6: Rosario city; 7: San Pedro city; 8: Paraná de las Palmas distributary; 9: Paraná Guazú distributary; 10: Río de la Plata Estuary.

and climatological regions that encompass the Paraná drainage basin influence both the river hydrology and sedimentology. The upper Paraná flows on the Brazilian Shield but most of the drainage network develops over landscapes of Jurassic–Cretaceous age, formed by tholeiitic basalts and siliceous sandstones (Iriondo, 1988). Thus, downstream of the mouth of the Iguazú River, the Paraná flows through a plateau border, although the channel is confined by rocky outcrops up to the west side of Posadas city (Argentina). Due to these geological features, some authors consider that the middle Paraná River begins downriver of Posadas (M. Iriondo, personal communication). Others divide the river into two segments, the upper Paraná upstream to the Paraguay River confluence, and the lower Paraná downstream of that point (Neiff, 1990). A more general opinion, however, is that the middle reach extends between the Paraguay River and Diamante city (Argentina) (Fig. 7.1). According to Iriondo (1979), the Paraná River flowed into this area during the Pliocene (two–three million years ago). Along the middle reach, the main channel flows from north to south along a tectonic lineation that is probably Pleistocene in age (Iriondo, 1988). The deltaic area in the lower Paraná was formed during the Holocene, with an estimated age of 900 years for the lower delta (Iriondo & Scotta, 1979).

The tropical rainfall in the Paraná headwaters of Brazil is drained by the Paraná and Paraguay rivers. The lower reach of the upper Paraná and a large segment of the middle Paraná (up to 30°15′ S) flow through a region of subtropical climate. Downriver, the middle and Lower Paraná flow through a region of temperate climate up to the Río de la Plata estuary. This large fluvial segment is bounded on the north by the 22°C annual isotherm, and on the south by the 17°C annual isotherm. The rainfall in the middle and lower reaches of the Paraná River decreases from east to west. Thus, at the beginning of the middle reach, the mean annual rainfall is 1500 mm and 300 km downriver at the Paraguay River outlet, it is 1100 mm. Here, the river flows from east to west. Downstream of the Paraná–Paraguay confluence, the river flows in a north–south direction, but rainfall progressively decreases from 1100 mm (at the confluence) to 950 mm at the Paraná mouth. The middle and lower reaches of the Paraná River flow through three biomes: the *Subtropical Parklands and Savannas*, the *Espinal*, and the *Pampean Grasslands* (Marchetti & Prutkin, 1982). These biomes are included within the large geographical region known as the Chaco-Pampa Plain.

GEOMORPHOLOGY AND BED SEDIMENTS

The upper Paraná River in its 347 km lower segment between the Iguazú River mouth and Posadas city (Argentina), flows in a single sinuous channel deeply incised in basalt, flowing on a rocky bed with large sand patches. These patches are composed largely of medium to very coarse sands (2–0 on the phi scale), with an average hydraulic gradient of 0.16 m km^{-1} (Drago & Amsler, 1998). As the river flows into the Chaco-Pampa Plain, its geomorphological and sedimentological characteristics gradually change. The Yacyretá Dam was constructed to the west of Posadas city. Downstream of Posadas city, there is a transition reach of 300 km that extends to the Paraguay River mouth (0.08 m km^{-1} mean slope). Here the upper Paraná River gradually changes into the middle Paraná. Thus, the river bed is increasingly covered by sand rather than rock and channel bars, and islands have begun to form. In this reach, the river begins to develop a braided channel. However, the largest channel islands are rock outcrop remnants of the basaltic plateau, for example Apipé and Yacyretá islands. Furthermore, the basaltic outcrops are present in the bed and along the banks almost as far downriver as the Paraguay River mouth. The bed sediments are mainly composed of medium sands (2 phi).

The 707-km middle Paraná reach extends from 27°15′S to 32°05′S, flowing between the 58°W and 61°W meridians. The main channel has a braided pattern with a fringing flood plain, showing a sequence of wide reaches with two or more anabranches and narrow and short reaches in which the river flows in a single channel (Fig. 7.2). Its sinuosity is 1.2 and the mean slope is 0.045 m km^{-1} (Drago, 1989, 1990b). The average main channel width is 2.8 km (range, 0.4–8 km), with a maximum depth of 45 m. The channel constrictions are nodal points, forming the most geomorphologically stable river habitats, some of them approaching 100 years in age (Drago, 1977a, 1990b). These channel constrictions do not exceed 1 km in width, and range between 25 and 45 m in depth. The braided reaches of the main channel are the most geomorphologically unstable riverine habitats of the Paraná valley. According to Drago (1977a, 1990b), in some cross-sections the flood-plain border moved eastward at an average rate of 45 m year^{-1}. The thalweg shifts are important, reaching maximum rates of 350 m year^{-1}, and the bar and island migration downstream and laterally can reach mean values of 60 and 35 m year^{-1} respectively. The maximum rate of bankline buildout was about 139 m year^{-1}, and the maximum retreat was found to be 90 m year^{-1}. Usually, the maximum bankline migrations are detected on the flood-plain border and the minimum on the high fault-raised cliffs composed of Tertiary sandstone. However, as much as 20 000 m^3 of sediment may be involved in a single bank failure of 200 m length, drastically changing the onshore-strip habitat on the cliffs (Drago, unpublished).

The middle Paraná has a predominantly sandy mobile bed with bars and islands, where dunes up to 12 m high have been observed (Drago, 1977b). Bed sediments almost uniformly consist of sand (mean percentages always over 98%), composed mainly of medium well-sorted sands (2 phi) that are transported principally by saltation (Bertoldi de Pomar, 1980;

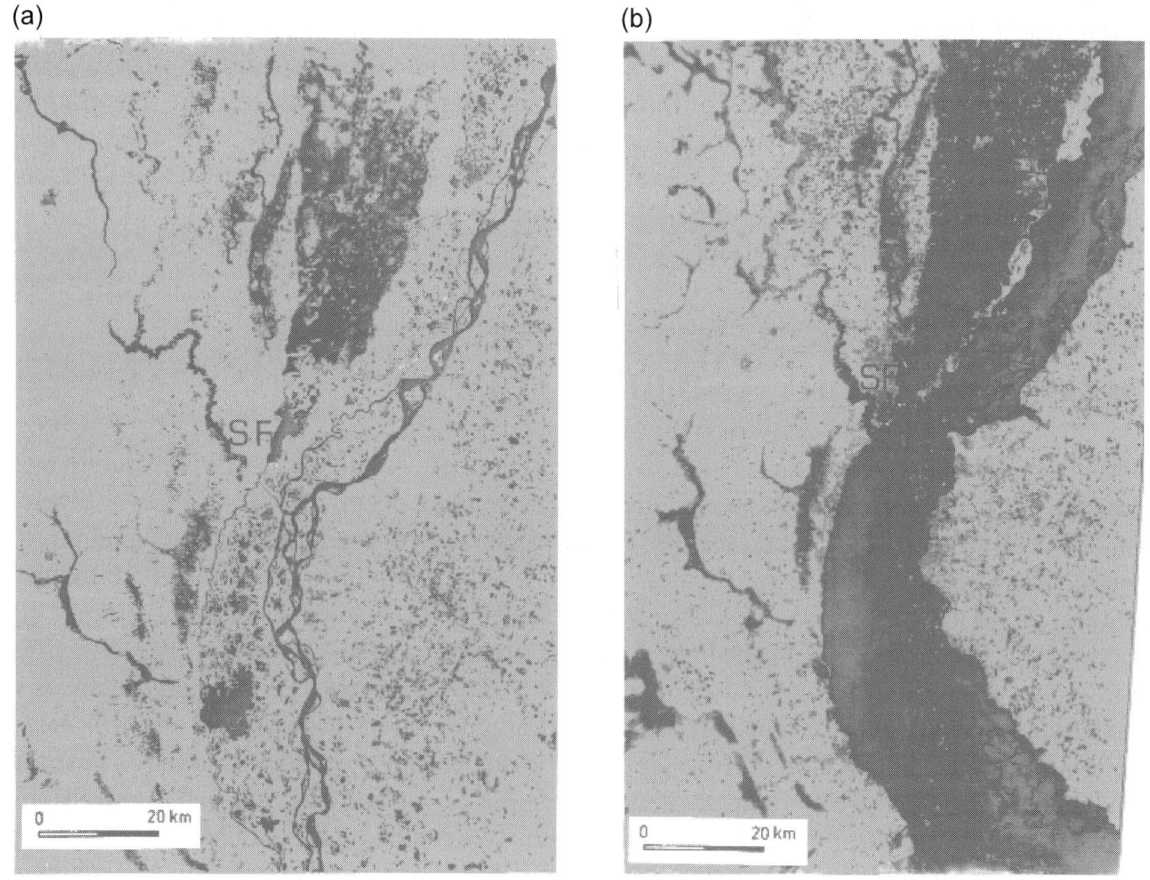

Fig. 7.2 Landsat images of the flood plain of the middle Paraná River near Santa Fe city (SF), Argentina (31°40′S, 60°30′W) during (a) mid-water level (8 November 1981) and (b) extraordinary flood (1 October 1983); water level difference: 3.16 m. The images cover an area of about 130 km × 95 km. (Images are from the CNIE Landsat Processing Centre, Buenos Aires.) (From Drago, 1977a,b, 1990a,b)

Drago & Amsler, 1998). Downriver, in the lower Paraná, the characteristic sandy bed is maintained, but the granulometry shows a shift to fine sands (3 phi). Therefore, there is a striking homogeneity in the bed sediments along the middle and lower reaches of the Paraná River. From the point of view of the substratum, it is possible to define a longitudinally homogeneous fluvial habitat of 1447 km in length, where the cross-sections along the main stem present the same pattern, with no less than 95% of the wetted perimeter composed of sand. Hereafter we will refer to this fluvial environment as the central strip environment. Thus, there are only narrow strips along the banklines where silt and clay percentages sometimes increase markedly, as is typical in channel areas where the water velocity is significantly reduced (slack-water habitat), allowing deposition of the finer grain sizes.

Bedforms range from ripples and dunes to sand waves; the latter bedforms show wave heights (wh) ranging from 7.5 to 15 m (Drago, 1977b). During high-water stages, the dunes move downriver at a mean velocity of 12 m day^{-1} and the average rate of movement of the superimposed ripples (wh = <0.30 m) and megaripples (wh = 0.30–1.50 m) reaches 37 m day^{-1}. During low water stages the rate of movement decreases to 3.3 m day^{-1} for dunes, and 9.9 m day^{-1} for the superimposed ripples and megaripples (Amsler & Gaudin, 1994; Lima *et al.*, 1990). Thus, the water depth plays an important role in controlling the rate of movement of the bedforms, mainly of dunes and sand waves. In the lower reach, the largest dunes are smaller than in the middle reach, with maximum and average wave heights of 7 m and 1.5 m, respectively. The current speeds in the main channel are always high, ranging from 0.5 to 2.6 m s^{-1} under low to moderate discharges and often exceeding 3 m s^{-1} during high flows. At a nodal point, the velocity of water on the river bed measured during a 5-year period, ranged from 0.44 to 1.11 m s^{-1}.

From 32°05′S to 34°29′S, the Paraná River has developed a large delta known as the lower Paraná reach, which extends for 345 km and occupies an area of 15 000 km^2. Most researchers (Soldano, 1947; Bonfils, 1962; Gentili & Rimoldi, 1979; Iriondo, 1972; Drago, 1990b) have designated the boundary between the middle Paraná flood plain and this deltaic area as just to the south of Diamante city (Argentina). Geomorphologically, the Paraná delta has been divided into two subregions known as the upper and lower delta. The upper delta lies between Diamante city and the Paraná Guazú-Paraná de las Palmas diversion point. The lower delta extends between this point and the Río de la Plata estuary. The multilobate shape of the delta reflects the sediment transport and deposition dynamics of the Paraná River, and the delta grows an average of 58 m year^{-1} (Drago & Amsler, 1998). The Paraná Guazú, the most important deltaic distributary, has been important from a hydro sedimentological standpoint since the Quaternary when neotectonic uplift increased its hydraulic gradient. This higher gradient allowed the river to capture 79% of the Paraná discharge, and most of its sediment load (which is now approximately six times greater than that of the Paraná de las Palmas). Consequently, the Paraná Guazú represents the morphologic and sedimentological continuation of the Paraná River at the delta (Drago & Amsler, 1998). It presents a predominantly sandy bed (99% of total weight) up to the Río de la Plata estuary. On the other hand, the Paraná de las Palmas has carved its channel through marine transgressions with silt–clay strata of Quaternary age. Therefore, the bed material along nearly all of its length shows high percentages of silt and clay.

HYDROLOGY

Despite its length, the Paraná River displays similar unimodal hydrographs along its upper, middle and lower reaches. During the first six months of the year the river is at high stage, with maximum peaks in March–April. Low waters occur in the second half of the year, with

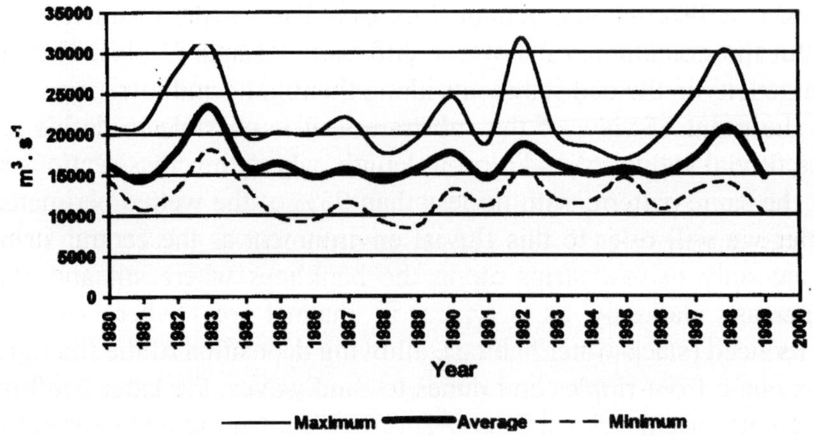

Fig. 7.3 Maximum, minimum and average discharges of the middle Paraná River at Paraná city, Argentina.

minimum flows in September–October. However, some climatic events as the ENSO may shift this regime resulting in protracted high or low water stages.

The upper reach of the Paraná, with an average discharge of 12 000 m^3 s^{-1}, governs the hydrological regime of the middle and lower reaches. The small discharges of the tributaries along the middle and lower reaches exert a negligible influence on the total discharge. The maximum and minimum water discharges of the middle Paraná measured over the last century were 60 000 and 4000 m^3 s^{-1}, respectively, with an average discharge of 20 000 m^3 s^{-1}. The single main channel carries an average discharge of 17 000 m^3 s^{-1}, or 85% of the total discharge (Fig. 7.3). The mean ratio of maximum and minimum discharge is especially low (1.8), because the upper Paraná and Paraguay Rivers differ in amplitude and phasing of discharge, and because storage of water in the flood plain attenuates the hydrograph. Interannual variation affects maximum and minimum discharge, the timing of the flood wave, the duration of the extremes of high and low water, and many other more subtle features of discharge. Furthermore, the middle and lower reaches receive tributaries that differ in amplitude and phasing of discharge. Although these tributaries are very low in discharge in comparison with the Paraná River, some of them affect the physical and chemical characteristics of the flood-plain channels flowing on the western border of the alluvial valley. Similarly, the effects of groundwater seepage are important along the western border. The contribution of both groundwater and surficial runoff from the Chaco-Pampa Plain, changes the hydrochemistry of the water flowing in secondary channels along the right margin of the middle and lower reaches, because waters of local origin tend to be more saline than those of the Paraná River (Depetris & Marchetti, 1973; Maglianesi, 1969; Stangenberg & Maglianesi, 1969).

The deltaic area has three reaches with distinct hydrological characteristics:

(a) a reach with fluvial characteristics (118 km in length) between Diamante and Rosario cities (Argentina),
(b) a fluvio-estuarine reach (139 km in length) between Rosario and San Pedro cities (Argentina), and
(c) an estuarine reach of 236 km in length between San Pedro city and Río de La Plata estuary.

The fluvial and fluvio-estuarine reaches are not directly influenced by seawater but they experience the hydrodynamic effects of the marine tides. These tidal effects together with the effects of strong south-to-southeast winds ("sudestadas") produce water level fluctuations that are detected occasionally as far upriver as Rosario city.

The fringing flood plain of the middle Paraná is mainly concentrated on the western margin of the main stem, whereas the deltaic flood plain has developed mainly on the eastern margin. The development of sand bars and channel islands and their further growth are the first steps in the formation of this flood plain (Drago, 1973, 1977a, 1990b), which covers an area of 13 063 km^2 with an upriver–downriver width ranging from 6 to 40 km. This surface is occupied by thousands of shallow lentic water bodies of different shapes and sizes, permanent and ephemeral, directly or indirectly connected with the river channels (Drago, 1976, 1980, 1981, 1989, 1990a).

The drainage network of the flood plain typically shows sinuous and irregular meander patterns. Sometimes, one single secondary channel changes in its planform features, ranging from sinuous to irregular meanders, with some straight reaches. The development of alternating shallows and deeps in the Paraná flood-plain channels is associated not only with the meander channels, but also with the stream junctions and in the bends of the low sinuosity channels. Mid-sand bars and islands are uncommon elements of the flood-plain streams. Dunes are strictly associated with the largest secondary channels, i.e. in channels with mean discharges over 500 $m^3 s^{-1}$. Usually, the dunes present wave heights less than 3 m. As in the main channel, the bed channel of these largest flood-plain streams shows high percentages of sands, always over 90% by weight. Furthermore, the cross-sections show that more than of 95% of their wetted perimeter is composed of sands. On the other hand, the lower hierarchy channels (mean discharges less than 100 $m^3 s^{-1}$) sometimes present ephemeral small ripples along the straight reaches. But the characteristic feature in the longitudinal profiles of these minor streams is the alternation of shallow reaches and scour holes. These scour holes are deep, from 5 to 15 m, with slow-water areas and up-river eddy flows, and may not have flow during the lowest river levels. The shallow and straight reaches are located between successive scour holes, showing maximum depths of 3–5 m, and sometimes these reaches become dry during severe droughts.

Flat and unconsolidated muddy beds are common in the smaller channels, where the banks offer support for rooted aquatic macrophytes that form floating meadows. These floating meadows sometimes cover the entire channel surface during high-water periods (summer–fall). During low waters, current velocity in the smallest channels becomes markedly reduced, until in some of them flow ceases, producing temporary lentic characteristics with a large decrease in suspended sediment concentration. Usually, the flood-plain channels with mean discharges over 500 $m^3 s^{-1}$ maintain a flow throughout the year. On the other hand, channels with less than 100 $m^3 s^{-1}$ show dry reaches during the lowest stages of severe droughts. The bed sediment of the minor channels contains large amounts of silt (mean average: 47% in weight) and clay (mean: 38% by weight), with lower percentages of sand (mean: 15% by weight). The predominant sand-grain size is very fine (4 phi), with a mean percentage of 69% of the total sand weight.

FLOOD-PLAIN LAKES

About 30% of the Paraná flood plain is covered by a wide variety of temporary and permanent lakes. During the largest floods, they tend to merge into a continuous sheet of water that covers the whole flood plain (Fig. 7.2(b)). However, at low river levels, processes such as the growth and decay of primary producers (mainly macrophytes) and the mixing and re-suspension of bottom sediments govern lake metabolism. Owing to the river and flood plain dynamics, the thousands of shallow lakes vary considerably in shape and size. The water bodies range from dendritic to circular, with lengths from less than 100 m to more than 10 km. They are shallow, rarely exceeding 5 m in depth at mid-water river stage and have a regular bottom topography. Usually, the maximum depths are found at the centre of the circular basins, while in the channel-shaped basins the maximum depths are located along the old thalwegs.

The bottom is composed mainly of clay–silt sediments (mean composition: 8% sand, 38% silt, 54% clay) with some patches of very fine sands (very fine sands always comprise over 65% of the total sand weight). Sandy substrata are found in the connections with the active channels or along the shoreline, with the largest sand percentages in the youngest basins. The evolution of these water bodies is related to geomorphological, hydrological, and biological (e.g. palustrine and aquatic vegetation) factors. Alluvial lakes may be classified in seven main types according to their origin (Drago, 1976, 1990a). Flood plain geomorphology and the origin and evolution of its water bodies determines the degree of connectivity between lotic and lentic environments (Drago, 1980, 1981). Usually, the connections between flood plain lakes and channels become more complex with increasing distance from the main channel. During mean water stage, two kinds of connections between rivers and lakes can be distinguished (Drago, 1981, 1990a):

(a) direct connections, such as connections through a mouth, erosion ditch, or a short channel, in which case the river level fluctuations are quickly reflected in the lakes, and the communication has a reversible flow; and
(b) indirect connections, such as connections through channels of greater length, with intermediate lentic basins, or by means of abandoned channels partially covered with sediments.

Flood plain lakes that have direct connections with active channels can be connected with the river all the year, while a lake with an indirect connection may be isolated from the river as much as 300 days of the year.

The hydraulic residence time varies according to the distance of the lakes from the active channels and the type and degree of connection and can also vary across microhabitats within a particular lake. This microhabitat variability is an important contributor to the aquatic habitat heterogeneity in the flood plain. The variability of microhabitats within a lake will depend of the origin and evolution of its basin. Usually, dendritic lakes with a high shoreline development (e.g. $D_L > 3$) will show a larger number of microhabitats compared to lakes with a low shoreline development (e.g. $D_L < 2$). The dish-shaped and especially the channel-shaped lake basins have the shortest water retention times. One dendritic lake with a direct connection had a residence time of 7 days in its main basin during the inundation phase, while its secondary basin (farther away from the river) had a residence time of 15 days. Two embayments had residence times of 30 and 50 days, respectively. The maximum retention time calculated in the lake was 230 days. However, other lakes show striking minimum and maximum residence times, ranging from 20 minutes to 365 days (Drago, unpublished).

STRUCTURE AND DISTRIBUTION OF BENTHIC MACROINVERTEBRATES

Benthic environments in the main river channels

The central strip of the main channel along 1447 km from Posadas city to the mouth of the Paraná River is longitudinally homogeneous in its benthic invertebrate community structure (Fig. 7.1). The dominant and sometimes the only representative species of the benthos is *Narapa bonettoi* (Oligochaeta) which attains maximum densities of 92 943 ind. m^{-2} (Marchese, 1984) and 112 000 ind. m^{-2} (Varela *et al.*, 1983). This is a small organism of 4 mm maximum length and 3 µg dw (dry weight) of biomass (Marchese, 1994). It is associated with lower densities of *Myoretronectes paranensis* (Turbellaria), *Haplotaxis aedeochaeta* (Oligochaeta) and *Tobrilus* sp. (Nematoda). Other species such as *Potamocaris* sp. (Crustacea

Harpacticoida) and *Parachironomus* sp. (Diptera Chironomidae) are occasionally detected. This species assemblage was also found in the upper Paraná by Varela *et al.* (1983), and Bonetto *et al.* (1986) and upstream of the Itaipú Dam by Takeda (1999) and Montanholi-Martins & Takeda (1999).

In the sandy bed of the Paraná main channel, where the substratum is 90–99% sand by weight with only small amounts of organic matter (0.003–0.008% C by weight), longitudinal patterns in species composition were not observed, although species abundance did decrease in the downriver direction. In the basaltic rock of the bed of the upper Paraná River the dominant species are the Porifera *Oncosclera navicella*, *O. schubarti*, *Trochospongilla repens*, *T. ringueleti y Corvospongilla seckti* (Ezcurra de Drago 1993) and the Mollusca *Byssanodonta paranensis*, and *Potamolithus* sp.

Principal components analysis of the species diversity in samples from the main channel central strip (Fig. 7.4), showed that component I explains 75.3% of the variance, which can account for the total density of *Narapa bonettoi.* Component II, accounting for water conductivity and the density of *Limnodrilus hoffmeisteri* (Oligochaeta), explains 9.9% of the variation. The analysis distinguished a group of samples from the Paraná de las Palmas distributary and the Río de la Plata estuary from those of the main distributary, the Paraná Guazú. The former environments have high densities of *Limnodrilus hoffmeisteri* and *Corbicula fluminea* (Bivalvia Corbiculidae) and *Limnoperna fortunei* (Bivalvia Mytilidae). *Limnoperna fortunei* is the last invader species among the Mollusca and was detected for the first time in South America in 1991 on the shores of the Río de La Plata estuary (Pastorino *et al.*, 1993). A few years later *L. fortunei* invaded the middle Paraná River flood plain and upper reaches of the Paraná River and Paraguay River on Asunción Harbour, Paraguay (Darrigran & Ezcurra de Drago, 2000). The Paraná Guazú, which is considered as the morphologic and sedimentological continuation of the Paraná main channel within the delta, has a benthic invertebrate species assemblage that resembles that of the main stem upriver, although with lower densities.

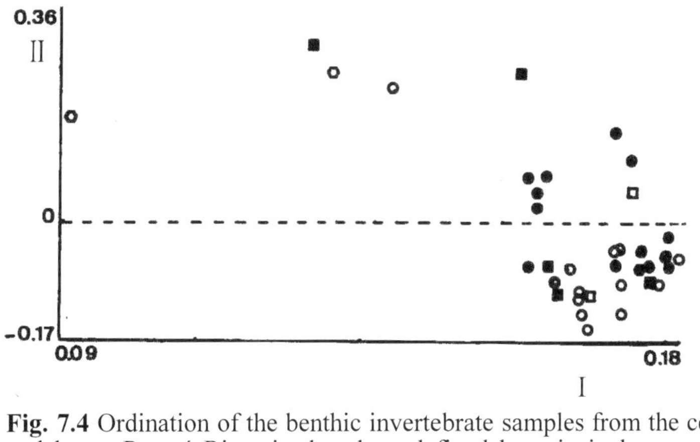

Fig. 7.4 Ordination of the benthic invertebrate samples from the central strips of the middle and lower Paraná River in the plane defined by principal components I and II. • middle Paraná River; □ Paraná Guazú distributary; ■ Paraná de las Palmas distributary; ○ lower Paraná River; O Río de la Plata Estuary.

The low benthic macroinvertebrate species diversity and biomass in the central strip of large rivers with sandy mobile beds and high discharges, has been mentioned by other authors throughout the world. Behning (1924) wrote about the scarce populations found in the sandy bed of the Volga River, and Mordukhai-Boltovskoi (1979) described for the same river a psammophilic biocoenosis of low species richness inhabiting the bed of the middle and lower reaches where the grain size is 85–90% sand. Monakov (1969) described a scarce and monotonous benthic population in the White Nile, represented by small larvae of Chironomidae (*Glyptotendipes* sp. and *Criptochironomus* sp.). In the dense sand sediments of the Amur River, Bogatov *et al.* (1995) reported low values of density (50–100 ind. m^{-2}) and biomass (0.002–0.01 g m^{-2}). Similar results have been recorded in the Ganges River with inter-site differences reflecting the sediment characteristics, i.e. their unstable sand substrata were an unsuitable habitat for most benthic animals and were unusually (for Asian waters) dominated by amphipods (Dudgeon, 1995). Lim (1987) also found lower species richness on sand substrata in two Malaysian rivers. The Sand River (Canada), with a mean annual discharge of 12–16 m^3 s^{-1} and moving sand ripples 6–15 cm high, showed macroinvertebrate production of at least 752 mg dw m^{-2} $year^{-1}$, based on measurements of two chironomid species which made up 81% of total biomass (50–490 mg m^{-2}). This was due to the small size of most organisms, whose densities were 12 000–78 000 ind. m^{-2}. Both dominant chironomid species were interstitial and maintained their position in shifting sand using silk strands (Mackay, 1995). Whereas the macroinvertebrate communities in these sand-bed rivers may appear to be sparse compared to other kinds of benthic habitat, Hynes (1970) pointed out the special microfauna that inhabit these environments.

In South America, Sioli (1975) estimated that the moving bed of the Amazon River with dunes up to 180 m long and 8 m high, would discourage the establishment of benthic animals. Subsequently, he recognized that nothing is known about this community in the Amazon main channel (Sioli, 1984). Lewis *et al.* (1995) also pointed out that the lack of benthic organisms in large rivers of northern South America is due to the instability of the fine-grained and shifting substrates. In the reviews of the research on benthic macroinvertebrate communities of the Orinoco and Amazon river–flood-plain systems, Vázquez *et al.* (1990), Reiss (1977), Junk & Robertson (1997) and Nessimian *et al.* (1998) mentioned only the benthos inhabiting different types of shallow lakes in the flood plains.

By comparison with the aforementioned references on the benthos of large flood plain rivers, we conclude that in relation to the influence of physical factors on the benthic community structure, the Paraná River presents the largest volume of information (Bertoldi de Pomar *et al.*, 1986; Bonetto, 1976; Bonetto *et al.*, 1986; Di Persia, 1986; Di Persia *et al.*, 1982; Ezcurra de Drago, 1980; Marchese, 1981, 1984, 1987; Marchese & Ezcurra de Drago, 1983, 1992; Varela *et al.*, 1983; Takeda, 1999; Montanholi-Martins & Takeda, 1999).

Considering the lateral dimension of large rivers, quantitative data have been obtained only in the middle Paraná River (Marchese & Ezcurra de Drago, 1992). There are no

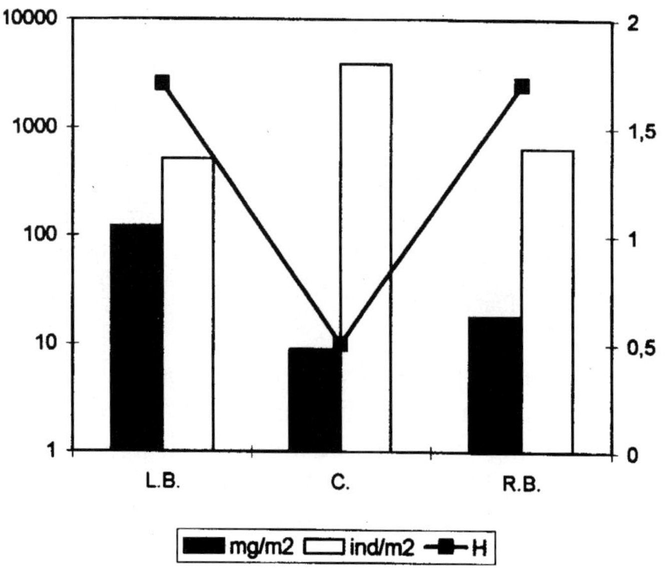

Fig. 7.5 Density (ind. m^{-2}), biomass (mg m^{-2}) and species diversity (H) near the left bank (L.B.), right bank (R.B.) and within the central strip (C.) of the middle Paraná River main channel.

publications on the subject for other large rivers of the world. Benthic samples collected along a lateral transect across the main channel showed strong differences between the benthos near the banks compared to the channel central strip (Fig. 7.5). The left bank, which has a higher percentage of clay than the right bank but has similar amounts of organic matter (0.543 and 0.524% C, respectively), showed a higher total benthic biomass with a dominance of Tubificidae (*L. hoffmeisteri*, *Paranadrilus descolei* and *Aulodrilus pigueti*). On the contrary, near the right bank the higher density is accounted for by Chironomidae (*Coelotanypus* sp., *Axarus* sp., *Criptochironomus* sp.). The highest density but lowest biomass was always found in the central strip (i.e. 95% of the wetted perimeter), where there was a very low biomass of *Narapa bonettoi*. Species diversity was always higher on the banks than at the channel centre and was similar near the left and right banks. In the channel bar habitat, some specimens of *Aulodrilus pigueti* were found, burrowing to 0.30 m in depth and to 2 m above the water line.

LATERAL AND TEMPORAL DIMENSIONS

Flood-plain drainage system

Most studies of river ecology emphasize the main channel and flood plain lakes, without presenting any information on the flood plain channels. However, these secondary channels together with the lentic water bodies enhance the spatial and temporal heterogeneity of aquatic environments of the flood plain (Ward, 1989). This spatio-temporal heterogeneity of the Paraná river–flood plain system is controlled by the alternating dry and wet phases that in turn, produce a high benthic invertebrate species richness. This variability contrasts with the monotonous benthos found in the longitudinal dimension of the Paraná main stem. Qualitative variations in some secondary channels with different bed sediment composition and macrophyte cover were described for the upper Paraná River by Takeda *et al.* (1999).

In the lateral dimensions of the middle Paraná River, variations in benthic community structure are related to the hierarchical gradient of the secondary channels and also the variable degree of hydrological connectivity between the shallow lakes and the river (Marchese & Ezcurra de Drago, 1992). For example, the study of several lotic environments revealed a decrease in diversity and biomass and an increase in density from the main channel to the secondary channels of decreasing hierarchy. Within the channels, there is a positive correlation between invertebrate biomass and the organic matter content of the sediment ($r = 0.73$, $p > 0.05$) and invertebrate density and discharge ($r = 0.89$, $p > 0.01$), and a negative correlation between invertebrate diversity and discharge ($r = -0.84$, $p > 0.01$).

Detrended analysis correspondence included monthly samples during one year for the central strip from the main channel and secondary channels of the Paraná River. These analyses using the density, species diversity, species richness, discharge, depth, organic matter contents and percentages of sand, silt, and clay reveal three groups: one formed by the sampling stations in the main channel, another formed by the secondary channels of higher-hierarchy ($Q > 300$ m^3 s^{-1}), and the third formed by secondary channels of lower-hierarchy (Fig. 7.6). Axes I and II explain 59.2% and 4.4% of the variance, with invertebrate density, discharge, clay percentage and organic matter content explaining most of the variance. Therefore benthic community heterogeneity increased in the secondary channels of lower-hierarchy. For example, the hydrological variability due to the inundation and drought phases in the low-hierarchy Correntoso River, result in spatio-temporal patches characterized by species assemblages that reflect different fluvial microhabitats. The rivers with fine bed sediments show spatio-temporal variations in their benthic invertebrate

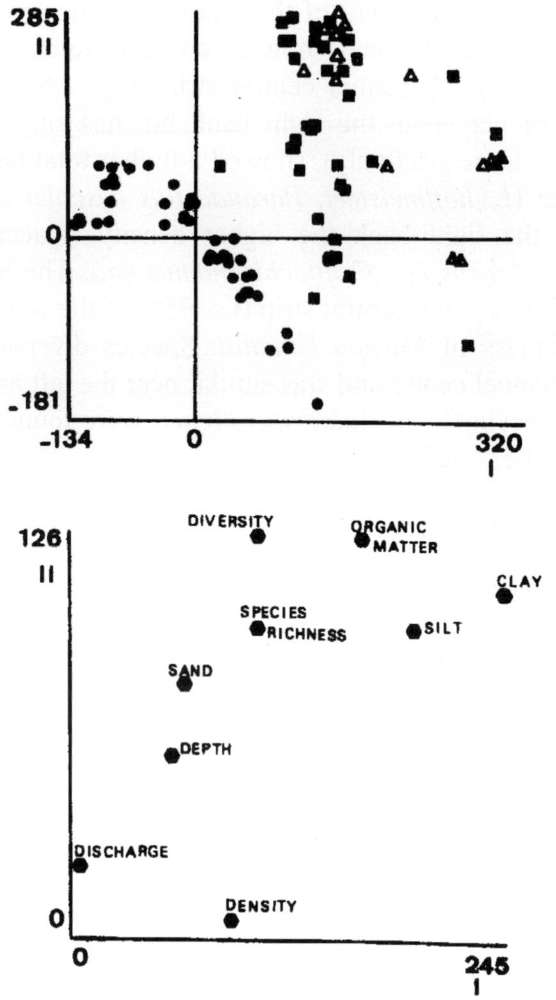

Fig. 7.6 Ordination of the benthic invertebrate samples from the middle Paraná River channel in the plane defined by detrended analysis correspondence. ● Main channel; ■ higher-hierarchy secondary channels; **Δ** lower-hierarchy secondary channels.

composition and structure according to the hydraulic characteristics of different riverine habitats.

Variations in channel morphology and hydrology produce microhabitats with particular hydraulic conditions. Minor secondary channels that have hydraulic behaviour similar to that of the main channel and major secondary channels during the highest water levels show species assemblages containing *Narapa bonettoi*, *H. aedeochaeta* and *M. paranensis.* (Ezcurra de Drago *et al.*, unpublished). With the change in hydraulic conditions during low water phases, this assemblage and that which had been found at high water disappeared. Two hypotheses explain this community shift: (a) these species found refuge in the hyporheos during low water stages, and (b) they were transported as drift and colonize these habitats temporarily during the high water stages (Ezcurra de Drago *et al.*, unpublished).

The spatial and temporal variation in the morphological and hydrological characteristics of the smaller flood-plain channels produce changes in the distribution of bed sediment sizes from the sand-silt-clay bed of the straight reaches to predominantly sandy patches with characteristic species assemblages. Localized variations in sediment grain sizes also result, due to the variable hydraulic conditions among microhabitats. Although the particle-size distribution of the substratum is an important determinant of the type of benthic fauna inhabiting a riverine habitat, from the studies of Statzner (1981), Statzner *et al.* (1988) and Marchese & Ezcurra de Drago (1992), we also consider the hydraulic–bed sediment relationship as a paramount factor.

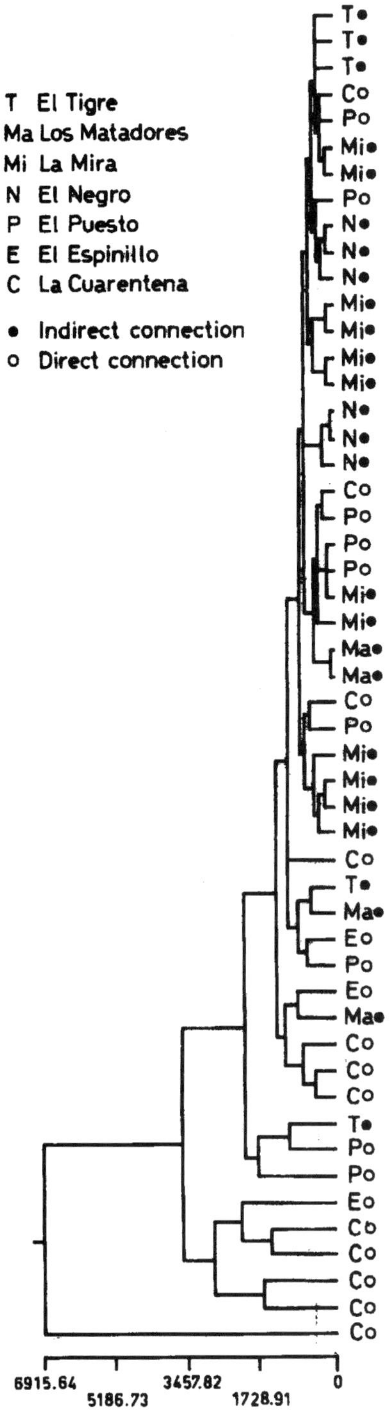

Fig. 7.7 Dendrogram of dissimilarity (Euclidian distance index) among benthic invertebrate samples from various flood-plain lakes during low and high water phases obtained by the unweighted pair group method.

In headwater streams, the importance of water velocity and substratum types in determining the micro-distribution of benthic macroinvertebrates has been stressed by several authors (e.g. Hynes, 1970; Minshall, 1984; Minshall & Minshall, 1977; Townsend, 1989). These authors document the influence of current velocity and particle-size heterogeneity on the macroinvertebrates in riffles and pools, emphasizing the importance of the variation in particulate organic matter in the different microhabitats. The spatio-temporal variations of benthic fauna in some reaches of small secondary channels in the Paraná River flood plain are comparable, within a particle size range of coarse sand to clay, to those found in small streams by numerous authors such as Minshall (1988), Doeg *et al.* (1989), and Jorde &

Bratrich (1998). Based on studies by Statzner & Higler (1986) and Statzner *et al.* (1988), it is possible to explain spatial variation in macroinvertebrate communities by a combination of substratum composition and hydraulic conditions. Thus, not only is high spatial diversity in invertebrate community structure found in higher-gradient, smaller streams that have mixtures of bedrock, boulders, cobbles and gravels (e.g. Cummins & Lauff, 1969; Minshall, 1984), but also in secondary flood-plain channels with fine-sediment substrata, where the channel morphology, hydrology and hydraulics determine the ecological variation.

Strong differences in the benthic community structure of flood-plain lakes are observed according to the lake's origin, age, degree of connection and water residence time. Multivariate cluster analysis applied to the attributes of the benthos and species more abundant of different flood-plain lakes during high-water and low-water phases reveals several different types of flood-plain lakes (Fig. 7.7). The youngest dendritic basins with direct connection to lotic waters, as for example La Cuarentena, El Puesto, Los Matadores, El Espinillo and La Mira lakes, are separated from lakes with indirect connections such as El Tigre pond or El Negrito pond, the latter with a longer residence time, 100% vegetation cover and very low oxygen concentration at the bottom (<2 mg l^{-1}). The representative species assemblage of lakes with direct connection with the river includes *Campsurus notatus* (Ephemeroptera), *Pisidium* sp.(Bivalvia), *Pristina* spp. and *Dero* spp.(Oligochaeta), whilst in the lakes with indirect connections and large vegetation cover the typical assemblage is formed by *Chironomus xanthus*, *Branchiura sowerbyi*, *L. hoffmeisteri*, *Aulodrilus pigueti* and *Chaoborus* sp. Dendritic lakes such as La Cuarentena and El Puesto show a very different qualitative and quantitative composition of the benthos that results from a patchy heterogeneity, with different hydraulic conditions due to their lotic connection and variable vegetation coverage. Some older flood-plain lakes with high percentages of aquatic macrophyte cover and long isolation phases from the river have the lowest benthic biomass and species diversity, sometimes reaching 0 bits. Typical organisms of these habitats include *C. xanthus*, *L. hoffmeisteri*, *B. sowerbyi*, *A. pigueti* and *Chaoborus* sp.

In the Amazonian lakes, Fittkau *et al.* (1975b) and Reiss (1977) found that the differences in the physico-chemical properties of the black waters and white waters as well as the type of bottom sediments influence the biomass and production of benthic fauna. Given that the chemistry of the Paraná River flood-plain lakes is not as variable as in the Amazon River system, the degree of connection and the hydraulic residence time are considered as the fundamental parameters influencing the benthic community structure. The benthic diversity, density and biomass are larger in lacustrine areas that lack vegetation cover, exhibit major water movements, maintain high oxygen concentrations and have a low organic matter content in the sediments. Thus, the bottom morphology and the hydrology of the different lake areas account for the variations of sediment sizes, amount of organic matter and plant cover percentage.

In the middle Paraná main channel and secondary channels of higher-hierarchy, the benthic biomass reaches an annual average of 107 mg m^{-2}. In these environments, *Narapa bonettoi* has a production of 9 mg m^{-2} dw $year^{-1}$ and a P/B relation of 1.5. In minor secondary channels such as the San Javier River, with a silt–clay bed and 0.607% C, the annual average biomass is 164 mg m^{-2} dw and the production of the dominant species *L. hoffmeisteri* is 47 mg m^{-2} $year^{-1}$ with a P/B = 1.12. In secondary channels and flood-plain lakes with larger amounts of organic matter in bed sediments, the annual average biomass exceeds the aforementioned values, ranging from 4 to 26 g m^{-2}. These values are given mainly by *C. notatus* and juvenile bivalves of Mycetopodidae, Hyriidae and Sphaeriidae (Fig. 7.8). According to Junk & Robertson (1997), in Amazonian lakes the greatest invertebrate biomass is measured in lentic water bodies with a lower vegetation cover. Thus, he found benthic biomass ranging from 0.230 g m^{-2} in lakes with greater plant cover to 2 g m^{-2} in lakes with less vegetation cover.

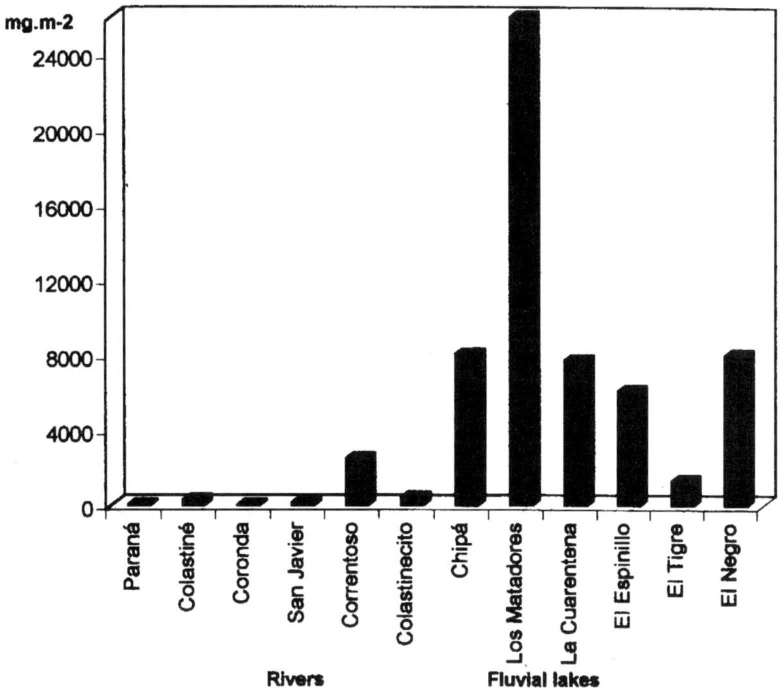

Fig. 7.8 Annual average benthic biomass (mg m^{-2} wet weight) in the main channel and flood-plain environments of the middle Paraná River.

In relation to the temporal dimension, the driving factor is the flood pulse (Junk *et al.*, 1989), which within flood-plain environments produces seasonal alternation between the potamophase and limnophase (Neiff, 1990, 1999). The inundation phase does not have a marked influence on benthic fauna in the main channel and flood-plain channels with larger discharges. On the contrary, in minor channels with lower discharges there is a strong influence of the flood pulse on benthos, because these environments become more lentic during the low-water phase (Drago, unpublished). This lentic phase in minor channels, during which there is very low or null water current and sediment transport, results in the impoverishment of benthic fauna. In flood-plain lakes seasonal differences are also observed, which are related to the hydraulic residence time and the degree of connection with the river (Fig. 7.9). If the amplitude and magnitude of the low water phase is large, it

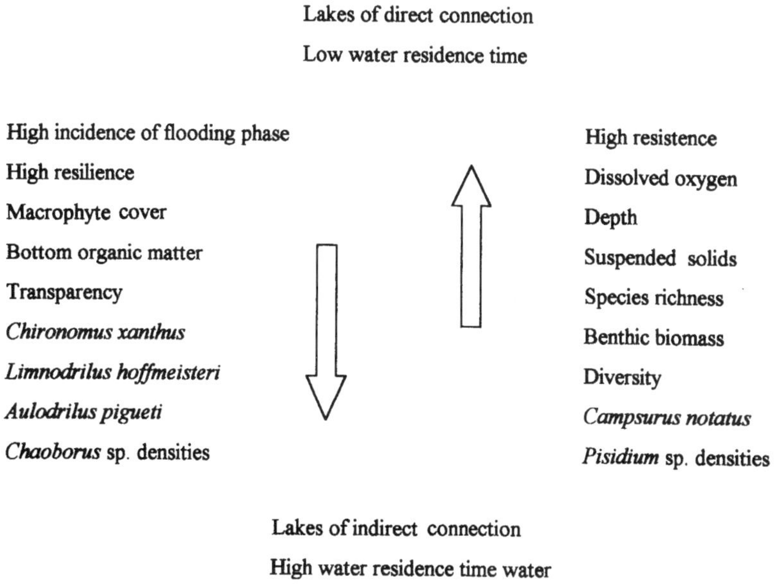

Fig. 7.9 Physical, chemical and biological linkages in flood-plain lakes spanning gradients between direct and indirect river connection and high and low water residence times.

has a strong stressing effect on the benthic community, mainly in lakes with strong isolation from the river and high macrophyte cover. Since these types of very isolated flood-plain lakes, even during the higher floods, have low possibilities of being reset, there is a great accumulation of dead organic matter and consequently a strong depletion of oxygen concentrations, usually reaching anoxia. Therefore, there is an impoverishment of the benthic species diversity and density with development of more tolerant species such as *Chironomus xanthus*, *Branchiura sowerbyi*, *Limnodrilus hoffmeisteri*, *Dero* spp. and *Chaoborus* sp. Reiss (1977) reported a qualitatively and quantitatively poor benthic fauna during the period of decreasing water level in central Amazon lakes. During the low water phase Chironomidae larvae were dominant in these environments. The benthic species diversity is higher in the minor secondary channels and flood-plain lakes during the inundation phase. The highest species richness at the rising water period was recorded by Nessimian *et al.* (1998) in the lower Rio Negro. When the river level begins to rise there is a rejuvenation of the environments, being colonized again by a large number of species. This was described by Sparks *et al.* (1990) as a reset of the system. By comparison, a large magnitude and duration of the inundation phase is less of a stress for benthic invertebrates than the low-water phase, which is similar to the findings of Neiff (1999) for aquatic macrophytes.

FUNCTIONAL FEEDING GROUPS

The particulate organic matter (POM) differences found among riverine habitats of the Paraná River flood plain are a consequence of hydraulic conditions (Fig. 7.10). The greatest amounts of litter occur in lower hierarchy secondary channels and flood-plain lakes where the physical fragmentation of the litter derived from macrophytes and forest is lowest. In the main channel and major secondary channels the mobile sandy beds contain low concentrations of POM. On the contrary, higher concentrations of coarse POM (CPOM) occur in the minor secondary channels and in flood-plain lakes, mainly in those with indirect river connections. Despite the high concentrations of CPOM found in several environments, the functional feeding groups in invertebrate communities are mainly collector-gatherers (Oligochaeta Narapidae, Tubificidae and Naididae, Ephemeroptera *Campsurus notatus* and some species of Chironomidae), filterer-collectors (the bivalve Mollusca Hyriidae, Mycetopodidae, Sphaeriidae, Corbiculidae and Mytilidae and the

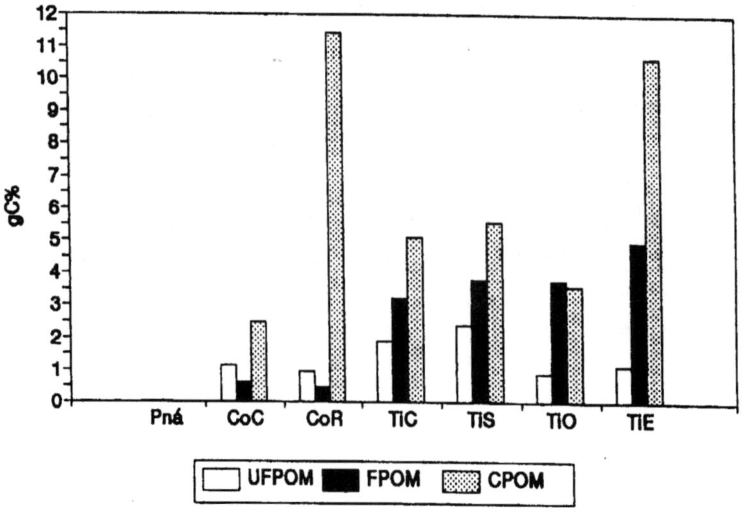

Fig. 7.10 Annual average particulate organic matter in the middle Paraná River system. Pná: Main channel central strip; CoC: Correntoso River centre; CoR: Correntoso River banks; TiC: central basin of Tigre lake; TiS, TiO, TiE: shoreline sampling sites at Tigre lake.

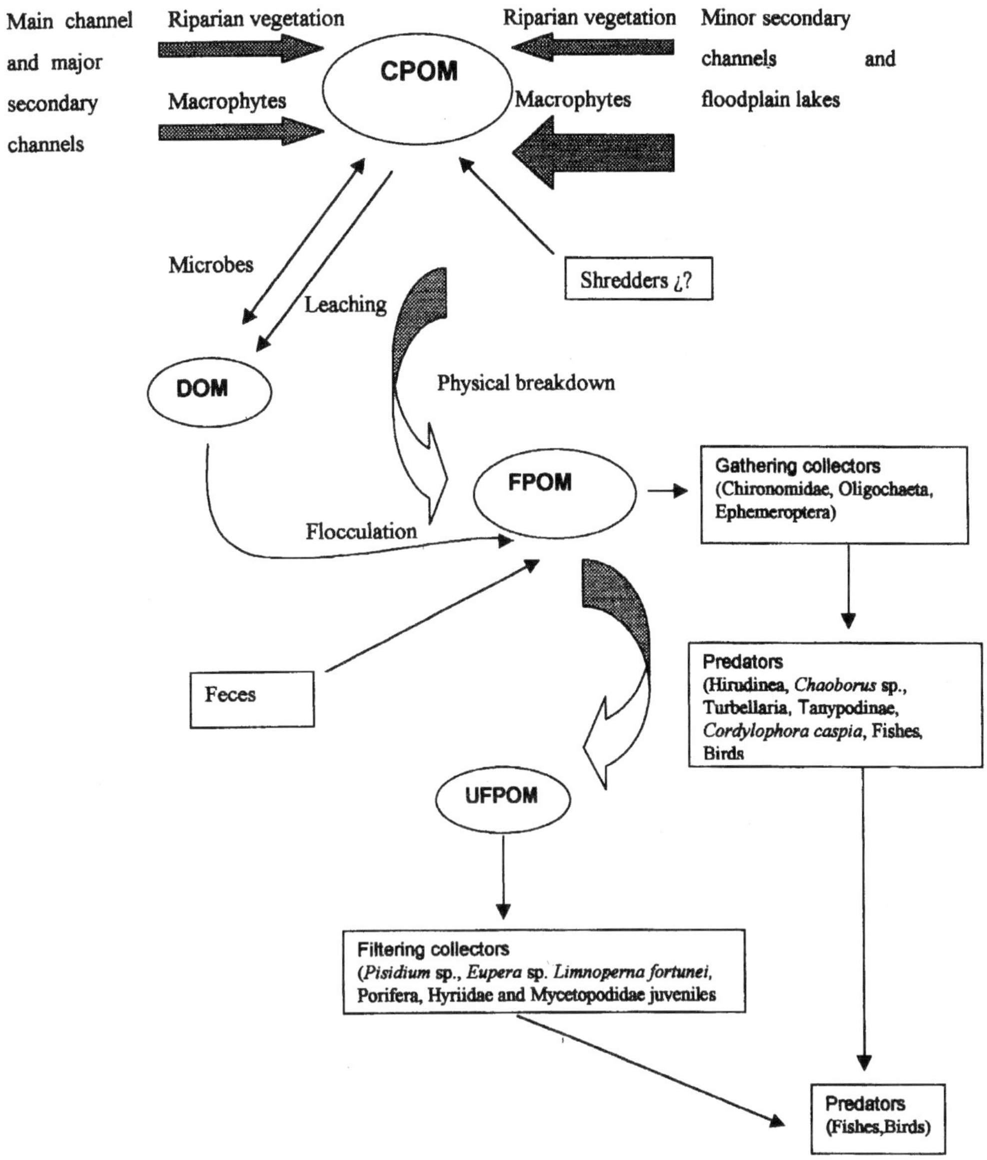

Fig. 7.11 Conceptual model of resources of particulate organic matter and their linkages with functional feeding groups for the middle Paraná River flood plain.

Porifera Spongillidae, Metaniidae and Potamolepidae), and predators (Hirudinea, Diptera Tanypodinae and Turbellaria). Shredders were not found (Ezcurra de Drago & Marchese, 1994) (Fig. 7.11). The fine POM (FPOM) and ultrafine POM (UFPOM) are the organic matter fractions used by benthic fauna. The lack of shredders explains the higher CPOM values found in the flood-plain lakes with indirect river connections and in the lower-hierarchy secondary channels during low-water phases. These lotic and lentic water bodies show very low rates of benthic invertebrate production. During the floods of largest, magnitude and amplitude, some channel-shaped flood-plain lakes located nearest to the active channels may be reset, eventually leading to greater secondary production of the benthic macro invertebrate community.

CONCLUSIONS

In its natural or seminatural state, the main channel–flood-plain complex of the Paraná River may be viewed as a mosaic of habitats undergoing succession. Controlling factors, such as flood and low-water phases, tributary inputs, riparian influences, groundwater seepage, hydraulic conditions, etc., interact to influence each aquatic habitat. The benthic macroinvertebrate communities respond to this spatial and temporal variability by changing their structure. Thus, the benthos of the aquatic habitats of the Paraná River flood plain vary according to the degree of connection with the main flow, showing the following characteristics:

- The benthic community structure varies along a compositional gradient from the main channel to the lower-hierarchy secondary channels and flood-plain lakes with less hydrological connectivity with the river.
- In the longitudinal dimension, the benthic community structure of the central strip of the main channel is constant spatially as well as temporally, because to the homogeneity of the hydrological, hydraulic and sedimentological characteristics of the main channel.
- In the central strip of the main channel as well as in the higher-hierarchy secondary channels, the typical benthic assemblage is formed by *Narapa bonettoi*, *Myoretronectes paranensis* and *Haplotaxis aedeochaeta*.
- In the main channel as well as in the higher-hierarchy secondary channels, there are marked differences in the benthic community composition of the central strip compared with the riverbed closer to the channel banks.
- The flood pulse produces more qualitative and quantitative changes in the benthic community structure of the lower-hierarchy secondary channels and in the flood-plain lakes than in the main channel and in the higher-hierarchy secondary channels.
- The variations in the benthic community structure of the lower-hierarchy secondary channels are more marked during the lowest water stages because of the transformation to more lentic conditions, and in the flood-plain lakes with an indirect connection with the river because there is a longer isolation phase.
- The lower-hierarchy secondary channels show a sequence of flat or uniform beds and scour holes, which undergo strong temporal variations in hydraulic characteristics.
- The flat bed reaches of the lower-hierarchy secondary channels show a characteristic benthic community structure, with a striking spatio-temporal homogeneity because of the regularity of the hydraulic parameters. The typical assemblage is formed by *Paranadrilus descolei*, *Limnodrilus hoffmeisteri*, *L. udekemianus*, *Pristina americana*, *Aulodrilus pigueti*, *Bothrioneurum americanus*, *Dero (Aulophorus) borelli*, *Campsurus notatus*, *Ablasbemyia* sp. and *Pisidium* sp., with *P. descolei* dominating. The quantitative structure of this assemblage changes only during the longer and lowest water stages, when the relative percentage of *P. descolei* decreases and *L. hoffmeisteri* and *L. udekemianus* become more abundant.
- On the contrary, the composition and structure of the benthos in the scour holes vary considerably during the flood phase as a result of the hydraulic variations of these habitats. The assemblage developed during the flood phase is the same as that of the central strip of the main channel and high-hierarchy secondary channels: *Narapa bonettoi*, *Myoretronectes paranensis* and *Haplotaxis aedeochaeta*.
- The species richness, species diversity and biomass are larger in the flood-plain lakes with direct connection to the river. Within these basins there can be a gradient of hydrological environments, as for example in the dendritic basins where mosaics of species assemblages exist according to the variable influence of the flood waters on different areas of the lake basin.
- In contrast to the aforementioned lacustrine basins, the smaller flood-plain lakes that show during the longest isolation phases undergo a progressive process of encroachment by aquatic vegetation which can eventually obliterate the open-water area until the successional process is reset by a large flood event.
- The typical species of the flood-plain lakes with a direct connection with the river are:

Aulodrilus pigueti, *Limnodrilus hoffmeisteri*, *Bothrioneurum americanus*, *Pristina americana*, *Slavina evelinae*, *Pristinella osborni*, *Campsurus notatus*, *Coelotanypus* sp., *Ablasbemyia* sp., *Polypedilum* sp., *Glyptotendipes* sp., juveniles of Hyriidae (*Diplodon parodizi*, *D. charruanus*, *D. delodontus delodontus*, *Castalia inflata*), Mycetopodidae (*Anodontites trapesialis susanae*, *A. ensiformis*, *A. trapezeus*, *A. trigonus georginae*, *Monocondylaea paraguayana*), Corbiculidae (*Neocorbicula limosa* and *Corbicula fluminea*), Sphaeriidae (*Pisidium* sp., *Eupera platensis*) and Mytilidae (*Limnoperna fortunei*).

- In the flood-plain lakes with indirect connection to the river or in some areas of the dendritic basins, which are highly dependent of the inundation phase for their "rejuvenation" process, the dominant species are *Limnodrilus hoffmeisteri*, *Eiseniella tetraedra*, *Dero (Aulophorus) borelli*, *Dero (Dero) multibranchiata*, *Trieminentia corderoi*, *Branchiura sowerby* and *Chironomus xanthus*.
- Some species of the main channel, secondary channels and flood-plain lakes, are endemic to the neotropical region, as for example several species of Porifera (*Corvospongilla seckti*, *Onscosclera navicella*, *O. schubarti*, *Trochospongilla repens*, *T. ringueleti*, *Drulia brownii* and *T. paulula)*, of Oligochaeta (*Narapa bonettoi*, *Haplotaxis aedeochaeta* and *Paranadrilus descolei)*, Turbellaria (*Myoretronectes paranensis*), Ephemeroptera (*Campsurus notatus*), Hirudinea (*Helobdella adiastola*), Bivalvia Hyriidae (*Diplodon parodizi*, *D. charruanus*, *D. delodontus delodontus*, *Castalia inflata*), Mycetopodidae (*Anodontites trapesialis susanae*, *A. ensiformis*, *A. trapezeus*, *A. trigonus georginae*, *Monocondylaea paraguayana)* and Sphaeriidae (*Byssanodonta paranensis*, *Eupera platensis*).
- Although many species inhabit lotic habitats as well as lentic habitats of the Paraná River system, it is possible to distinguish a potamobenthos (benthic community of river beds) mainly formed by *N. bonettoi*, *H. aedeochaeta*, *Myoretronectes paranensis*, *B. americanus*, *P. descolei*, *P. americana* and *Tobrilus* sp. and in some reaches *Oncosclera navicella*, *Corvospongilla seckti*, *Trochospongilla repens*, *T. ringueleti*, and *T. paulula.*
- It is very difficult to distinguish a characteristic limnobenthos (benthic community of standing-water habitats). However, in the flood-plain lakes the following species are more consistently abundant: *C. notatus*, *Pisidium* sp., *Eupera platensis*, *Dero (D.) nivea*, *Dero (D.) borellii*, *B. sowerbyi*, *Axarus* sp., *Glyptotendipes* sp., *Polypedilum* sp., *Chironomus xanthus*, *Ablasbemyia* sp. and *Chaoborus* sp.
- The benthic community of the main channel, the higher-hierarchy secondary channels and flood-plain lakes with direct connection to the river are characterized by their higher resistance to the marked variability of the hydrological conditions, while the communities found in lower-hierarchy secondary channels and flood-plain lakes with indirect connection have lower resistance and higher resilience.
- The benthic species of the central strip of the main channel and higher-hierarchy secondary channels are *r* strategists while those of the lower-hierarchy secondary channels and flood-plain lakes are mainly *K* strategists.

Acknowledgements We thank Dr Stephen K. Hamilton and Dr Wolfgang J. Junk for critically reading the manuscript and offering many helpful comments. Special thanks to Dr Stephen K. Hamilton for improving the English and to the Editor Dr Michael McClain for the encouragement to prepare this chapter. We are grateful to Aldo R. Paira for field and laboratory assistance.

REFERENCES

Amsler, M. & Gaudin, H. (1994) La superposición de dunas y el transporte de la carga de fondo en el Río Paraná. In: *Anales del XV Congreso Nacional* del *Agua (La Plata, Argentina)*. Universidad Nacional de La Plata, La Plata, Prov. de Buenos Aires, Argentina.

Behning, A. L. (1924) K. Izucheniu pridonnoi zhizni r. Volgi (Study of the benthic life in the Volga). *Monographie Volzhskoi Biology*. In: *The River Volga and its Life* (ed. by D. Mordukhai-Boltovskoi). W. Junk, London, UK.

Bertoldi de Pomar, H. (1980) Campaña "Keratella I" en el Río Paraná Medio: sedimentos de fondo. *Ecología* (Argentina) **4**, 31–43.

Bertoldi de Pomar, H., Copes, C. Ezcurra de Drago, I. & Marchese, M. (1986) Características limnológicas del Río Paraná y sus principales tributarios en el tramo Goya-Diamante. Los sedimentos y su fauna. *Revista Asociación Ciencias Naturales del Litoral* **17**, 79–97.

Bonetto, A. A. (1976) *Calidad de las Aguas del Río Paraná*. Introducción a su estudio ecológico. Dir. Nac. Constr. Porturarias y Vías Navegables, INCYTH, PNUD, ONU, Argentina.

Bonetto, A. A., Varela, M. E. & Bechara, J. A. (1986) El bentos del Paraná medio en el tramo Corrientes-Esquina. *Ecosur* (Argentina) **12/13**, 37–57.

Bonfils, C. (1962) Los suelos del delta del Río Paraná. Factores generadores, clasificación y uso. *Revista Investigaciones Agrícolas* **16**, 257–370.

Bogatov, V., Sirotsky, S. & Yuriev, D. (1995) The ecosystem of the Amur River. In: *Ecosystem of the World. River and Stream Ecosystems* (ed. by C. E. Cushing, K. W. Cummins & G. W. Minshall), 601–614. Elsevier Science, Amsterdam.

Cummins, K. W. & Lauff, G. H. (1969) The influence of substrate on the microdistribution of stream macrobenthos. *Hydrobiologia* **34**, 145–181.

Darrigran, G. & Ezcurra de Drago, I. (2000) Invasion of the exotic freshwater mussel *Limnoperna fortunei* (Dunker, 1857) (Bivalvia: Mytilidae) in South America. *The Nautilus* **114**(2), 69–73.

Depetris, P. & Marchetti, P. J. (1973) Turbiedad y conductividad de las aguas del Río Paraná Medio y su relación con las principales unidades fitogeográficas. *Boletin Paranaense de Geociencias (Brasil)* **31**, 135–140.

Di Persia, D. H. (1986) Zoobenthos of the Paraná system. In: *The Ecology of River Systems* (ed. by B. R. Davies & K. F. Walker), 589–598. Dr W. Junk, Dordrecht, The Netherlands.

Di Persia, D. H., Poledri, J. C. & D'Angelo, R. (1982) El zoobentos del arroyo Yacarecito, Prov. de Santa Fe, Argentina. *Revista Asociación Ciencias Naturales del Litoral* **13**, 13–24.

Doeg, T. J., Marchant, R., Douglas, M. & Lake, P. S. (1989) Experimental colonization of sand, gravel and stones by macroinvertebrates in the Acheron River, southeastern Australia. *Freshwater Biology* **22**, 57–64,

Drago, E. C. (1973) Caracterización de la llanura aluvial del Paraná Medio y de sus cuerpos de agua. *Boletim Paranaense de Geociencias (Brasil)* **31**, 31–44.

Drago, E. C. (1976) Origen y clasificación de ambientes leníticos en llanuras aluviales. *Revista Asociación Ciencias Naturales del Litoral* **7**, 123–137.

Drago, E. C. (1977a) Erosión y sedimentación en un tramo del cauce del Río Paraná Medio (República Argentina). *Asociación Geológica Argentina, Revista* **32**, 277–290.

Drago, E. C. (1977b) Campaña "Keratella I" a lo largo del Río Paraná Medio. II: Formas del lecho en su cauce principal. *Revista Asociación Ciencias Naturales del Litoral* **8**, 57–62.

Drago, E. C. (1980) Aspectos limnológicos de ambientes próximos a la ciudad de Santa Fe (Paraná Medio): comportamiento hidrológico y sólidos suspendidos. *Ecología* (Argentina) **5**, 31–48.

Drago, E. C. (1981) Grados de conexión y fases hidrológicas en ambientes leníticos de la llanura aluvial del Río Paraná (Argentina). *Ecología* (Argentina) **6**, 27–33.

Drago, E. C. (1989) Morphological and hydrological characteristics of flood-plain ponds in the middle Paraná River (Argentina). *Revue Hydrobiologie Tropicale* **22**, 183–190.

Drago, E. C. (1990a) Hydrological and geomorphological characteristics of flood-plain ponds in the middle Paraná River. *Acta Limnológica Brasiliense* **3**, 907–930.

Drago, E. C. (1990b) Geomorphology of large alluvial rivers: lower Paraguay and middle Paraná. *Interciencia* **15**, 378–387.

Drago, E. C. & Amsler, M. (1998) Bed sediment characteristics in the Paraná and Paraguay rivers. *Water International* **23**, 174–183.

Dudgeon, D. (1995) The ecology of rivers and streams in tropical Asia. In: *Ecosystem of the World. River and Stream Ecosystems* (ed. by C. E. Cushing, K. W. Cummins & G. W. Minshall), 615–658. Elsevier Science, Amsterdam.

Elwood, J. W., Newbold, J. D., O'Neil, R. V. & Van Winkle, W. (1983) Resource spiraling: an operational paradigm for analyzing lotic ecosystems. In: Dynamics of Lotic Ecosystems (ed. by T. D. Fontaine & S. M. Bartell), 3–28. Ann Arbor Science Publishers, Ann Arbor, Michigan, USA.

Ezcurra de Drago, I. (1980) Campaña limnológica "*Keratella I*" en el Río Paraná Medio: complejo bentónico del río y ambientes leníticos asociados. *Ecología* (Argentina) **4**, 89–101.

Ezcurra de Drago, I. (1993) Distribución geográfica de las esponjas argentinas (Porifera: Spongillidae, Potamolepidae y Metaniidae). Relaciones zoogeográficas, vías de poblamiento. In: *Conferences on Limnology* (La Plata,

Argentina) (ed. by A. Boltovkoy & H. López), 115–125. Instituto de Limnología "Dr R. A. Ringuelet", La Plata, Prov. Buenos Aires, Argentina.

Ezcurra de Drago, I. & Marchese, M. (1994) Zoobentos del Río Paraná en sus tramos medio e inferior. Grupos funcionales. *Tankay* (Argentina) **1**, 197–198.

Fittkau, E. J., Junk, W., Klinge, H. & Sioli, H. (1975a) Substrate and vegetation in the Amazon region. *Ber. Int. Symposium Internationale Vereinigung Vetat.* In: *The Central Amazon Floodplain, 1997*, 73–90. Ecological Studies vol. 126, Springer, Berlin, Germany.

Fittkau, E. J., Irmler, W., Junk, W., Reiss, F. & Schmidt, W. (1975b) Productivity, biomass and population dynamics in Amazonian water bodies. In: *Tropical Ecological Systems* (ed. by F. B. Golley & E. Medina), 289–311. Springer, Berlin.

Gentili, C. & Rimoldi, H. (1979) Mesopotamia. In: *Segundo Simposio de Geología Regional Argentina, Academia Nacional de Ciencias de Córdoba* **I**, 185–223. Universidad Nacional de Córdoba, Córdoba, Argentina.

Hynes, H. B. N. (1970) *The Ecology of Running Waters*. University Press, Liverpool, UK.

Iriondo, M. (1972) Mapa gemorfológico de la llanura aluvial del Río Paraná entre Helvecia y San Nicolás (República Argentina). *Asociación Geológica Argentina, Revista* **27**, 155–160.

Iriondo, M. H. (1979) Origen y evolución del Río Paraná. *II Jornadas del Paraná Medio, Paraná, Entre Ríos* (Argentina) **1**, 1–5.

Iriondo, M. H. (1988) A comparison between the Amazon and Paraná river systems. *Mitteilungen aus dem Geologisch-Palaeontologischen Institut der Universitaet Hamburg, SCOPE-UNEP, Sonderband* **66**, 77–92.

Iriondo, M. H. & Scotta, E. (1979) The evolution of the Paraná River delta. In: *Proceedings of the* 1978 *Symposium on Coastal Evolution in the Quaternary* (São Paulo, Brazil) (ed. by K. Suguío, T. Fairchild, L. Martin & J. Flexor), 405–418. Instituto Geociencias, USP, São Paulo, Brazil.

Jorde, K. & Bratrich, C. (1998) River bed morphology and flow regulations in diverted streams effects on bottom shear patterns and hydraulic habitat. In: *Advances in River Bottom Ecology* (ed. by G. Bretschko & J. Helesic), 47–63. Backhuys Publishers, Leiden, The Netherlands.

Junk, W. J. & Robertson, B. A. (1997) The benthos. In: *The Central Amazon Floodplain* (ed. by W. J. Junk), 286–289. Springer, Berlin.

Junk, W. J., Bayley, P. B. & Sparks, R. E. (1989) The flood pulse concept in river–flood-plain systems. In: *Proceedings of the International Large River Symposium (LARS)* (ed. by D. P. Dodge). *Canadian Special Publication Fisheries Aquatic Sciences 106*, 110–127. NRC Research Press, Ottawa.

Lewis, W., Jr, Hamilton, S. K. & Saunders III, J. F. (1995) Rivers of northern South America. In: *Ecosystem of the World. River and Stream Ecosystems* (ed. by C. E. Cushing, K. W. Cummins & G. W. Minshall), 219–256. Elsevier Science, Amsterdam.

Lim, R. P. (1987) Water quality and faunal composition in the streams and rivers of the Ulu Endau area, Johore, Malaysia. *Malaysia Naturalist Journal* **41**, 337–347.

Lima, D., Campana, N., Amsler, M., Schreider, M. & Gaudin, H.(1990) Desplazamiento de dunas y carga de fondo en un tramo del Río Paraná. *Memorias XIV Congreso Latinoamericano de Hidráulica (Montevideo, Uruguay)*, vol. 3, 1203–1214. Universidad Nacional de la República, Montevideo, Uruguay.

Mackay, R. J. (1995) River and stream ecosystems of Canada. In: *Ecosystem of the World. River and Stream Ecosystems* (ed. by C. E. Cushing, K. W. Cummins & G. W. Minshall), 33–60. Elsevier Science, Amsterdam.

Maglianesi, R. (1969) Fenómenos de salinización en algunos brazos de la margen derecha del Paraná Medio. *Physis* (Argentina) **28**, 1–13.

Marchese, M. (1981) Contribución al conocimiento del complejo bentónico del Río Paraná medio. *Ecología* (Argentina) **6**, 55–65.

Marchese, M. (1984) Estudios limnológicos en una sección transversal del tramo medio del Río Paraná. XI: Zoobentos. *Revista Asociación Ciencias Naturales del Litoral* **15**, 157–174.

Marchese, M. (1987) The ecology of some benthic Oligochaeta from Paraná River, Argentina. *Hydrobiologia* **155**, 209–214.

Marchese, M. (1994) Population dynamics of *Narapa bonettoi Righi and Varela,* 1983 (Oligochaeta: Narapidae) from the main channel of the middle Paraná River, Argentina. *Hydrobiologia* **278**, 303–308.

Marchese, M. & Ezcurra de Drago, I. (1983) Zoobentos de los principales tributarios del Río Paraná medio en el tramo Goya-Diamante. Su relación con el cauce principal y cauces secundarios. *Revista Asociación Ciencias Naturales del Litoral* **14**, 97–107.

Marchese, M. & Ezcurra de Drago, I. (1992) Benthos of the lotic environments in the middle Paraná River system: transverse zonation. *Hydrobiologia* **237**, 1–13.

Marchetti, B. & Prutkin, N. (1982) Los Biomas. In: *Atlas de la República Argentina* (ed. by E. Chiozza E. & R. Figueira R.), 422–432. Centro Editor de América Latina SA, Buenos Aires, Argentina.

Minshall, G. W. (1984) Aquatic insect-substratum relationships. In: *The Ecology of Aquatic Insects* (ed. by V. H. Resh & D. M. Rosenberg), 358–400. Praeger Special Studies, Praeger Scientific, New York.

Minshall, G. W. (1988) Stream ecosystem theory: a global perspective. *Journal North American Benthological Society* **7**(4), 263–288.

Minshall, G. W. & Minshall, J. N. (1977) Microdistribution of benthic invertebrates in a rocky mountain (USA) stream. *Hydrobiologia* **55**, 231–249.

Monakov, A. V. (1969) The zooplankton and zoobenthos of the White Nile and adjoining waters in the Republic of the Sudan. *Hydrobiologia* **33**, 161–185.

Montanholi-Martins, M. C. & Takeda, A. M. (1999) Communities of Benthic Oligochaetes in relation to sediment structure in the upper Paraná River, Brazil. *Studies Neotropical Fauna and Environments* **34**, 52–58.

Mordukhai-Boltovskoi, D. (1979) *The River Volga*. Dr W. Junk, The Hague.

Neiff, J. J. (1990) Ideas para la interpretación ecológica del Paraná. *Interciencia* **15**, 424–441.

Neiff, J. J. (1999) El régimen de pulsos en ríos y grandes humedales de Sudamérica. In: *Tópicos Sobre Humedales Subtropicales y Templados de Sudamérica* (ed. by A. I. Malvárez), 97–146. UNESCO, Montevideo, Uruguay.

Nessimian, J. L., Dorvillé, L. F. M., Sanseverino, A. M. & Baptista, D. F. (1998) Relation between flood pulse and functional composition of the macroinvertebrate benthic fauna in the lower Rio negro, Amazonas, Brazil. *Amazoniana* **XV**(1/2), 35–50.

Newbold, J. D., Elwood, J. W., O'Neill, R. V. & Van Winkle, W. (1981) Measuring nutrient spiralling in streams. *Canadian Journal of Fisheries and Aquatic Sciences* **38**, 860–863.

Newbold, J. D., O'Neill, R. V., Elwood, J. W. & Van Winkle, W. (1982) Nutrient spiralling in streams: implications for nutrient limitation and invertebrate activity. *American Naturalist* **120**, 628–652.

Pastorino, G., Darrigran, G., Martin, S. & Lunaschi, L. (1993) *Limnoperna fortunei* (Dunker, 1857) (Mytilidae), nuevo bivalvo invasor en aguas del Río de La Plata. *Neotropica* **39**(101–102), 34.

Pringle, C. M., Naiman, R. J., Bretschko, G., Karr, J. R., Oswood, M. A., Webster, J. R., Welcomme, R. L. & Winterbourn, M. J. (1988) Patch dynamics of lotic systems: the stream as a mosaic. *Journal North American Benthological Society* **7**, 503–524.

Reiss, F. (1977) Qualitative and quantitative investigations on the macrobenthic fauna of Central Amazon lakes. I. Lago Tupé, a black water lake on the lower Río Negro. *Amazoniana* **VI**(2), 203–235.

Resh, V. H., Brown, A. V., Covich, A. P., Gurtz, M. E., Li, H. W., Minshall, G. W., Reice, S. R., Sheldon, A. L., Wallace, J. B. & Wissmar, R. (1988) The role of disturbance in stream ecology. *Journal North American Benthological Society* **7**, 433–455.

Sioli, H. (1975) Tropical river: the Amazon. In: *River Ecology* (ed. by B. A. Whitton), 461–488. University of California Press, Berkeley, California.

Sioli, H. (1984) *The Amazon. Limnology and Landscape Ecology of a Mighty Tropical River and its Basin* (ed. by H. Sioli). Dr W. Junk, Dordrecht.

Soldano, F. (1947) *Régimen y Aprovechamiento de la Red Fluvial Argentina* I: *El Río Paraná y sus Tributarios*. Editorial CIMERA, Buenos Aires, Argentina.

Sparks, R., Bayley, P., Kohler, S. & Osborne, L. (1990) Disturbance and recovery of large floodplain rivers. *Environmental Management* **14**(5), 699–709.

Stangenberg, M. & Maglianesi, R. (1969) Composición química de las aguas de la cuenca del Río Paraná Medio. Segunda parte: el Río Colastiné. *Physis* (Argentina) **28**, 111–121.

Statzner, B. (1981) The relation between "hydraulic stress" and microdistribution of benthic macroinvertebrates in a lowland running water system, the Schierenseebrooks (north Germany). *Archiv für Hydrobiologie* **91**, 192–218.

Statzner, B. & Higler, B. (1986) Stream hydraulics as a major determinant of benthic invertebrate zonation patterns. *Freshwater Biology* **16**, 127–139.

Statzner, B., Gore, J. A. & Resh, V. H. (1988) Hydraulic streams ecology: observed patterns and potential applications. *Journal North American Benthological Society* **7**(4), 307–360.

Strahler, A. N. (1957) Quantitative analysis of watershed geomorphology. *American Geophysical Union Transactions* **38**, 913–920.

Takeda, A. M. (1999) Oligochaete community of alluvial upper Paraná River, Brazil: spatial and temporal distribution (1987–1988). *Hydrobiologia* **412**, 35–42.

Townsend, C. R. (1989) The patch dynamics concept of stream community ecology. *Journal North American Benthological Society* **8**, 36–50.

Vannote, R. L., Minshall, G. W., Cummins, R. W., Sedell, J. R & Cushing, C. E. (1980) The river continuum concept. *Canadian Journal of Fisheries and Aquatic Sciences* **37**, 130–137.

Varela, M. E., Bechara, J. & Andreani, N. L. (1983) Introduccion al estudio del bentos del Alto Paraná. *Ecosur* (Argentina) **19**(19/20), 103–126.

Vázquez, E., Sanchez, L., Perez, L. E. & Blanco, L. (1990) Estudios hidrobiológicos y piscicultura en algunos cuerpos de agua (rios, lagunas y embalses) de la cuenca baja del Río Orinoco. In: *El Río Orinoco como Ecosistema* (ed. by F. H. Weibezahn, H. Alvarez & W. M. Lewis Jr), 301–335. Impresos Rubel CA, Venezuela.

Ward, J. W. (1989) The four dimensional nature of lotic ecosystems. *Journal North American Benthological Society* **8**(1), 2–8.

Ward, J. V. & Stanford, J. A. (1983) The serial discontinuity concept of lotic ecosystems. In: *Dynamics of Lotic Ecosystems* (ed. by T. D. Fontaine & S. M. Bartell, 29–42. Ann. Arbor Science Publishing, Ann Arbor, Michigan.

CHAPTER 8

HYDROLOGICAL CONTROLS OF ECOLOGICAL STRUCTURE AND FUNCTION IN THE PANTANAL WETLAND (BRAZIL)

Stephen K. Hamilton

The Pantanal is one of the world's largest tropical wetlands, occupying an area of approximately 450 by 250 km (140 000 km^2) in the Paraná River basin of South America (Fig. 8.1). The Pantanal is defined for the present purposes as the contiguous flood plains and wetlands of the upper Paraguay River and its major tributaries, above the confluence of the Apa River. Most of the region is in Brazil, with smaller areas in Bolivia and Paraguay.

Scientific studies have been carried out mostly in the Brazilian Pantanal. The Bolivian and Paraguayan portions are generally more remote and remain much less studied. A rich body of research on the Pantanal exists in the Brazilian scientific literature, written largely in Portuguese. The English-language literature contains reviews of scientific information on the Pantanal that vary in their emphasis, including plant and animal life (Heckman, 1998; Junk & da Silva, 1995; Por, 1995), environmental problems and management (Alho *et al.*, 1988; da Silva, 2000; Swarts, 2000), and the potential impacts of navigation projects in the Paraguay River system (Hidrovia Panel of Experts, 1997; Ponce, 1995). The increasing international attention paid to the Pantanal has accelerated scientific progress through multinational research programmes, and has also led to important conservation initiatives by governmental and nongovernmental organizations. The recent designation of the Pantanal by UNESCO as both a Biosphere Reserve and World Heritage Site should help to further focus international attention on the region.

The objective of this chapter is to provide an overview of the Pantanal and to discuss our emerging understanding of the critical linkages between hydrology and ecological structure and function in this vast flood-plain region. Flood-plain ecosystems such as the Pantanal are defined by their seasonal inundation and desiccation, and shift seasonally between phases with standing water and phases in which the soil surface is dry and the subsurface water table may fall to well below the rooting zone. The depth and duration of maximum inundation as well as the degree of desiccation during the dry season vary considerably across the Pantanal, and this variability in the hydrological regime of the flood-plain ecosystems has numerous ramifications for ecological structure and function, as articulated in the Flood Pulse Concept (Junk *et al.*, 1989; Junk & da Silva, 1995). This chapter also discusses the potential consequences of anthropogenic alterations in the hydrological regimes of the Pantanal flood plains, particularly in regard to the recently proposed navigation project known as the Paraguay–Paraná Waterway (or Hidrovía).

GEOMORPHOLOGY AND HYDROLOGY

The Pantanal lies in an immense isostatically subsided depression that was formed as a result of tectonic uplift associated with the Andes, which caused basement-rock reactivation and formed a back-bulge basin that now contains up to 500 m of unlithified sediments

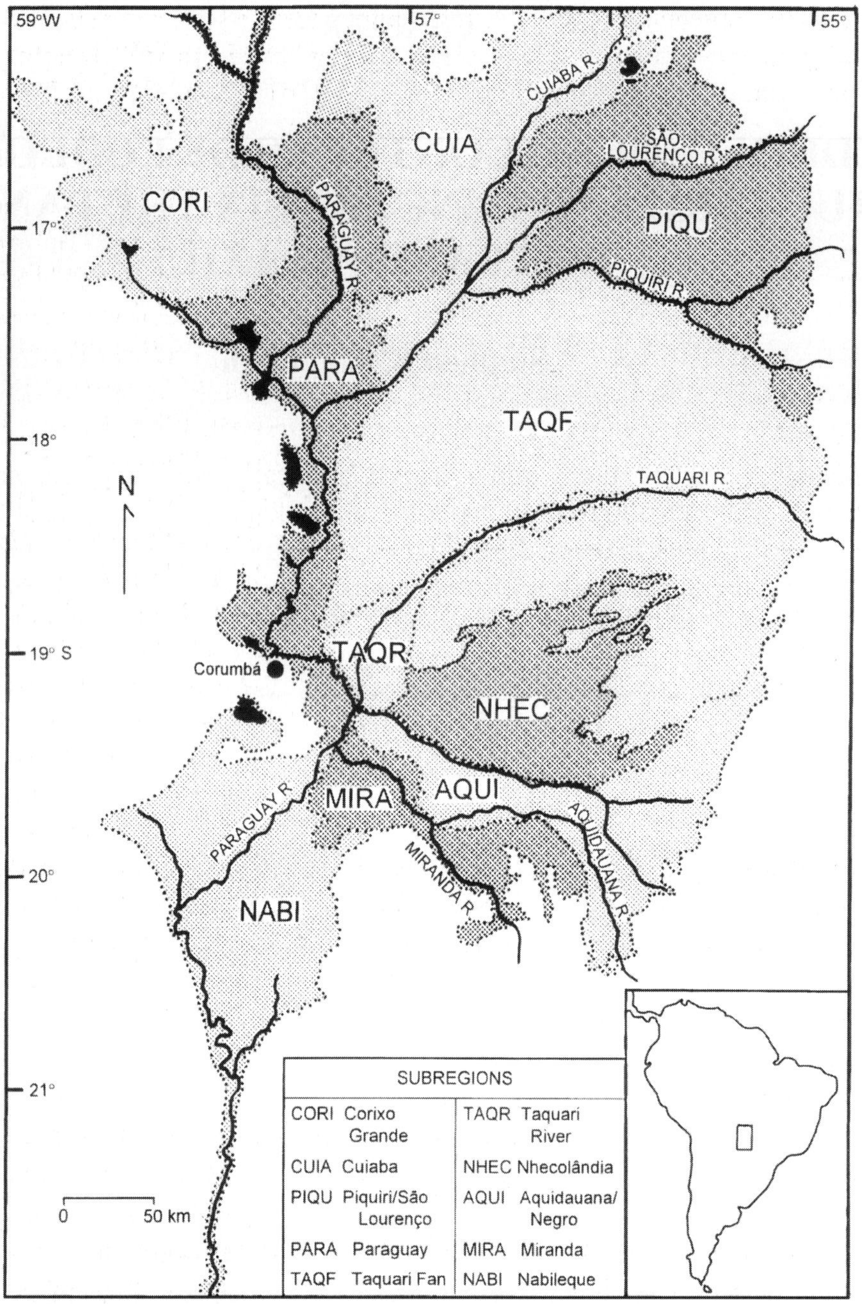

Fig. 8.1 Map of the Pantanal wetland, showing the 10 subregions delineated by Hamilton *et al.* (1996). Most of the wetland area is in Brazil with smaller areas along the west in Bolivia and Paraguay. Figure from Hamilton *et al.* (1996).

(Ussami *et al.*, 1999). Much of the alluvial material that fills the Pantanal depression was deposited during the late Cenozoic, when the climate was drier than today and the sandstone rocks in the surrounding uplands underwent intense denudation (Ab'Saber, 1988; Klammer, 1982). Since the Pleistocene, the relief of the Pantanal has been modified under the influence of a more humid climate. The Pantanal is still in a process of active aggradation, although this is occurring at much slower rates now, and the landforms are more stabilized by vegetation cover.

The surface geomorphology of the Pantanal is structured by a mosaic of coalesced alluvial fans built by inflowing rivers (Klammer, 1982). These fans are particularly visible in side-looking airborne radar imagery collected during the RADAMBRASIL Project (RADAMBRASIL, 1982). The extensive alluvial fan of the Taquari River is an especially predominant feature that covers *c.* 60 000 km^2 (Short & Blair, 1986; Hamilton *et al.*, 1998).

Elevation within the Pantanal flood plains ranges from 80 to 150 m above sea level, and elevational gradients within the flood plains average *c.* 25 cm km^{-1} from east to west, and *c.* 2 cm km^{-1} from north to south along the Paraguay River (Carvalho, 1986). Some isolated mountains occur along the Paraguay River axis (RADAMBRASIL 1982); their geological composition is variable, and includes crystalline rock, limestones, and iron and manganese oxides (Klammer, 1981; Urban *et al.*, 1992; Ussami *et al.*, 1999).

The upland drainage basin surrounding the Pantanal occupies 356 000 km^2 and consists of elevated plateaus and low mountains to the north and east (250–1200 m elevation) and flat plains to the west. The Brazilian uplands are locally referred to as the *Planalto*. Rivers in the Brazilian uplands have much steeper gradients with relatively little flood plain, and they deliver most of the water and sediment to the Pantanal. In contrast, the much flatter uplands to the west in Bolivia and Paraguay contribute little water to the Pantanal; they receive less rainfall, and a larger proportion of this water is lost by evaporation and infiltration.

Soils in the Pantanal are variable but generally contain more silt and clay in areas subject to riverine overflow and tend to be sandy on the higher parts of the alluvial fans, including areas subject to flooding by local rainfall. Alluvial deposits distant from the main rivers tend to be older and highly weathered. Soils within the Pantanal are described by Amaral Filho (1986), EDIBAP (1979), PCBAP (1997) and RADAMBRASIL (1982). Peat accumulations are generally not abundant in tropical flood-plain soils, probably because of high decomposition rates and seasonal desiccation (Junk, 1983). Soils of the Brazilian uplands surrounding the Pantanal are typical of the central Brazilian *cerrado*, and are described by Furley & Ratter (1988).

The climate of the Pantanal is tropical with a marked wet season, and is influenced by the subtropical southern Atlantic anticyclone (EDIBAP, 1979; IBGE, 1989; Tarifa, 1986). Annual rainfall is 1000–1500 mm across the basin, with most rain falling between November and March (Cadavid Garcia & Castro, 1986). Higher rainfall occurs in some parts of the surrounding uplands, particularly in the north. Rainfall patterns are less seasonal in the uplands to the south of the Pantanal (Serra da Bodoquena), which often receive significant rainfall during the austral winter as well. Due to its low elevation and continental location, the region is subject to penetration of polar air masses during the winter, which can cause temperatures to fall to freezing during brief episodes and can produce light frosts (IBGE, 1989).

The hydrology of the main river systems remains largely natural, although a large dam on the Manso River (a tributary of the Cuiabá River) has recently been completed. The general direction of water flow across the flood plains is from east to west until it reaches the Paraguay River, where it flows southward (UNESCO, 1973). The Paraguay River runs from north to south along the western side of the Pantanal, collecting water from the various tributaries and non-channelized flood-plain flow paths (Fig. 8.1). Water levels in the Paraguay River fluctuate seasonally by 2–5 m; water-level fluctuations in most other parts of the Pantanal are less than this.

Major tributaries of the Paraguay River in the Pantanal include the Jaurú, Cuiabá-São Lourenço, Taquari, Abobral, and Miranda rivers (Fig. 8.12). The Jaurú drains uplands to the northwest of the Pantanal. The Cuiabá-São Lourenço river system drains uplands to the north and northeast, as well as substantial flood-plain areas. The Taquari and Miranda rivers drain upland areas to the east and southeast of the Pantanal. The Abobral River drains a flood-plain area located between the Taquari and Miranda rivers, and its discharge is highly seasonal. Several other flood-plain water courses can carry substantial discharges during inundation and are locally denominated as rivers. For example, the Paraguai-mirim and Nabileque rivers collect water primarily from overbank flow of the Paraguay River, although they also receive water from adjacent flood-plain areas.

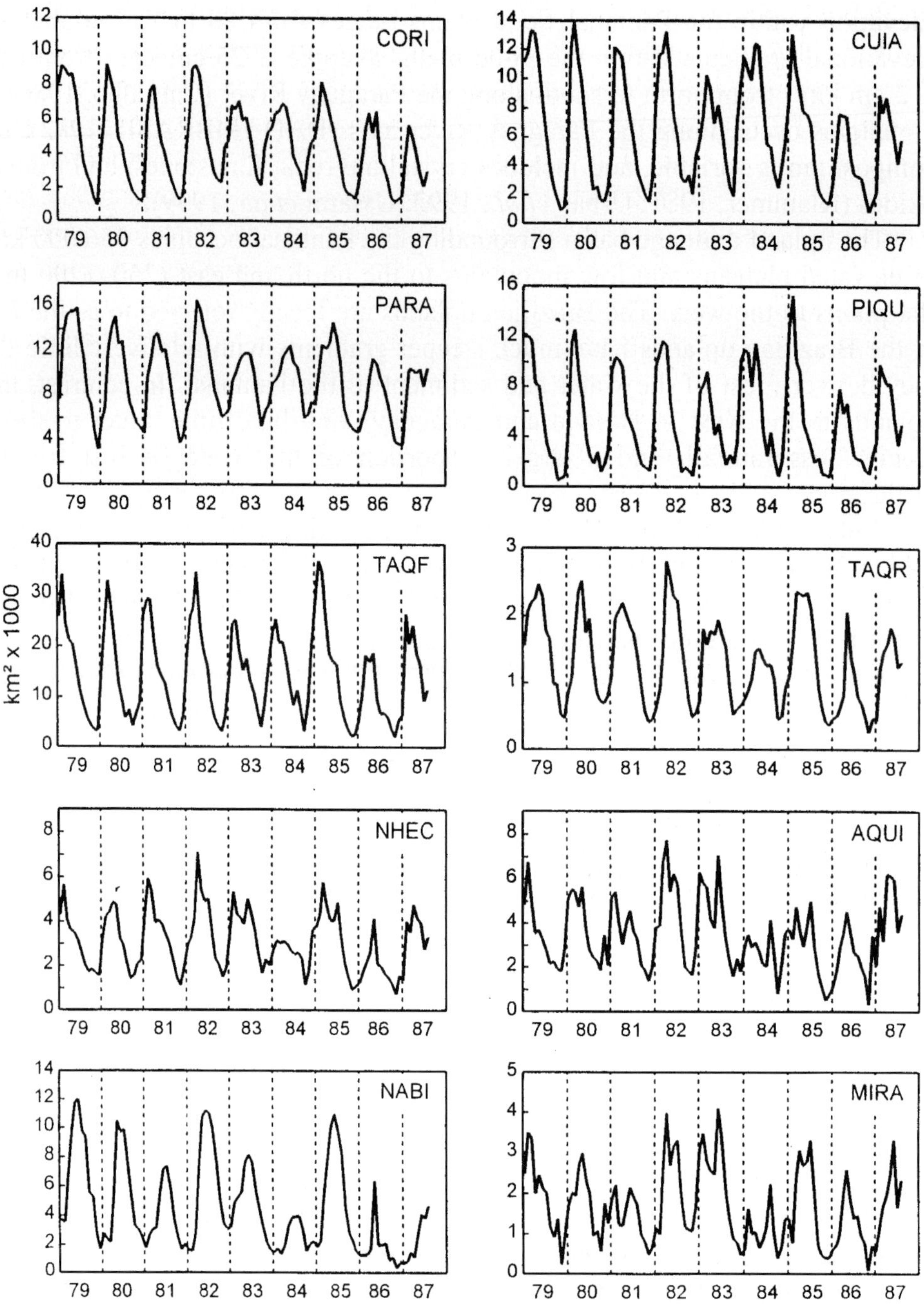

Fig. 8.2 Inundation patterns in the 10 subregions of Fig. 8.1, determined from satellite observations of passive microwave emission. Figure from Hamilton *et al.* (1996).

The Pantanal contains a variety of flood-plain subregions with distinct geomorphological, hydrological and ecological characteristics. Various authors have distinguished subregions within the Pantanal based on traditional perceptions and political jurisdictions (e.g. Adámoli, 1982) or on geomorphological and hydrological characteristics (EDIBAP, 1979; Hamilton *et al.*, 1996; RADAMBRASIL, 1982). Figure 8.1 shows the 10 subregions distinguished by Hamilton *et al.* (1996). The nomenclature of these subregions reflects the major rivers that flow through them, and although they are not necessarily entirely subject to flooding from those rivers, the boundaries have been chosen to reflect hydrological and geomorphological differences. Alternatively, the Pantanal can be subdivided by depth and duration of inundation, as in a map found in EDIBAP (1979). In another approach, 12 distinct wetland units were identified on the basis of radar images and hydrological characteristics (RADAMBRASIL, 1982).

Flooding in much of the region is distinctly seasonal, although the flooding period may be delayed for as long as 6 months after the rains due to slow passage of flood waters through the Pantanal. Many areas are flooded by riverine overflow and are thus true flood plains, while other areas normally flood with local rainfall, although their geomorphological origin may be alluvial. Hamilton *et al.* (1996) analysed passive microwave remote sensing observations made by satellite to reveal inundation patterns throughout the Pantanal (Fig. 8.2). The flood pulse is typically unimodal in most of the Pantanal, although more erratic patterns are often observed in the flood plains of the southern Pantanal, particularly those affected by the Miranda and Aquidauana rivers. Areas flooded by the Paraguay River show the most regular, predictable flood pattern. The total area inundated at a given time fluctuated between 10 000 and 110 000 km^2 during 1979–1987.

Analysis of the regional water balance shows that the sum of the discharges of inflowing rivers is approximately equal to the outflow from the Pantanal via the Paraguay River, on an annual basis (EDIBAP, 1979). A representative water balance is depicted in Table 8.1 (Hamilton *et al.*, 1997). Thus the water lost by evapotranspiration is roughly balanced by direct precipitation inputs, which are typically 1000–1300 mm year^{-1} within much of the Pantanal. The regional water balance can vary substantially from one year to the next.

Table 8.1 Regional water balance for the Pantanal above the Nabileque subregion (Hamilton *et al.*, 1997). This average annual balance is based on data from relatively wet years. The sum of the 11 principal riverine inflows to the Pantanal as well as the outflow discharge at Porto Esperança are based on monthly means for 3 years (1974–1976); inflow data are from EDIBAP (1979) and outflow data were calculated from daily stage records using the rating curve from that period. Rainfall and Thornthwaite potential evapotranspiration (PET) are based on means for four sites within the Pantanal, each of which has 3–6 years of data during 1976–1982 (data from Cadavid Garcia, 1984). The area of flooded land was estimated from the mean seasonal cycle of flooded area above Porto Esperança, as determined from satellite observations of passive microwave emission between 1979 and 1987 (Hamilton *et al.*, 1996).

	Flux ($\times 10^6$ m^3 year^{-1})
Riverine inflows	78 221
Rain on flooded land	62 455
PET from flooded land	–69 607
Balance of above	71 069
Riverine outflow	65 804

Flood waters throughout most of the region tend to flow in a characteristic direction, albeit slowly (often 2–10 cm s^{-1}: Hamilton *et al.*, 1995) because of the low gradients and the resistance offered by the dense vegetation. Flow paths across the flood plain are often indicated by poorly-defined channels, even though much of the flow may occur outside of the channels at higher water levels. The more defined channels are known as *corixos* or *córregos*. Less defined channels that are reminiscent of the sloughs of the Florida Everglades are known locally as *vazantes*; these are usually filled with emergent plants and often dry completely for part of the year.

Rainfall shows considerable interannual variability, which in turn causes variability in the flooding patterns (EDIBAP, 1979; Robertson & Mechoso, 1998). Multiyear series of higher or lower rainfall dramatically affect the flooding in the region (Hamilton *et al.*, 1996). A comparative study of inundation patterns showed that interannual variation in flooding is greater in the Pantanal than in five of the other major flood plains of South America (Hamilton *et al.*, in press). The river level at Ladário provides a good indication of overall flooding in the region because of its central location along the Paraguay River axis (Hamilton *et al.*, 1996). Hamilton *et al.* (1996) used the relationship between stage of the Paraguay River and area inundated throughout the Pantanal during the 9 years of satellite observations (1979–1987) to extend the inundation record over the period of stage records (1900–1995); that record is further extended through January 2000 in Fig. 8.3. Occasional

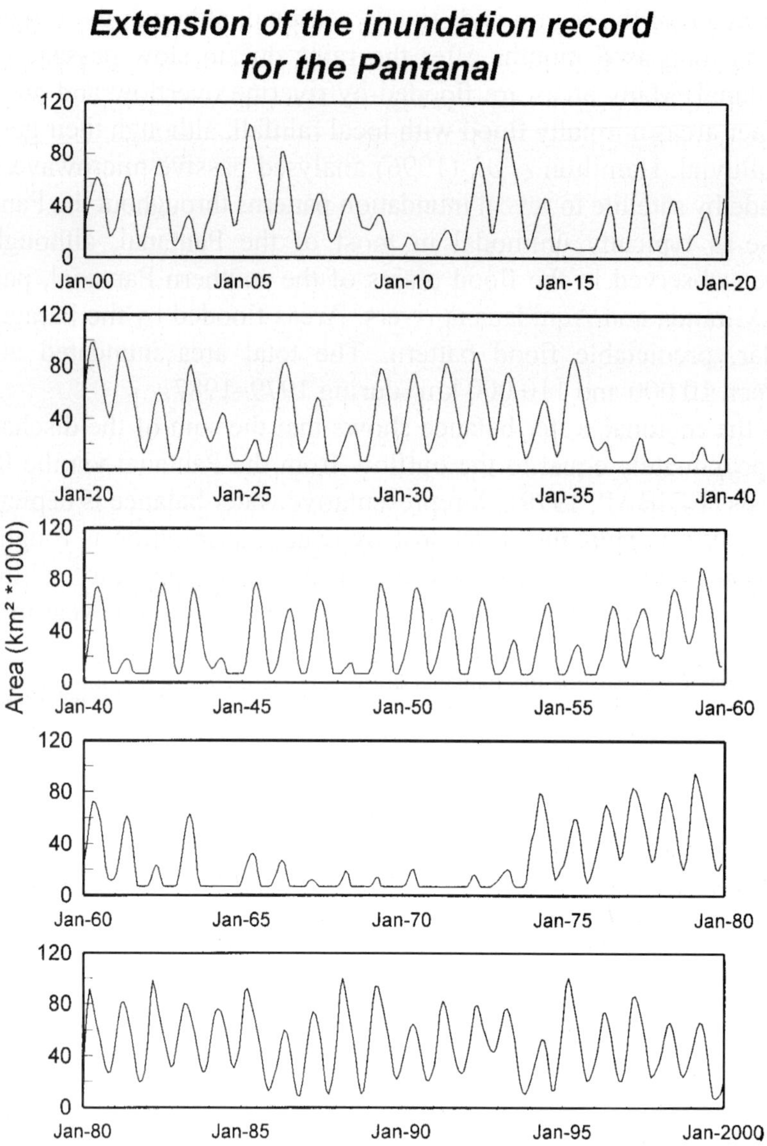

Fig. 8.3 Extension of the inundation record over the century of stage observations at Ladário (near Corumbá). This record was produced using a predictive equation that describes the relationship between river stage and inundated area during 1979–1987, when passive microwave observations of inundation area were available (Hamilton *et al.*, 1996).

prolonged dry periods are apparent from this inundation record, with the longest in the record occurring between 1960–73, but the annual flooding has been relatively predictable during the last quarter of the century.

The progress of flood waters through the Paraguay River system is likely to be regulated by occasional geomorphological control points (EDIBAP, 1979; Hamilton, 1999; Ponce, 1995). These control points are probably composed of resistant bedrock, gravel, clay lenses, or conglomerates that confine the river channel, preventing further excavation and/or widening of the channel to accommodate excess discharge. Although the locations of these control points remain uncertain, their existence can be inferred from the water surface profiles of the major rivers (EDIBAP, 1979).

Because of the low elevational gradients within the Pantanal, backwater effects control river water levels and thus flood-plain inundation (EDIBAP, 1979; Hamilton, 1999; Ponce, 1995). The occasional control points affect water levels not only in the river channels but also in adjacent flood plains and tributary courses. Water levels of the tributaries are controlled by the Paraguay River near their confluences. The peak discharge of these tributaries tends to occur earlier than that of the Paraguay River, and the later rise of the

Paraguay impounds the lower courses of the tributaries, decreasing their current and sometimes even reversing their flow temporarily. Thus the highest water levels in the lower courses of these tributaries do not necessarily correspond with their peak discharges. This backwater effect often extends for considerable distances upriver. Backwater effects are also important in other large flood-plain rivers, such as the Amazon River system (Meade *et al.*, 1991).

The topographic profiles of the flood plain of the Paraguay River and its major tributaries are typical of large flood-plain rivers, in which the highest elevations tend to occur on the levees along the river, and the land slopes downward behind the levees to extensive backswamps (Junk, 1983). These levees support forest along most of the rivers, and although the levees may be only tens of metres wide they often form a nearly continuous strip along the river banks. The backswamps are generally not densely forested, resembling savannas or grasslands in their vegetation. The inundation hydroperiod is longest in the backswamps, and the levees are generally flooded at lower depths for less time, or remain dry throughout the high-water phase.

Lakes with open-water surfaces commonly represent the deepest waters outside of the river channels, and are the last flood-plain waters to dry. The most important lakes of the Pantanal can be grouped into three major types: (a) the large lakes in the vicinity of the Paraguay River, (b) the small lakes and ponds of the Nhecolândia subregion on the Taquarí River fan, and (c) the small lateral levee lakes of the riverine flood plains. Lakes are defined here as basins that contain open water throughout the year and that do not exhibit substantial flow (e.g. not greater than *c.* 5 cm s^{-1}). Lakes in the Pantanal comprise a smaller proportion of the overall flood-plain area than do lakes in the fringing flood plains of the Amazon and Orinoco rivers (Sippel *et al.*, 1992; Hamilton & Lewis, 1990a). Large lakes are locally called *lagoas* or *baías*, smaller lakes are known as *baías*, and saline lakes of the Nhecolândia subregion are called *salinas*.

Large lakes that are hydrologically connected to the Paraguay River include Uberaba, Guaíba, Mandioré, Vermelha, and Cáceres. Each of these lakes has recently had an open-water surface of >100 km^2 and, based on the limited information available, they all appear to be shallower than the Paraguay River. The proportion of river discharge that flows through these lakes is unknown. A few large lakes are also found in the northeastern Pantanal, along the Cuiabá River near Barão do Melgaço, and these have been extensively studied by researchers from the Federal University of Mato Grosso (e.g. da Silva, 1990; Pinto-Silva, 1991).

Several thousand small lakes and ponds are found in Nhecolândia, the southwestern part of the Taquarí alluvial fan, where they occupy depressions that are generally <1 km^2 in area, interspersed among aeolian sand formations. In general, such lakes are hydrologically independent of the Paraguay River system. These water bodies are chemically and biologically diverse (Mourão *et al.*, 1988). Most lack surface water connections and appear to contain water originating from local rainfall, although some coalesce at high water to form drainage paths. Vascular plants commonly cover the water surface in these lakes. Some relatively saline lakes are found in the more topographically isolated basins, and these lack vascular plants but contain high concentrations of microalgae. Smaller areas with similar lakes and ponds are found in several other places in the Pantanal (Klammer, 1982).

The small lateral-levee lakes are most common along the Paraguay River, and include channel-shaped lakes in meander scrolls as well as more circular depressions. At high water these lakes are hydrologically continuous with the sheet flooding of the vegetated flood-plain areas, and they often have high rates of hydraulic through-flow. They are similar to lakes of the fringing flood plains of other large South American rivers, such as the Amazon (Lesack & Melack, 1995; Melack & Fisher, 1990; Sippel *et al.*, 1992) and Orinoco (Hamilton & Lewis, 1990b; Lewis *et al.*, 2000).

WATER QUALITY

The waters of the Pantanal are, in general, of low to moderate salinity. Chemical characteristics of the major rivers during 1992–1993 are summarized in Hamilton *et al.* (1997, 1999). The Paraguay River at Porto Esperança has a mean specific conductance of 56 $\mu S\ cm^{-1}$, and the hydrochemical composition of this river is summarized in Fig. 8.4. Most of the tributary and flood-plain waters range in conductance between 10 and 300 $\mu S\ cm^{-1}$ (Junk, 1993; Hamilton *et al.*, 1999). Pantanal waters are generally slightly acidic to circumneutral (pH 5.7–7.9), with high partial pressures of dissolved carbon dioxide (Hamilton *et al.*, 1995). Salinity and pH tend to be higher in the waters in the southern Pantanal, particularly in areas that receive drainage from the Serra da Bodoquena, where limestones are abundant. The Paraguay River increases significantly in salinity after receiving the Miranda River, which drains the aforementioned areas.

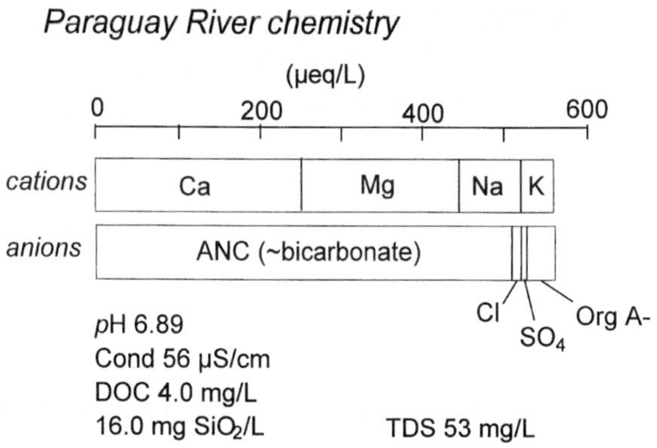

Fig. 8.4 Major solute chemistry of the Paraguay River at Porto Esperança. Concentrations are discharge-weighted means for 11 samples taken during 1992–1993 (Hamilton *et al.*, 1997). ANC = acid-neutralizing capacity or total alkalinity; Org A- = organic acid anions (estimated from pH and DOC); Cond = specific conductance (25°C); DOC = dissolved organic carbon; TDS = total dissolved solids (calculated from individual solute measurements).

In comparison with industrialized nations, nutrient concentrations are low in waters of the region, reflecting the high biological demand and the low levels of anthropogenic eutrophication (da Silva, 1990; Hamilton *et al.*, 1997). Within the Pantanal, the Paraguay River carries lower inorganic nutrient concentrations than the Amazon and Orinoco rivers during most of the year, probably because of its greater contact with the flood plain, where nutrients are rapidly removed from the water. In general, nutrient fluxes are very high in tropical flood plains despite the low concentrations, and nutrient recycling processes support the high biological productivity (Melack & Fisher, 1990).

Suspended sediment concentrations are moderately high in many of the rivers outside of the Pantanal, but most of the sediment load of these rivers is lost within the Pantanal before reaching the Paraguay River (Semmelman *et al.*, 1999). This is due to passage of river waters through flood-plain environments as well as to reduced current velocities within the river channels. At the lowest river stages, when there is little contact of river waters with adjacent flood-plain environments, suspended sediments do reach the Paraguay River, which becomes moderately turbid throughout the Pantanal. There is thus an inverse relationship between the Paraguay River stage and suspended sediment concentration, at least for the fine fraction. The larger lakes can develop significant turbidity by wind-induced re-suspension of sediments during low-water periods (da Silva, 1990), as has been observed in other flood plains (e.g. Hamilton & Lewis, 1990b).

Low concentrations of dissolved oxygen are characteristic of waters in the region (Fig. 8.5; Hamilton *et al.*, 1995). This results from the high biological productivity of the flood-plain ecosystems, and because respiration greatly exceeds photosynthesis beneath the

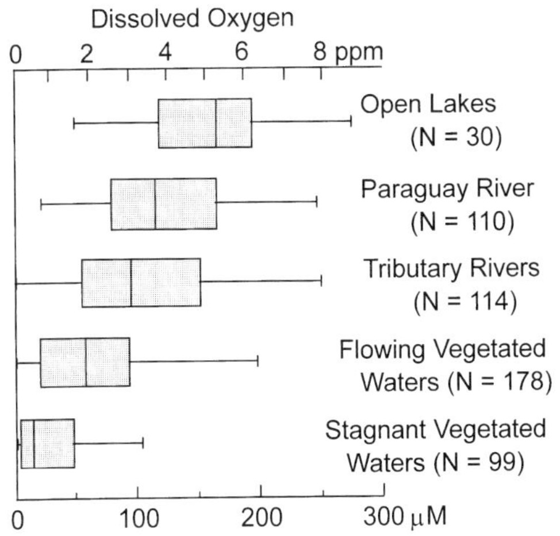

Fig. 8.5 Dissolved oxygen concentrations in samples from various aquatic environments sampled throughout the Pantanal (Hamilton *et al.*, 1995). The boxes extend from the 25th to 75th percentiles, with the median marked as a vertical bar inside the box, and the horizontal lines indicate the range of values falling within the interquartile range times 1.5 (outliers are excluded). All samples are from water columns with little thermal stratification (i.e. that mixed daily). Vegetated waters are those that contained emergent aquatic plants. Numbers of samples are given in parentheses.

water surface. This excess of respiration is partly explained by the root metabolism of emergent vascular plants. The oxygen concentrations tend to be highest in the largest rivers, but even in the Paraguay River the median concentration is only *c.* 45% saturation. Flood-plain waters are frequently anoxic, even though they may be shallow and in direct contact with the atmosphere. Stagnant flood-plain waters tend to be anoxic more frequently than waters with slight flow.

Oxygen depletion of flood-plain and river waters is exceptionally strong during the rising water phase (Calheiros & Hamilton, 1998; Hamilton *et al.*, 1995). In some years, particularly after a severe dry season, the Paraguay River becomes anoxic for a few weeks or months. This phenomenon, known locally as the *dequada*, typically occurs when river waters first contact previously dry land (Hamilton *et al.*, 1997; Calheiros *et al.*, 2000). A flood wave of anoxic water travels progressively across the flood plain. The river can become anoxic where the flood-plain waters return to the main channel. This anoxic flood wave originates in the northern Pantanal around January and travels down the Paraguay River, normally passing Corumbá around March. Even the largest lakes along the Paraguay River can be affected, although they are undoubtedly important sites of re-oxygenation of flood-plain waters because of their exposure to wind-induced turbulence. The lakes are too shallow at this time to develop anoxic bottom waters that could be suddenly mixed to the surface by the wind. Massive natural fish kills can result when there is no refuge of more oxygenated waters (see below). These fish kills are popularly believed to be caused by leaching of ashes from dry season fires on the flood plain, but so far there is no scientific evidence to support that assertion (Calheiros & Hamilton, 1998; Calheiros *et al.*, 2000). Heckman (1998) speculates that a bacterial toxin rather than oxygen deficiency is the likely cause of such fish kills in the northern Pantanal, since chronic oxygen deficiency is normal in many flood-plain waters containing fish.

In general, the quality of waters of the region is evidently not strongly affected by human activities. Some significant point sources of pollution affect the rivers outside the Pantanal, particularly where agroindustrial plants are situated (EMBRAPA, 1990; Ferreira *et al.*, 1994), and at Cuiabá, the largest city in the drainage basin (Mittermeier *et al.*, 1990). There is still little sewage treatment in the region, and even the largest urban area in and around Cuiabá is only partially served by a treatment plant, but new facilities are planned for most of the cities of the region. Mercury pollution by gold-mining operations centred around Poconé in the northern Pantanal has been the most studied water-quality problem, and mercury contamination has been shown in sediments as well as in the biota of the region around Poconé (Lacerda *et al.*, 1991; Leady & Gottgens, 2001).

PLANT LIFE

Vegetation of the Pantanal is often described as the "Pantanal complex", a mixture of plant communities from the surrounding biomes: the moist forest of the Amazon basin, the *cerrado* savannas of the central Brazilian Shield, and the *Chaco* savannas of Bolivia and Paraguay (Prance & Schaller, 1982; Adámoli & Pott, 1999). The Bolivian and Paraguayan portions of the upper Paraguay River basin resemble the *Chaco* savannas to the south more than the *cerrado* savanna formation (Prado *et al.*, 1992). The northernmost parts of the upland drainage basin contain elements of tropical moist forest as well as *cerrado*. Restricted areas of low-elevation calcareous soils in the southern Pantanal around Corumbá support a deciduous woodland that contains species typical of the semiarid northeast of Brazil (Ratter *et al.*, 1988). Pott & Pott (1994) describe the most important plant species of terrestrial environments in the Pantanal, and Pott & Pott (1997, 2000) provide checklists and descriptions of the aquatic plants of the region.

Seasonal flooding and desiccation strongly affect the flood-plain vegetation, and terrestrial or woody communities often exhibit patchy distributions with sharp interfaces (Dubs, 1992a). Savanna and mixtures of grassland with semi-deciduous forest are the most common formations (Allem & Valls, 1987), although gallery forest occurs along river levees. In many parts of the Pantanal, forest occurs on the more elevated ground and grassland covers areas subject to seasonal inundation (Dubs, 1992a, Ponce & Cunha, 1993). Narrow strips of forest on elevated ground surrounded by flood plain are locally called *cordilheiras*, and more circular forest patches on elevated ground are known as *capões*. During the dry season, water stress is a limiting factor for plant growth in many parts of the flood plains and probably rivals inundation as the primary determinant of the species composition and distribution of the vegetation.

Several common forest types are subject to seasonal inundation (RADAMBRASIL, 1982). Dense gallery forests (*mata ciliar*) with a distinct species composition often occur on levees along the major rivers (Haase, 1999; da Silva 1990, Prance & Schaller, 1982; Ratter *et al.*, 1988). Forests dominated by particular tree species also occur in the region. Extensive stands of the Carandá palm (*Copernicia australis*) are found in some parts of the southern Pantanal, and the Cambará (*Vochysia divergens*) commonly forms dense stands throughout the Pantanal. Open, park-like stands of the Paratudo tree (*Tabebuia aurea*) are found in the Miranda subregion as well as in some parts of the Nabileque subregion, and reach the Paraguay River in the vicinity of Porto Esperança.

Human impact on vegetation in the Pantanal has occurred mainly through the introduction of cattle and the use of fire during the last 300 years, which have influenced the nature and distribution of vegetation (Prance & Schaller, 1982). Cattle density remains low (*c.* 3 ha/head: Wilcox, 1992), in part because of the extent of land subject to inundation. Deforestation has recently increased but is most common in the least inundated parts of the Pantanal (Coutinho *et al.*, 1994), where exotic grasses (particularly *Brachiaria decumbens*) are planted for pasture (Allem & Valls, 1987).

Aquatic or semi-aquatic vascular plants (macrophytes) occur almost everywhere surface water persists for more than a few weeks. Floating and rooted emergent forms are most abundant (da Silva & Esteves, 1993; Pott & Pott, 1997; Prance & Schaller, 1982). Floating emergent plants such as *Oxycaryum cubense* (=*Scirpus cubensis*), *Eichhornea azurea*, *E.*

crassipes, *Azolla* spp. and *Salvinia auriculata* tend to dominate in more deeply flooded environments; the latter three species are free-floating while the former are often anchored to the bottom by roots. Rooted emergent species with more vertical, erect stems colonize areas of shallower water (< *c.* 1-m depth), and sometimes develop immense stands of one or a few species (Penha *et al.*, 1999). The most abundant rooted emergent species include *Cyperus giganteus*, *Thalia geniculata*, *Typha domingensis*, *Pontederia parviflora*, *Ipomoea fistulosa*, *Polygonum* spp., *Oryza subulata* and *Paspalum* spp. Most of these species produce large quantities of adventitious roots in the water column during inundation.

The floating macrophyte mats are particularly abundant along the Paraguay River, and patches are frequently found floating freely in the river. These may break off from the river banks, where such plants are ubiquitous, or be carried into the river via waters flowing from flood-plain and lake environments (Bonetto, 1975). They frequently present problems for navigation, and can completely block channels less than about 150-m wide. The ecology of these plants is detailed in Junk (1983, 1993).

Submersed vascular plants are relatively less abundant in most environments, but *Utricularia* spp. and *Cabomba piauhyensis* are often present. Some areas that receive shallow flooding by local rain waters are dominated by submersed vascular plants; such areas are particularly common in the higher parts of the alluvial fan of the Taquari River. The more deeply flooded environments along the Paraguay River have few submersed vascular plants, which evidently cannot cope with the fluctuating water levels and the poor light penetration (Junk, 1993).

Algae have been studied in only a few sites in the Pantanal (Calheiros & Oliveira, 1999), but some generalizations are possible. Many of the aquatic environments contain limited phytoplankton biomass because of the slowly flowing waters, which strain suspended particulates through dense vegetation, as has been observed in other flood plains (Garcia de Emiliani, 1993; Hamilton *et al.*, 1990). The large lakes along the Paraguay River have hydraulic residence times of sufficient length for phytoplankton to grow, and they are a significant source of algae to the river when they drain into the river (Espíndola *et al.*, 1992a; Oliveira & Calheiros, 2000). Smaller flood-plain lakes and pools accumulate high phytoplankton biomass when they are hydrologically isolated (De-Lamonica-Freire & Heckman, 1996). Phytoplankton productivity has been studied in some large lakes in the vicinity of the Cuiaba River in the northern Pantanal (da Silva, 1990; Pinto-Silva, 1991), and in some of the small lakes of Nhecolândia (Mourão, 1988; Mourão *et al.*, 1988). The former lakes are most similar to the large lakes along the Paraguay River. Species composition of algae in waters of the northern Pantanal has been documented by De-Lamonica-Freire & Heckman (1996).

Although phytoplankton production may be limited in many flood-plain environments by water flow, attached algae (periphyton or epiphyton) flourish on the abundant submersed surfaces of vascular plants (De-Lamonica-Freire & Heckman, 1996; Engle & Melack, 1990). The rates of primary productivity attained by attached algae in flood-plain waters of the Pantanal are not known (Calheiros & Oliveira, 1999), but studies in floating macrophyte mats of the Amazon have shown that their productivity can be very high (e.g. Doyle, 1991), and they are probably important in supporting aquatic food webs leading to fish (see below).

Waters containing emergent vascular plants are by far the most important aquatic environment of the Pantanal by area. Most of these environments dry seasonally, although some have been flooded nearly continuously during the present series of relatively wet years. During the dry phase, some aquatic plants die while others persist, mixed with terrestrial plants (Junk, 1983). When water is present, the degree of coverage of the water surface by vegetation canopies ranges from complete in the case of dense emergent marshes, dense floating mats of herbaceous plants, and the denser flooded forests, to sparse in the case of flooded shrublands where only an occasional shrub or palm tree emerges from the water, as can be observed in the Nabileque subregion.

Occasional large floods or particularly severe dry seasons can dramatically modify the composition of flood-plain vegetation, changing the relative abundance of herbaceous and woody (shrub) forms. Long-term changes in depth and duration of inundation, even changes that appear subtle, can also invoke major changes in vegetation. In many areas in the Pantanal, forests that had expanded during the dry period of 1960–1973 died during the ensuing series of wetter years. Such cyclic fluctuations were likely common prior to that time as well, since the long-term river stage record shows several other low-water episodes.

AQUATIC INVERTEBRATES

Ecological studies of aquatic invertebrates in the Pantanal have only recently been carried out (Espíndola *et al.*, 1992b, Kretzschmar *et al.*, 1993, Turner & da Silva 1992), although similar communities have been more studied in the Amazon (Goulding *et al.*, 1988; Junk, 1993; Robertson & Hardy, 1984) and Orinoco (Lewis *et al.*, 2000; Vásquez, 1989) river systems. As with phytoplankton, zooplankton growth occurs mainly in situations where water flow does not present a limiting factor (Hamilton *et al.*, 1990a,b; Saunders & Lewis, 1989). Zooplankton in the Paraguay River presumably originates from the larger lakes (Espíndola *et al.*, 1992b), or in temporarily stagnant side channels (Saunders & Lewis, 1989), and could be significant in riverine food webs (Lundberg *et al.*, 1987).

Studies in other tropical flood plains indicate that the highest production of invertebrate fauna occurs in the emergent vascular plants rather than in the open waters or in benthic environments (Junk, 1993). The productivity and density of invertebrates in the floating vegetation is high; densities of macroinvertebrates are known to exceed 100 000 organisms m^{-2}. Among the macroinvertebrates, the most common taxa feed by filtration or feed on fine particles associated with submersed plant surfaces.

Benthic invertebrates may be limited by low dissolved oxygen in flood-plain environments (Junk, 1993), and by unstable bottom sediments and high fish predation pressure in the larger rivers (Junk *et al.*, 1989). However, there have been few studies of benthos in the main river channels, in part because of the difficulty in sampling large deep rivers (Lewis *et al.*, 1995); abundant benthic invertebrates have been found in similar river bottoms in some North American rivers (e.g. Soluk, 1985).

Molluscs, crabs and prawns are particularly important invertebrates in tropical flood-plain environments (Goulding *et al.*, 1988; Heckman, 1998; Welcomme, 1985). Prawns appear to be less abundant in the Pantanal than in the Orinoco and Amazon river systems, but the paleomonid *Macrobrachium amazonicum* is common in flood-plain waters. In the Pantanal, large ampulariid snails of the genus *Pomacea* spp. (*Ampullaria*) as well as *Marisa planogyra* are abundant and important in the diets of several vertebrates, including the snail kite (*Rostrhamus sociabilis*) and the caiman lizard (*Dracaena paraguayensis*).

FISH AND FISHERIES

The fish fauna of the Pantanal is rich in species and highly productive (Bonetto, 1986; Catella, 1992; Heckman, 1998; Lima, 1986; Paiva, 1984), as in other South American flood plains (Bayley & Petrere, 1989; Welcomme, 1990). Unfortunately, there are only a few detailed studies of the life histories and ecology of fish communities in the Pantanal. Studies of fish communities in other large flood-plain rivers have shown that the flood plain is critical for the maintenance of high fish production and diversity (Bayley & Petrere, 1989; Goulding *et al.*, 1988; Welcomme, 1990). Even fish that spend most of their lives in the main river channels are often dependent on the flood plain for spawning, or as a source of food. Conversely, fish that inhabit the flood plain are often dependent on the main river channels as migration routes (Lima, 1986; Novoa, 1982) and as refuges during the dry season (Welcomme, 1985).

The majority of the biomass of the fish communities in the Paraguay–Paraná river system generally comprises a few species whose mode of feeding is classed as detritivory because they ingest principally fine particles, often from sediment or plant surfaces (Bonetto, 1975, 1986; Bowen *et al.*, 1984; Catella, 1992). This dominance by detritivores has also been found in other South American flood-plain rivers (Araujo-Lima *et al.*, 1986).

The energy sources that sustain aquatic food webs leading to fish in tropical flood-plain rivers remain a subject of scientific debate. The dominance of fish communities by detritivorous species has inspired inquiries into the ultimate sources of their nutrition. Analyses of stomach contents and observations of feeding behaviour have led some investigators to assume that vascular plants support much of the fish production directly, through consumption of nutritious fruits and seeds, or via a detrital pathway, in which vascular-plant organic matter is first processed by microbial communities before being ingested as fine detrital particles (Bayley, 1989; Goulding, 1980). Detritivorous fish actually consume live and detrital algae as well as vascular-plant detritus, and this algal organic matter may be much more nutritious than the vascular-plant detritus. Recently, stable isotope studies of fish in the Orinoco and Amazon river systems have provided compelling evidence for the importance of algae in supporting fish production (Araujo-Lima *et al.*, 1986; Forsberg *et al.*, 1993; Hamilton *et al.*, 1992; Lewis *et al.*, 2000). Similar investigations are currently underway in the Pantanal. Regardless of whether vascular plants or algae are most important in sustaining aquatic food webs leading to fish, however, the importance of the flood plain to fish production is well-established, even for those species that spend most of their lives in the main channels.

Two kinds of major fish migrations are known to occur in the Pantanal (da Silva, 1990; Paiva, 1984; Lima, 1986; Resende *et al.*, 1995), and similar migrations have been documented in other flood-plain rivers (Bonetto *et al.*, 1988; Welcomme, 1985). The *piracema* is a reproductive migration coinciding approximately with the rainy season (October–March, depending on location in the Pantanal and on the timing of flooding). In the *piracema*, various species use the river channels to spawn, or to reach spawning areas on the flood plains. In recent years, commercial and sport fishing have been banned in Mato Grosso do Sul during the most important part of the *piracema*. The *lufada* is a migration from the flood-plain areas to the main river channels, and occurs toward the end of the flood season (especially April–June), when the inundation area on the flood plain begins to contract. The fish generally migrate upriver after leaving the flood plain.

The prevalence of oxygen-depleted waters on the flood plain must present a limiting factor to fish, at least for larger species. Tropical fish show a variety of anatomical, physiological, and behavioural adaptations to survive in oxygen-depleted waters. A number of fish species are able to breathe air from the surface, although these species are generally not dominant, even in hypoxic waters (Kramer *et al.*, 1978). Many others utilize the relatively oxygenated surface layer for branchial respiration, and may have morphological adaptations that allow them to efficiently channel a surface layer of a few millimetres depth across the gills. Other adaptations involve increased physiological efficiency of oxygen uptake and utilization, and behavioural changes to reduce oxygen demand. Smaller fish may flourish in oxygen-depleted waters where larger piscivorous fishes are excluded, although the need to surface exposes them to predation by birds.

As in other tropical flood-plain rivers (Junk, 1982; Novoa, 1982; Welcomme, 1990), exploitation of fisheries in the Pantanal is focused on a few species and current levels of harvest are probably not sustainable (da Silva, 1990; Mittermeier *et al.*, 1990). Sport fishing by national and international tourists is growing in popularity, and increasingly conflicts with commercial fishing interests. The Paraguay River and its major tributaries are the focus of most of the fishing activity, with greatest concentrations of infrastructure in the vicinity of Corumbá, Porto Murtinho, and Cáceres.

The most important species from a fisheries perspective are the following (Paiva, 1984; PCBAP, 1997): Pintado (*Pseudoplatystoma corruscans*), Cachara (*P. fasciatum*), Dourado (*Salminus maxillosus*), Pacu (*Piaractus mesopotamicus*), Curimbata (*Prochilodus lineatus*), Jau (*Paulicea luetkeni*), Jeripoca (*Hemisorubim platyrhynchos*), and piranhas (*Serrasalmus* spp.). The first four of these are the most utilized. A small but significant industry exists to collect smaller bait fish (*isca*) to sell to sport fishermen; these are generally obtained from floating vegetation or from residual flood-plain water bodies at falling water. The flood plain and running waters support a great diversity of other fish species, including some that have not been described (Willink *et al.*, 2000).

OTHER VERTEBRATE FAUNA

The Pantanal is renowned for its abundance and diversity of wildlife, which includes terrestrial as well as aquatic animals (Alho *et al.*, 1987; Dubs, 1992b; Heckman, 1998; Magalhães, 1992; Mittermeier *et al.*, 1990; Quigley & Crawshaw, 1992; Strussmann & Sazima, 1993; Willink *et al.*, 2000), although there are few endemic species of vertebrates in the region. The Pantanal is a stronghold for a number of threatened and endangered mammals and birds, which are described in Mittermeier *et al.* (1990). The Paraguayan caiman (*Caiman crocodylus yacare*) and the capybara (*Hydrochaeris hydrochaeris*) are found at high densities throughout the region (Mourão, 1997; PCBAP, 1997). Overall, however, population densities are not extraordinary but the wildlife is readily visible to visitors because of the open savanna vegetation and the tendency for animals to concentrate at residual waters during the dry season.

Several of the threatened and endangered vertebrate species are not aquatic or even necessarily dependent on aquatic environments for food, and were formerly abundant over a much wider range. Examples include the hyacinthine macaw (*Anodorhyncus hyacinthinus*), maned wolf (*Chrysocyon brachyurus*), tapir (*Tapirus terrestris*), giant anteater (*Myrmecophaga tridactyla*), and the jaguar (*Panthera onca*). Their survival in the Pantanal while their populations have been decimated in other non-flood-plain environments is likely explained by the fact that seasonal inundation has historically impeded development and restricted access to the region. In addition, there is not a cultural tendency among the human inhabitants of the Pantanal to hunt wildlife more than necessary for subsistence, and fish and beef are generally preferred foods despite the abundance of wildlife (Emory, 1985).

Many of the birds and some of the mammals and reptiles are directly or indirectly dependent on aquatic environments. A great variety of waterfowl occurs in the region, including several migratory species (Dubs, 1992b; Bucher et al., 1993). Among the rarest mammals that inhabit aquatic environments are the giant river otter (*Pteroneura brasiliensis*) and the marsh deer (*Blastocerus dichotomus*). In addition to caiman, abundant reptiles include the anaconda (*Eunectes notaeus*), tegu lizard (*Tupinambis teguixin*), jabuti tortoise (*Geochelone* sp.) and iguana (*Iguana iguana*). Aquatic turtles are present but not frequently observed in the waters of the Pantanal. Amphibians are poorly known and have not been systematically surveyed (Willink *et al.*, 2000), but many species of frogs are commonly observed.

HYDROLOGICAL CONTROLS ON ECOLOGICAL STRUCTURE AND FUNCTION

Much attention has recently focused on the importance of natural flow regimes, including particularly their range of variability, in maintaining or restoring river and stream ecosystems (Poff *et al.*, 1997; Richter *et al.*, 1997). River regulation projects such as dams and navigation works have tended to reduce the range of variability in natural flow regimes, resulting in long-term changes in channel geomorphology, river–flood-plain connectivity, and riparian and flood-plain vegetation. Research on relatively unaltered tropical flood plains such as those of the mainstem Amazon River in Brazil has underscored the importance of the natural regime of seasonal inundation, referred to as the flood pulse, as the primary force controlling the structure and function of flood-plain ecosystems (Junk *et al.*, 1989; Junk, 1997).

The principal hydrological processes that are likely to control ecological structure and function in the flood plains of the Pantanal can be deduced from work on other fluvial systems as well as the existing body of knowledge for the Pantanal. The salient hydrological characteristics of the seasonal flood pulse include frequency (generally once per year in most subregions of the Pantanal: Fig. 8.2), depth of inundation (ranges from <10 cm to 2 m or more; overall the mean depth of maximum inundation is probably around 0.5 m), and the duration of inundation (variable but usually continuous over 1–2 months or longer). The sources of flood waters are also ecologically important, as river overflow tends to be richer in dissolved and suspended materials compared with flood waters originating as rain that falls directly on the flood plain (Junk & da Silva, 1995). Flow paths across the flood plain are important because river waters can lose their nutrients and suspended material as they flow across the flood plain. Finally, the severity of the dry season is important to consider as well; this is dictated not only by the duration of inundation but also by the timing of inundation relative to local rainfall, and by the nature of the soils. For example, in the northern Pantanal inundation coincides approximately with the wet season, while in the southernmost flood plains, inundation occurs during the local dry season (Hamilton *et al.*, 1996).

The high species diversity and biological productivity of the tropical flood-plain ecosystem is explained by the seasonal inundation, which maintains a spatially and temporally variable environment with both aquatic and terrestrial characteristics (Junk *et al.*, 1989; Junk, 1993, 1997; Lewis *et al.*, 2000; Neiff, 1990; Welcomme, 1985). Periodic inundation also brings nutrients to the flood-plain ecosystem, stimulates nutrient fluxes, and maintains plant communities in early stages of community succession. Measurements of the primary productivity of aquatic vascular plants in the Amazon flood plain reveal that these plant communities rival the productivity of the most intensively managed agricultural crops in spite of the seasonally changing conditions (Piedade *et al.*, 1992).

Flood plains such as the Pantanal seem to be in a constant state of disturbance, and many of the plant and animal species must be resilient to flourish in such an environment. However, the predictability and extended duration of the seasonal flood pulse have allowed organisms to adapt to it (Junk *et al.*, 1989). The regular occurrence of seasonal inundation invokes a pulse stability to the flood-plain ecosystem (*sensu* Odum, 1969; see also Mitsch & Gosselink, 2000). The greatest diversity and abundance of wildlife occur in areas subject to partial inundation, probably because these areas offer habitat for both terrestrial and aquatic animals throughout the year, and tend to support a mosaic of forest, savanna and marsh environments.

For many terrestrial animal species, the existence of dry refuges during inundation is critical, and exceptional floods are known to cause high mortality of these species. Dry corridors are provided by levees along rivers (particularly along the Paraguay River), and these are especially important for mammals with large home ranges such as the jaguar (*Panthera onca*; Quigley & Crawshaw, 1992). These dry areas also serve as refuges for many smaller animals, and for arboreal wildlife such as primates, since the dry patches are often the only dense forests on the flood plain.

Conversely, for many aquatic animals the persistence of some flooded areas during the dry season is critical. The larger lakes and river channels can serve as refuges for many aquatic animals, although they are exposed to much higher risks of predation at this time. Fish yields are often improved after a series of years in which the flood plain retains water (Quirós, 1990), and the duration of flooding is correlated with success of caiman reproduction (Coutinho & Campos, 1996). Many water birds are dependent on the annual drainage phase, when their prey becomes concentrated in remnant water bodies on the flood plain.

In addition to its role as a dry-season refuge, the main river channel in a river–flood-plain system such as that of the Paraguay River serves as a migration and dispersion route for aquatic animals that depend on the flood plain for sustenance and shelter (Junk *et al.*, 1989). The fish migrations discussed above are one example; many fish also take advantage of the river to disperse their larvae into downstream flood-plain habitats. Thus it is essential to maintain the environmental integrity of the river in order to preserve the high productivity and species diversity of associated flood-plain ecosystems.

During extended series of dry years such as the 1960–1973 period (Fig. 8.3), many wildlife populations must have been greatly reduced. Unfortunately, few scientific studies of the wildlife populations were performed before 1973. Populations of rare species have likely come closest to local extinction during these extended dry periods. Evidence of the changes in the distribution of forest vegetation that took place during that period is still visible as extensive stands of dead trees in some parts of the Pantanal, which long-time residents say were killed by the long series of wet years after 1974.

ENVIRONMENTAL PROBLEMS RESULTING FROM HYDROLOGICAL ALTERATIONS

The most important economic activities within the Pantanal include cattle ranching, commercial and sport fishing, and eco-tourism; the environmental problems associated with each of these activities are beyond the scope of this chapter but have been discussed elsewhere (Alho *et al.*, 1987; da Silva, 2000; Mittermeier *et al.*, 1990). The following discussion summarizes the most important environmental problems that result from hydrological alterations, which include erosion and sedimentation, construction of raised roads and dikes, and river channel alterations to facilitate navigation.

As a result of changing land use in the uplands, accelerated erosion and sedimentation of rivers that enter the Pantanal have evidently already occurred (EDIBAP, 1979; Ferreira *et al.*, 1994; Godoy *et al.*, 1998; Tricart, 1982). A large proportion of the upland basin has been converted from natural *cerrado* savanna to managed pasture or intensive agriculture during the past 25 years, which greatly increases erosion rates (Salati *et al.*, 1989). Many areas in the Planalto have been completely deforested. There is considerable anecdotal evidence that sedimentation of the rivers has reduced fish populations and made the rivers less navigable (e.g. Bucher *et al.*, 1993; Emory, 1985). The problems seem to be particularly acute in the Taquari, Piquiri, and São Lourenço river systems. In the case of the Taquari basin, the soil erosion is evidently most severe in areas used for pasture, as opposed to those converted to row-crop agriculture. In the lower part of the Taquari River fan, occasional changes in the river course in recent decades have considerably altered the distribution of flooding, and these changes may be related to the increased sediment load of the river.

Construction of raised roads within the Pantanal can have serious ecological consequences, and as recently as 1981–1983 regional development plans included an extensive network of *c.* 800 km of roads to be built within the Pantanal (Emory, 1985). In addition to increasing access to the region, raised roads often alter the natural hydrological flow patterns during inundation, particularly if they do not run parallel to the direction of water flow. Even when occasional culverts or bridges are present, a difference in water level between the upstream and downstream sides of roads is often visible, and eventually the vegetation may develop differences. Decreased flow of flood-plain waters reduces the nutrient supply and often leads to greater probability of oxygen depletion, particularly where dissolved oxygen concentrations are low under normal conditions (Hamilton *et al.*, 1995). On the other hand, the numerous borrow pits created for construction of raised roads can be important permanent water bodies during the dry season for aquatic and terrestrial wildlife. The attraction of wildlife to the vicinity of roads has positive and negative consequences; animals are more readily viewed by tourists but are also more susceptible to mortality by vehicles and poaching.

Raised dikes have been constructed in a few places in the Pantanal to exclude flood waters, including a 60 000-ha area at Ilha Camargo in the Cuiabá subregion (Campos Filho & Machado, 1999; Emory, 1985) and a dike around the small city of Porto Murtinho. The Ilha Camargo project is generally regarded to have caused negative environmental effects, resulting in problems with undesirable plant species invading the diked area, diminished carrying capacity of the land for cattle, and possibly increasing flood levels outside of the dikes. Another diked area was constructed along the Paraguay River downriver of Ladário; the dike was to be part of the Transpantaneira Highway and the flood plain inside the diked area was to be used for agriculture. This project is also regarded as a failure; the land was not found to be useable, and the dike is now open to allow exchange of water with the river. However, the dike still interferes with natural flow patterns, and waters within the diked area are typically stagnant compared with natural flood-plain areas.

A number of dams have been proposed for the tributary rivers entering the Pantanal, including the Sepotuba, Cuiabá, Itiquira, and Taquari river systems. So far the only significant dam was completed in 1998 on the Manso River, a major tributary of the Cuiabá River. This and the other dams were originally proposed in the 1970s for the primary purpose of flood control, although the Manso Dam is also a hydroelectric facility with 210 MW of installed generating capacity. At the time of this writing (2001) the Manso reservoir was still filling; the dam was closed in late 1999. Downstream impacts of the Manso Dam have yet to be studied but will certainly include altered flow and flood regimes in the Cuiabá subregion of the Pantanal, where flood waters across much of the area are supplied by overbank flow from the Cuiabá River. There is no indication of active plans to build any other significant dams in the region.

A recently proposed navigation project known as the Paraguay–Paraná Waterway (or Hidrovía) would modify the Paraguay River channel to facilitate year-round navigation of barge trains through the Pantanal. The river channel alterations required for this project have aroused concerns about the potential environmental impact on adjacent flood plains (summarized by Gottgens *et al.*, 2001; Hidrovía Panel of Experts, 1997). The potential direct impacts of navigation projects within the Pantanal may be grouped as follows: (a) degradation of the river environment; (b) degradation of riparian areas; and (c) alteration of river–flood plain exchanges of water, materials, and aquatic animals. Secondary effects of development associated with the Hidrovia have been reviewed by Bucher *et al.* (1993) and are therefore not considered further here.

Perhaps the greatest risk inherent in any modification of the channel of a large river involves changes in the geomorphological controls that regulate the river's velocity, discharge, water surface elevation, and sediment load. For example, enlargement of the channel at its control points may lead to faster flow, resulting in more rapid drainage of water and lower river levels. The Paraguay River is evidently regulated by such control points, which confer a more equitable distribution of discharge over the course of the year than would otherwise occur, and result in the backwater effects that inundate lateral flood plain areas (EDIBAP, 1979; Hamilton, 1999; Ponce, 1995). Once the control points are altered, the river channel is likely to become unstable as it attempts to reach a new equilibrium. Sediment deposits in the river bed and along the edges will be increasingly mobilized, and the present-day configuration of islands and banks may change in unpredictable ways, leading to threats to buildings and property and accelerated erosion of gallery forests. Lateral channel migration rates are likely to increase (Drago, 1990).

The river environment will also be subject to direct impacts of its use as a navigation corridor (Wantzen *et al.*, 1999). The effects of barge traffic on sediment re-suspension, on fishes (particularly larval fishes and migrating adults), and on bank erosion have been discussed by Bucher *et al.* (1993). Chemical pollution by petroleum products, ballast water, and accidental spills of cargo will be likely as the use of the river is increased. Contamination of the river will be difficult to contain, and will be readily transported into adjacent flood-plain environments because there is so much hydrological contact between the river and flood plain in the Pantanal. Pollution of the Paraguay River will thus affect extensive areas of flood plain in addition to the main river channel. The flood-plain environment will tend to retain pollutants through filtration, sedimentation, and biotic uptake, thereby exacerbating their ecological impacts.

The role of the river as habitat for aquatic animals is particularly critical during the dry season and during fish migrations, as mentioned above. During the dry season, inundated flood plain may persist only in certain areas close to the main river channel, such as the vicinity of the larger lakes and around the confluences of the Cuiabá-São Lourenço and Paraguay rivers (Hamilton, 1999; UNESCO, 1973). These areas must serve as refuges for fish, water birds, caiman and other animals that depend on aquatic environments, including endangered species such as the giant river otter. Given that the inundation area in the Pantanal can be reduced by a factor of 10 between high and low water (Hamilton *et al.*, 1996), aquatic animals can become highly concentrated in these refuges. The survival of these animals is critical to provide populations to disperse into the newly flooded environments in the following high-water season (Welcomme, 1990). Some terrestrial animals habitually stay near permanent water bodies as well, such as the jaguar and the marsh deer (Quigley & Crawshaw, 1992). During extended dry periods, these refuges must become even more important.

The riparian areas, which are often the highest ground on the flood plain and generally support gallery forest, are critical components of the flood-plain landscape, as discussed above. They are also the terrestrial environment that would be most severely impacted by dredging activities, spoils deposits, construction of navigation infrastructure, and enhanced erosion of river margins. The gallery forests contain many plant species that are not found in the more lowlying areas, and their species composition is just beginning to be documented (da Silva, 1990; Prance & Schaller, 1982; Ratter *et al.*, 1988). Flood-plain forests contain a unique combination of species that tolerate the seasonal alternation between soil saturation and desiccation (Worbes, 1985), and any long-term changes in either the inundation or dry phase could alter their species composition.

Destruction of the gallery forests would represent a serious loss of habitat for terrestrial animals and for water birds. Their importance as a dry refuge during high water has been discussed above. They are also used as roosting and nesting sites by many species of water birds that forage in the river or backswamp environments. Fragmentation of the gallery forests would interrupt an important wildlife corridor used by animals such as capybaras and large cats, which need to migrate to avoid seasonal flooding (Mittermeier *et al.*, 1990).

The gallery forests are also recognized as important sources of food to riverine fish, many of which are adapted to feed on fruits and seeds from trees and shrubs, particularly during high water (Gottsberger, 1978; Goulding, 1980). A strip of floating aquatic plants such as water hyacinths (*Eichhornea* spp.) typically grows between the gallery forest and the river channel, providing protection to fishes and other aquatic animals and buffering the erosive action of waves and currents; in many reaches, the forest may be necessary to anchor this vegetation along the bank.

The gallery forests are also important in stabilizing the river banks, greatly increasing the resistance of the banks to erosion by anchoring the levee soils with their root systems and by reducing current velocities over the levees (Drago, 1990; Neiff *et al.*, 1985). Removal or degradation of the forest vegetation is likely to result in more rapid rates of lateral channel migration, such as those observed in the largely unforested middle Paraná River flood plain in the vicinity of Santa Fe, Argentina (Drago, 1990).

River channel alterations for navigation could potentially alter river–flood-plain exchanges of water, materials, and aquatic animals. The Paraguay River in the Pantanal is distinct from other large rivers of the world because of the very high proportion of its discharge that contacts the flood plain, either before or after entering the main channel (EDIBAP, 1979; Hamilton *et al.*, 1997; UNESCO, 1973). Exchanges of water between the river and the flood plain occur through channels that pass through occasional breaks in the levees (tie channels) or as non-channelized flow across lower spots along the levees. These exchanges are complex and often display seasonal changes in flow direction at a particular point, depending on the relative levels of the river and flood plain (Hamilton & Lewis, 1987; Lesack & Melack, 1995; UNESCO, 1973). The tributary channels commonly reverse their flow during rapid rises of the Paraguay River, acting as temporary conduits to carry river water onto the flood plain.

Significant alterations in river levels and discharge patterns will affect these river–flood-plain exchanges (Hamilton, 1999). The net effect of such changes is likely to be negative for flood-plain environments if riverine through-flow on the flood plain is reduced, if drainage of flood-plain waters back to the river is accelerated, or if the annual flood pulse becomes less predictable (Junk *et al.*, 1989; Petts, 1990). Reduced riverine through-flow would result in reduced nutrient supply to and oxygenation of flood-plain waters (Engle & Melack, 1990; Hamilton & Lewis, 1990b; Hamilton *et al.*, 1995). Stagnant flood-plain waters are frequently noticeably less productive, containing sparser and more stunted aquatic plants and fewer large fish. Accelerated drainage of the flood plain would reduce the hydroperiod, thereby invoking changes in flood-plain vegetation and increasing the severity of the dry season for biota dependent on aquatic environments (Junk *et al.*, 1989). Invasion of the flood plain by woody plants may occur more commonly, decreasing the value of these areas as natural pasture and encouraging more disruptive land management practices by ranchers.

By deepening and straightening the main channel of the Paraguay River, navigation projects such as the Hidrovia could result in lower water levels with respect to the adjacent flood plain (Hamilton, 1999). Even a seemingly small decrease of 0.2–0.5 m could dramatically affect the amount of river water exiting the main channel, and might reduce the total area of flood plain that is inundated at a given river discharge. The backwater effects that are presently propagated far up the tributary courses might be reduced or eliminated. Large-scale changes in flood plain vegetation, wildlife populations, and fish production could ensue.

Once a navigation project such as the Hidrovia is established, economic interests will be dependent on year-round navigability of the Paraguay River. In unusually dry years, there will be strong pressure to restrict outflows from the main channel to maintain better conditions for navigation. This has already been attempted in the Pantanal, when the tie channels feeding the flood plain leading to the Paraguai-mirim River system were blocked during the 1960s to reduce water losses from the main channel. Such actions are easily implemented but could result in untold ecological damage to downstream flood-plain ecosystems.

The extent of flood plain that might be influenced by channel alterations of the Paraguay River is difficult to determine precisely, but it is likely that the entire Paraguay River and Nabileque subregions would be affected (Fig. 8.1; Hamilton, 1999). Significant portions of the Corixo Grande, Taquari River, lower Cuiabá, and lower Negro–Aquidauana subregions would also be affected if backwater effects are reduced or eliminated. Together these areas comprise much of the total flood-plain area in the region that is subject to deeper, longer lasting inundation (EDIBAP, 1979; Hamilton *et al.*, 1996). Areas that are least likely to be affected include the Taquari fan, Piquiri-São Lourenço, and Miranda subregions, and upper portions of the Corixo Grande, Cuiabá, and Negro-Aquidauana subregions.

CONCLUSIONS

The ecological structure and function of Pantanal ecosystems are intimately tied to the hydrological regime of seasonal inundation and desiccation. The considerable spatial variability in hydrological regimes explains many of the ecological differences among subregions. Conservation and management of these flood-plain ecosystems requires consideration of the natural hydrological regime and its range of variability. Projects resulting in hydrological alterations, such as dams, raised roads and dikes, and river channel alterations for navigation, require careful analysis of their potential environmental impacts. Alternatives that cause the least perturbation of the natural flood regime should be considered whenever possible.

In the United States and Europe, large flood plain rivers have been extensively altered for navigation, hydropower, and flood plain "reclamation", and the consequent losses in natural resources and ecosystem services are only now being fully appreciated. For example, using the Illinois River (a major tributary of the Mississippi River) as a case study, Sparks *et al.* (2000) have demonstrated how the historical re-engineering of the river channel for navigation and flood control led to the degradation of important river fisheries and ultimately created a host of environmental problems. Throughout the Mississippi River system, alternatives to structural flood control are now being seriously considered as evidence accumulates to suggest that the overall effect of engineering modifications of the river system has been to exacerbate the magnitude of large floods and their resultant costs to society. Some degree of restoration of the original hydrological connectivity between the river channels and their flood plains is now considered a viable and attractive option. In the case of the Pantanal and other large flood-plain river systems that still retain their natural hydrology, economic development must be planned with ample consideration of the mistakes of the past in other great rivers of the world, lest that unfortunate history be repeated.

Acknowledgements This research was based upon work supported by the US National Aeronautics and Space Administration under grants NAGW-2724 and NAGW-4352, and by the National Science Foundation under grant DEB-9701714. The Pantanal research centre operated by the Empresa Brasileira de Pesquisa Agropecuária facilitated the field research. Comments from Richard Sparks and Johan Gottgens improved the manuscript. This is contribution 966 of the W. K. Kellogg Biological Station.

REFERENCES

Ab'Saber, A. N. (1988) O Pantanal Mato-grossense e a teoria dos refúgios. *Revista Brasileira de Geografia* **50**, 9–57.

Adámoli, J. (1982) O Pantanal e suas relações fitogeográficas com os cerrados. Discussão sobre o conceito de "Complexo do Pantanal". In: *Anais do XXXII Congresso Nacional de Botânica*, 109–119. Universidade Federal de Piaui, Teresina, Brazil.

Adámoli, J. & Pott, A. (1999) Las fuentes de diversidad en el Pantanal. In: *Biodiversidad y uso de la tierra: Conceptos y ejemplos de Latinoamérica* (ed. by S. D. Matteucci, O. T. Solbrig, J. Morello & G. Halffter), 317–360. University of Buenos Aires, Argentina.

Alho, C. J. R., Campos, Z. M. S. & Gonçalves, H. C. (1987) Ecologia de capivara (*Hydrochaeris hydrochaeris*, Rodentia) do Pantanal: II—Atividade, Sazonalidade, uso de espaço e manejo. *Revista Brasileira de Biologia* **47**(1/2), 99–110.

Alho, C. J. R., Lacher, T. E. & Gonçalves, H. C. (1988) Environmental degradation in the Pantanal ecosystem. *BioScience* **38**, 164–171.

Allem, A. C. & Valls, J. F. M. (1987) *Recursos Forrageiros Nativos do Pantanal Mato-Grossense*. Brazilian Corporation for Agricultural Research (EMBRAPA), Brasília, Brazil.

Amaral Filho, Z. P. (1986) Solos do Pantanal Matogrossense. In: *Anais do Simpósio sobre Recursos Naturais e Sócio-econômicos do Pantanal*, 29–42. Brazilian Corporation for Agricultural Research (EMBRAPA), Brasília, Brazil.

Araujo-Lima, C., Forsberg, B., Victoria, R. & Martinelli, L. (1986) Energy sources for detritivorous fishes in the Amazon. *Science* **234**, 1256–1258.

Bayley, P. B. (1989) Aquatic environments in the Amazon basin, with an analysis of carbon sources, fish production, and yield. In: *Proceedings of the International Large Rivers Symposium* (ed. by D. P. Dodge). *Canadian Special Publication Fisheries Aquatic Sciences 106*, 399–408. NRC Research Press, Ottawa.

Bayley, P. B. & Petrere, M., Jr (1989) Amazon fisheries: assessment methods, current status and management options. In: *Proceedings of the International Large Rivers Symposium* (ed. by D. P. Dodge). *Canadian Special Publication Fisheries Aquatic Sciences 106*, 385–398. NRC Research Press, Ottawa.

Bonetto, A. A. (1975) Hydrologic regime of the Parana River and its influence on ecosystems. *Ecological Studies* **10**, 175–197.

Bonetto, A. A. (1986) The Paraná River system. In: *The Ecology of River Systems* (ed. by B. R. Davies & K. F. Walker), 541–588. W. Junk, Dordrecht, The Netherlands.

Bonetto, A. A., Wais, I. R. & Arquez, G. S. (1988) Ecological considerations for river regulation of the Del Plata basin according to flatland characteristics. *Water International* **13**, 2–9.

Bowen, S. H., Bonetto, A. A. & Ahlgren, M. O. (1984) Microorganisms and detritus in the diet of a typical neotropical riverine detritivore, *Prochilodus platensis* (Pisces: Prochilodontidae). *Limnology and Oceanography* **29**(5), 1120–1122.

Bucher, E. H., Bonetto, A., Boyle, T., Canevari, P., Castro, G., Huszar, P. & Stone, T. (1993) *Hidrovia: an Initial Environmental Examination of the Paraguay–Paraná Waterway*. Wetlands for the Americas Publ. no. 10, Manomet, Massachusetts.

Cadavid Garcia, E. A. (1984) O clima no Pantanal do Mato Grosso. *Brazilian Corporation for Agricultural Research (EMBRAPA), Brasília, Technical Circular 14*.

Cadavid Garcia, E. A. & Castro, L. H. R. (1986) Análise de frequência de chuva no Pantanal Mato-Grossense. *Pesquisa Agropecuária Brasileira* **21**(9), 909–925.

Calheiros, D. F. & Hamilton, S. K. (1998) Limnological conditions associated with natural fish kills in the Pantanal wetland (Baía do Castelo, Paraguay River, Brazil). *Proceedings of the International Association of Applied and Theoretical Limnology* **26**, 2189–2193.

Calheiros, D. F. & Oliveira, M. D. (1999) Pesquisa limnológica no Pantanal: uma revisão. In: *Proceedings of the II Symposium on Natural and Socioeconomic Resources of the Pantanal: Management and Conservation*, 115–125. Brazilian Corporation for Agricultural Research, Brasília, and the Federal University of Mato Grosso do Sul, Campo Grande, Brazil.

Calheiros, D. F., Seidl, A. F. & Ferreira, C. J. A. (2000) Participatory research methods in environmental science: local and scientific knowledge of a limnological phenomenon in the Pantanal wetland of Brazil. *Journal of Applied Ecology* **37**, 684–696.

Campos Filho, L. V. & Machado, M. F. R. (1999) Quando a modernidade não é adequada a todos os lugares: O caso dos diques no Pantanal de Poconé—MT. In: *Proceedings of II Symposium on Natural and Socioeconomic Resources of the Pantanal: Management and Conservation*, 487–495. Brazilian Corporation for Agricultural Research, Brasília, and the Federal University of Mato Grosso do Sul, Campo Grande, Brazil.

Carvalho, N. de O. (1986) Hidrologia da Bacia do Alto Paraguai. In: *Anais do Primeiro Simpósio sobre Recursos Naturais e Socio-econômicos do Pantanal* (Corumbá, Mato Grosso do Sul, 28 November–4 December 1984), 43–49. Brazilian Corporation for Agricultural Research, Brasília, Brazil.

Catella, A. C. (1992) Estrutura de comunidade e alimentação dos peixes da Baía da Onça, uma algoa do Pantanal do rio Aquidauana, MS. Master's Thesis, University of Campinas (UNICAMP), São Paulo, Brazil.

Coutinho, M. E. & Campos, Z. (1996) The effects of habitat and seasonality on the densities of caiman in southern Pantanal, Brazil. *Journal of Tropical Ecology* **12**, 741–747.

Coutinho, M. E., Mourao, G. M., Silva, M. P. & Campos, Z. (1994) The sustainable use of natural resources and the conservation of the Pantanal wetland, Brazil. *Acta Limnologica Brasilensia* **5**, 165–176.

Da Silva, C. J. (1990) Influência da variação do nível d'água sobre a estrutura e funcionamento de uma área alagável do Pantanal Matogrossense (Pantanal de barão de Melgaço, Município de Santo Antônio de Leverger e Barão de Melgaço—MT). Doctoral Dissertation, Universidade Federal de São Carlos, Departamento de Ciências Biológicas, São Carlos, Brazil.

Da Silva, C. J. (2000) Ecological basis for the management of the Pantanal—upper Paraguay River basin. In: *New Approaches to River Management* (A. J. M. Smits, P. H. Nienhuis & R. S. E. W. Leuven), 97–117. Backhuys Publishers, Leiden, The Netherlands.

Da Silva, C. J. & Esteves, F. A. (1993) Biomass of three macrophytes in the Pantanal of the Mato Grosso, Brazil. *International Journal of Ecology and Environmental Science* **19**, 11–23.

De-Lamonica-Freire, E. & Heckman, C. W. (1996) The seasonal succession of biotic communities in wetlands of the tropical wet-and-dry climatic zone: III. The algal communities in the Pantanal of Mato Grosso, Brazil, with a comprehensive list of the known species and revision of two desmid taxa. *Internationale Revue der gesamten Hydrobiologie* **81**, 253–280.

Doyle, R. D. (1991) Primary production and nitrogen cycling within the periphyton community associated with emergent aquatic macrophytes in an Amazon flood plain lake. Doctoral Dissertation, University of Maryland.

Drago, E. C. (1990) Geomorphology of large alluvial rivers: lower Paraguay and middle Parana. *Interciencia* **15**(6), 378–387.

Dubs, B. (1992a) Observations on the differentiation of woodland and wet savanna habitats in the Pantanal of Mato Grosso, Brazil. In: *Nature and Dynamics of Forest–Savanna Boundaries* (P. A. Furley, J. Proctor & J. A. Ratter), 431–449. Chapman and Hall, London.

Dubs, B. (1992b) *Birds of Southwestern Brazil: a Catalogue and Guide to the Birds of the Pantanal of Mato Grosso and its Border Areas*. Betrona-Verlag, Switzerland.

EDIBAP (1979) *Estudo de Desenvolvimento Integrado da Bacia do Alto Paraguai. Relatorio de 1ª. Fase*. Brazilian Government–United Nations Development Programme–Organization of American States, Brasília.

EMBRAPA (1990) *Avaliação da Contaminação Ambiental da Bacia Hidrográfica do Río Miranda. Relatório Final.* Centro de Pesquisa Agropecuária do Pantanal, Brazilian Corporation for Agricultural Research (EMBRAPA), Brasília.

Emory, J. C. (1985) Perception and development of the Brazilian Pantanal. MS Thesis, University of California, Berkeley.

Engle, D. L. & Melack, J. M. (1990) Floating meadow epiphyton: biological and chemical features of epiphytic material in an Amazon flood plain lake. *Freshwater Biology* **23**, 479–494.

Espíndola, E. L. G. E., Matsumura-Tundisi, T. & Moreno, I. H. (1992a) Composicão, dominância, e variação do "standing crop" da communidade de fitoplâncton em função das alterações hidrológicas da Lagoa Albuquerque, Pantanal Matogrossense. In: *Abstracts of the II Congresso Latino-Americano de Ecologia and I Congresso de Ecologia do Brasil* (Minas Gerais, Brazil, 6–11 December 1992), p. 506.

Espíndola, E. L. G. E., Matsumura-Tundisi, T. & Moreno, I. H. (1992b) Efeitos da dinâmica hidrológica do sistema Pantanal Matogrossense sobre a estrutura da communidade zooplanctônica da Lagoa Albuquerque. In: *Abstracts of the II Congresso Latino-Americano de Ecologia and I Congresso de Ecologia do Brasil* (Minas Gerais, Brazil, 6–11 December 1992), p. 448.

Ferreira, C. J. A., Soriano, B. M. A., Galdino, S. & Hamilton, S. K. (1994) Factors of anthropogenic origin affecting waters of the Pantanal Wetland and associated rivers in the upper Paraguay River basin of Brazil. *Acta Limnologica Brasiliensia* **5**, 135–148.

Forsberg, B. R., Araujo-Lima, C. A. R. M., Martinelli, L. A., Victoria, R. L. & Bonassi, J. A. (1993) Autotrophic carbon sources for fish of the central Amazon. *Ecology* **74**(3), 643–652.

Furley, P. A. & Ratter, J. A. (1988) Soil resources and plant communities of the central Brazilian cerrado and their development. *Journal of Biogeography* **15**, 97–108.

Garcia de Emiliani, M. O. (1993) Seasonal succession of phytoplankton in a lake of the Paraná River flood plain, Argentina. *Hydrobiologia* **264**, 101–114.

Godoy, J. M., Padovani, C. R., Pereira, J. C. A. & Vieira, L. M. (1998) Aplicabilidade da geocronologia da deposição de sedimento com ^{210}Pb como ferramenta na avaliação do assoreamento do rio Taquari, Pantanal, MS. *Geochimica Brasiliensis* **12**(1/2), 113–121.

Gottgens, J. F., Perry, J. E., Fortney, R. H., Meyer, J. E., Benedict, M. & Rood, B. E. (2001) The Paraguay–Paraná Hidrovía: protecting the Pantanal with lessons from the past. *BioScience* **51**(4), 301–308.

Gottsberger, G. (1978) Seed dispersal by fish in the inundated regions of Humaita, Amazonia. *Biotropica* **10**(3), 170–183.

Goulding, M. (1980) *The Fishes and the Forest: Explorations in Amazonian Natural History.* Univ. California Press, Berkeley.

Goulding, M., Carvalho, M. L. & Ferreira, E. G. (1988) *Rio Negro, Rich Life in Poor Water.* SPB Academic Publishing, The Hague, The Netherlands.

Haase, R. (1999) Litterfall and nutrient return in seasonally flooded and non-flooded forest of the Pantanal, Mato Grosso, Brazil. *Forest Ecology and Management* **117**(1–3), 129–147.

Hamilton, S. K. (1999) Potential effects of a major navigation project (the Paraguay–Paraná Hidrovía) on inundation in the Pantanal flood plains. *Regulated Rivers: Research and Management* **15**(4), 289–299.

Hamilton, S. K. & Lewis, W. M. Jr (1987) Causes of seasonality in the chemistry of a lake on the Orinoco River floodplain, Venezuela. *Limnology and Oceanography* **32**, 1277–1290.

Hamilton, S. K. & Lewis, W. M., Jr (1990a) Physical characteristics of the fringing flood plain of the Orinoco River, Venezuela. *Interciencia* **15**(6), 491–500.

Hamilton, S. K. & Lewis, W. M., Jr (1990b) Basin morphology in relation to chemical and ecological characteristics of lakes on the Orinoco River flood plain, Venezuela. *Archiv für Hydrobiologie* **119**(4), 393–425.

Hamilton, S. K., Lewis, W. M., Jr & Sippel, S. J. (1992) Energy sources for aquatic animals in the Orinoco River flood plain: evidence from stable isotopes. *Oecologia* **89**(3), 324–330.

Hamilton, S. K., Sippel, S. J., Lewis, W. M., Jr & Saunders, J. F. (1990) Zooplankton abundance and evidence for its reduction by macrophyte mats in two Orinoco flood plain lakes. *Journal of Plankton Research* **12**(2), 345–363.

Hamilton, S. K., Sippel, S. J. & Melack, J. M. (1995) Oxygen depletion and carbon dioxide and methane production in waters of the Pantanal wetland of Brazil. *Biogeochemistry* **30**(2), 115–141.

Hamilton, S. K., Sippel, S. J. & Melack, J. M. (1996) Inundation patterns in the Pantanal wetland of South America determined from passive microwave remote sensing. Archiv für Hydrobiologie **137**(1), 1–23.

Hamilton, S. K., Sippel, S. J., Calheiros, D. F. & Melack, J. M. (1997) An anoxic event and other biogeochemical effects of the Pantanal wetland on the Paraguay River. *Limnology and Oceanography* **42**, 257–272.

Hamilton, S. K., Corrêa de Souza, O. & Coutinho, M. E. (1998) Dynamics of flood plain inundation in the alluvial fan of the Taquari River (Pantanal, Brazil). *Proceedings of the International Association of Applied and Theoretical Limnology* **26**, 916–922.

Hamilton, S. K., Sippel, S. J., Calheiros, D. F., & Melack, J. M. (1999) Chemical characteristics of Pantanal waters. In: *Proceedings of II Symposium on Natural and Socioeconomic Resources of the Pantanal: Management and Conservation*, 89–100. Brazilian Corporation for Agricultural Research, Brasília, and the Federal University of Mato Grosso do Sul, Campo Grande, Brazil.

Hamilton, S. K., Sippel, S. J. & Melack, J. M. (in press) Comparison of inundation patterns in South American floodplains. *Journal of Geophysical Research* **D20** (available in electronic form; doi 10.1029/2000JD000306).

Heckman, C. W. (1998) *The Pantanal of Poconé: Biota and Ecology of the Northern Section of the World's Largest Pristine Wetland.* Monographiae Biologicae 77, Kluwer Academic Publishers, Dordrecht.

Hidrovía Panel of Experts (1997) *The Hidrovía Paraguay–Paraná Navigation Project: Report of an Independent Review*. The Environmental Defense Fund (Washington, DC) and the Fundação Centro Brasileiro de Referência e Apoio Cultural (Brasília, DF, Brazil).

IBGE (1989) *Geografia do Brasil*, vol. 1: *Região Centro–Oeste*. Fundação Instituto Brasileiro de Geografia e Estatística, Diretoria de Geociências, Rio de Janeiro.

Junk, W. J. (1982) Amazonian flood plains: their ecology, present and future use. *Revue d'Hydrobiologie Tropicale* **15**(4), 285–301.

Junk, W. J. (1983) Ecology of swamps on the Middle Amazon. Chapter 9 in: *Mires: Swamp, Bog, Fen and Moor* (ed. by A. J. P. Gore). Ecosystems of the World 4B, Elsevier, Amsterdam.

Junk, W. J. (1989) The use of Amazonian flood plains under an ecological perspective. *Interciencia* **14**(6), 317–322.

Junk, W. J. (1993) Wetlands of tropical South America. In: *Wetlands of the World: Inventory, Ecology and Management* (ed. by D. F. Whigham, D. Dykyjová & S. Hejný), vol. I, 679–739. Kluwer, Dordrecht.

Junk, W. J. (1997) *The Central Amazon Flood Plain: Ecology of a Pulsing System.* Ecological Studies, vol. 126, Springer, New York.

Junk, W. J. & da Silva, C. J. (1995) Neotropical flood plains: a comparison between the Pantanal of Mato Grosso and the large Amazonian River flood plains. In: Limnology in Brazil (J. G. Tundisi, C. E. M. Bicudo & T. M. Tundisi), 195–217. Brazilian Academy of Sciences and Brazilian Limnological Society.

Junk, W. J., Bayley, P. B. & Sparks, R. E. (1989) The flood-pulse concept in river–flood plain systems. In: *Proceedings of the International Large Rivers Symposium* (ed. by D. P. Dodge). *Canadian Special Publication Fisheries Aquatic Sciences 106*, 110–127. NRC Research Press, Ottawa.

Klammer, G. (1981) Landforms, cyclic erosion and deposition, and late Cenozoic changes in climate in southern Brazil. *Zeitschrift für Geomorphologie N.F.* **25**(2), 146–165.

Klammer, G. (1982) The paleodesert of the Pantanal of Mato Grosso and the Pleistocene climatic—history of the central Brazilian tropics (in German). *Zeitschrift für Geomorphologie N.F.* **26**, 393–416.

Kramer, D. L, Lindsey, C. C., Moodie, G. E. E. & Stevens, E. D. (1978) The fishes and the aquatic environment of the central Amazon basin, with particular reference to respiratory patterns. *Canadian Journal of Zoology* **56**, 717–729.

Kretzschmar, A. U., Ferreira, S. A., Hardoim, E. L. & Heckman, C. W. (1993) Peak growth of the *Asplanchna sieboldi* (Leydig 1854) rotifer aggregation in relation to the seasonal wet and dry cycle in the Pantanal, Mato Grosso, Brazil. In: *Wetlands and Ecotones: Studies in Land–water Interactions*, 293–301. International Science Publishers, New Delhi.

Lacerda, L. D., Salomons, W., Pfeiffer, W. C. & Bastos, W. R. (1991) Mercury distribution in sediment profiles from lakes of the high Pantanal, Mato Grosso State, Brazil. *Biogeochemistry* **14**, 91–97.

Leady, B. S. & Gottgens, J. F. (2001) Mercury accumulation in sediment cores and along food chains in two regions of the Brazilian Pantanal. *Wetlands Ecology and Management* **9**(4), 349–361.

Lesack, L. F. W. & Melack, J. M. (1995) Flooding hydrology and mixture dynamics of lake water derived from multiple sources in an Amazon flood plain lake. *Water Resources Research* **31**(2), 329–345.

Lewis, W. M., Jr, Hamilton, S. K. & Saunders, J. F. III (1995) Rivers of northern South America. In: *River and Stream Ecosystems* (C. E. Cushing, K. W. Cummins & G. W. Minshall), 219–256. Ecosystems of the World, vol. 22. Elsevier, Amsterdam.

Lewis, W. M., Jr, Hamilton, S. K., Lasi, M. A., Rodríguez, M. A. & Saunders, J. F. III (2000) Ecological determinism on the Orinoco flood plain. *BioScience* **50**, 681–692.

Lima, J. A. F. de (1986). A pesca no Pantanal de Mato Grosso (rio Cuiaba: importancia dos peixes migradores). *Acta Amazonica* **16/17**, 87–94 (1986/1987).

Lundberg, J. G., Lewis, W. M., Jr, Saunders, J. F. & Mago-Leccia, F. (1987) A major food web component in the Orinoco River channel: evidence from planktivorous fishes. *Science* **237**, 81–83.

Magalhães, N. W. (1992) *Conheça o Pantanal*. Terragraph, Sao Paulo, Brazil.

Meade, R. H., Rayol, J. M., da Conceição, S. & Natividade, J. R. G. (1991) Backwater effects in the Amazon River basin of Brazil. *Environmental Geology and Water Sciences* **18**(2), 105–114.

Melack, J. M. & Fisher, T. R. (1990) Comparative limnology of tropical flood plain lakes with an emphasis on the central Amazon. *Acta Limnologica Brasiliensia* **3**, 1–48.

Mitsch, W. J. & Gosselink, J. G. (2000) *Wetlands*, third edn. Wiley, Chichester, UK.

Mittermeier, R. A., Camara, I. G., Padua, M. T. J. & Blanck, J. (1990) Conservation in the Pantanal of Brazil. *Oryx* **24**, 103–112.

Mourão, G. M. (1988) Limnologia comparativa de três lagoas (duas "baías" e uma "salina") do Pantanal da Nhecolândia, MS. Unpublished Master's Thesis, Universidade Federal de São Carlos, Departamento de Ciências Biológicas, São Carlos, Brazil.

Mourão, G. M. (1997) Uso de levantamentos aéreos para estudo da distribuição e abundância de grandes vertebrados no Pantanal Mato-Grossense. Doctoral Dissertation, Instituto Nacional de Pesquisas da Amazônia/Universide do Amazonas, Manaus, Brazil.

Mourão, G. M., Ishii, I. & Campos, Z. (1988) Alguns fatores limnológicos relacionados com a ictiofauna de baías e salinas do Pantanal da Nhecolândia, Mato Grosso do Sul, Brasil. *Acta Limnologica Brasiliensia* **11**, 181–198.

Neiff, J. J. (1990) Ideas para la interpretación ecológica del Paraná. *Interciencia* **15**, 424–441.

Neiff, J. J., Reboratti, H. J., Gorlero, M. C. & Basualdo, M. (1985) Impacto de las crecientes extraordinarias sobre los bosques fluviales del Bajo Paraguay. *Boletin de la Comisión Especial del Río Bermejo* (Cámara de Diputados de la Nación (Buenos Aires, Argentina)) **4**, 13–31.

Novoa, D. (1982) *Los Recursos Pesqueros del Río Orinoco e su Explotación*. Corporación Venezolana de Guayana (Editorial Arte), Caracas, Venezuela.

Odum, E. P. (1969) The strategy of ecosystem development. *Science* **164**, 262–270.

Oliveira, M. D. & Calheiros, D. F. (2000) Flood pulse influence on phytoplankton communities of the south Pantanal flood plain, Brazil. *Hydrobiologia* **427**(1–3), 101–112.

Paiva, M. P. (1984) *Aproveitamento de Recursos Faunísticos do Pantanal de Mato Grosso: Pesquisas Necessárias e Desenvolvimento de Sistemas de Produção mais Adequados à Região.* Brazilian Corporation for Agricultural Research (EMBRAPA), Brasília.

PCBAP (1997) *Plano de Conservação da Bacia do Alto Paraguai (Pantanal)*. Ministério do Meio Ambiente, dos Recursos Hídricos e da Amazonia Legal, Brasília, Brazil.

Penha, J. M. F., da Silva, C. J. & Bianchini-Junior, I. (1999) Productivity of the aquatic macrophyte *Pontederia lanceolata* Nutt. (Pontederiaceae) on flood plains of the Pantanal Mato-grossense, Brazil. *Wetlands Ecology and Management* **7**, 155–163.

Petts, G. E. (1990) Regulation of large rivers: problems and possibilities for environmentally sound river development in South America. *Interciencia* **15**(6), 388–395.

Piedade, M. T. F., Junk, W. J. & de Mello, J. A. N. (1992) A flood plain grassland of the central Amazon. Chapter 5 in: *Primary Productivity of Grass Ecosystems of the Tropics and Sub-tropics* (ed. by S. P. Long, M. B. Jones & M. J. Roberts). Chapman and Hall, London.

Pinto-Silva, V. (1991) Variação diurna dos principais parâmetros limnológicos nos lagos Recreio e Buritizal—Pantanal Mato-Grossense, Barão de Melgaço, MT. Doctoral Dissertation, Universidade Federal de São Carlos, Centro de Ciências Biológicas e da Saude, São Carlos, Brazil.

Poff, N. L., Allan, J. D., Bain, M. B., Karr, J. R., Prestegaard, K. L., Richter, B. D., Sparks, R. E. & Stromberg, J. C. (1997) The natural flow regime: a paradigm for conservation. *BioScience* **47**, 769–784.

Ponce, V. M. (1995) *Hydrologic and Environmental Impact of the Paraná–Paraguay Waterway on the Pantanal of Mato Grosso, Brazil: a Reference Study*. San Diego State University.

Ponce, V. M. & da Cunha, C. N. (1993) Vegetated earthmounds in tropical savannas of central Brazil: a synthesis. *Journal of Biogeography* **20**, 219–225.

Por, F. D. (1995) *The Pantanal of Mato Grosso (Brazil): World's Largest Wetlands*. Kluwer Academic Publishers, Dordrecht.

Pott, A. & Pott, V. J. (1994) *Plantas do Pantanal*. Empresa Brasileira de Pesquisa Agropecuária, Brasília, Brazil.

Pott, A. & Pott, V. J. (1997) Checklist das macrófitas aquáticas do Pantanal, Brasil. *Acta Botanica Brasilica* **11**, 215–227.

Pott, V. J. & Pott, A. (2000) *Plantas Aquáticas do Pantanal*. Empresa Brasileira de Pesquisa Agropecuária, Brasília, Brazil.

Prado, D. E., Gibbs, P. E., Pott, A. & Pott, V. J. (1992) The Chaco-Pantanal transition in southern Mato Grosso, Brazil. In: *Nature and Dynamics of Forest–savanna Boundaries* (ed. by P. A. Furley, J. Proctor & J. A. Ratter), 451–470. Chapman and Hall, London.

Prance, G. T. & Schaller, G. B. (1982) Preliminary study of some vegetation types of the Pantanal, Mato Grosso, Brazil. *Brittonia* **34**(2), 228–251.

Quigley, H. B. & Crawshaw P. G., Jr (1992) A conservation plan for the jaguar *Panthera onca* in the Pantanal region of Brazil. *Biological Conservation* **61**, 149–157.

Quirós, R. (1990) The Paraná River basin development and the changes in the lower basin fisheries. *Interciencia* **15**(6), 442–451.

RADAMBRASIL (1982) *Folha SE.21 Corumbá e Parte da Folha SE.20: Geologia, Geomorfologia, Pedologia, Vegetação e Uso Potencial da Terra*. Ministério das Minas e Energia, Rio de Janeiro, Brazil.

Ratter, J. A., Pott, A., Pott, V. J., da Cunha, C. N. & Haridasan, M. (1988) Observations on woody vegetation types in the Pantanal and at Corumba, Brazil. *Notes of the Royal Botanical Garden of Edinburgh* **45**(3), 503–525.

Resende, E. K., Catella, A. C., Nascimento, F. L., Palmeira, S. S., Pereira, R. A. C., Lima, M. S. & Almeida, V. L. L. (1995) *Biologia do Curimbatá (Prochilodus lineatus), Pintado (Pseudoplatystoma corruscans) e Cachara (Pseudoplatystoma fasciatum) na Bacia Hidrografica do Río Miranda, Pantanal do Mato Grosso do Sul, Brasil*. Brazilian Corporation for Agricultural Research (EMBRAPA), Brasília.

Richter, B. D., Baumgartner, J. V., Wigington, R. & Braun, D. P. (1997) How much water does a river need? *Freshwater Biology* **37**, 231–249.

Robertson, B. A. & Hardy, E. R. (1984) Zooplankton of Amazonian lakes and rivers. In: *The Amazon: Limnology and Landscape Ecology of a Mighty Tropical River and its Basin* (ed. by H. Sioli), 337–352. Dr W. Junk Publishers, The Hague.

Robertson, A. W. & Mechoso, C. R. (1998) Interannual and decadal cycles in river flows of southeastern South America. *Journal of Climate* **11**, 2570–2581.

Salati, E., Victoria, R. L., Martinelli, L. A. & Richey, J. E. (1989) Deforestation and its role in possible changes in the Brazilian Amazon. In: *Global Change and Our Common Future: Papers from a Forum* (ed. by R. S. DeFries & T. S. Malone), 159–171. National Academy Press, Washington, DC.

Saunders, J. F. & Lewis W. M., Jr (1989) Zooplankton abundance in the lower Orinoco River, Venezuela. *Limnology and Oceanography* **34**(2), 397–409.

Semmelman, F. R., Borges, A., Risso, A., Bordas, M. & Lopes, M. S. (1999) Balanço preliminar de sedimentos no Pantanal da bacia do Alto Paraguai—BAP—Brasil. In: *Proceedings of the II Symposium on Natural and Socioeconomic Resources of the Pantanal: Management and Conservation*, 79–87. Brazilian Corporation for Agricultural Research, Brasília, and the Federal University of Mato Grosso do Sul, Campo Grande, Brazil.

Short, N. M. & Blair, R. W. (1986) *Geomorphology from Space: a Global Overview of Regional Landforms*. NASA SP-486, Washington, DC.

Sippel, S. J., Hamilton, S. K. & Melack, J. M. (1992) Inundation area and morphometry of lakes on the Amazon River flood plain, Brazil. *Archiv für Hydrobiologie* **123**(4), 385–400.

Soluk, D. A. (1985) Macroinvertebrate abundance and production of psammophilous Chironomidae in shifting sand areas of a lowland river. *Canadian Journal of Fisheries and Aquatic Sciences* **42**(7), 1296–1302.

Sparks, R. E., Braden, J. B., Demissie, M., Mitra, P., Schneider, D. W., White, D. C. & Xia, R. (2000) Technical support of public decisions to restore flood plain ecosystems: a status report on the Illinois River Project, USA. In: *New Approaches to River Management* (ed. by A. J. M. Smits, P. H. Nienhuis & R. S. E. W. Leuven), 225–247. Backhuys, Leiden, The Netherlands.

Strussmann, C. & Sazima, I. (1993) The snake assemblage of the Pantanal at Pocone, western Brazil: faunal composition and ecological summary. *Studies on Neotropical Fauna and Environment* **28**(3), 157–168.

Swarts, F. A. (ed.) (2000) *The Pantanal: Understanding and Preserving the World's Largest Wetland* (Selected papers from the First and Second World Conference on Preservation and Sustainable Development in the Pantanal, Washington, DC, February and October 1999). Paragon House, St Paul, Minnesota.

Tarifa, J. R. (1986) O sistema climático no Pantanal: Da compreensão do sistema a definição de prioridade de pesquisa climatológica. In: *Anais do Simpósio sobre Recursos Naturais e Socio-econômicos do Pantanal*, 9–27. Brazilian Corporation for Agricultural Research (EMBRAPA), Brasília.

Tricart, J. (1982) El Pantanal: Un ejemplo del impacto de la geomorfología sobre el medio ambiente. *Geographia* **7**, 37–50.

Turner, P. N. & da Silva, C. (1992) Littoral rotifers from the state of Mato Grosso, Brazil. *Studies on Neotropical Fauna and Environment* **27**, 227–241.

UNESCO (1973) *Hydrological Studies of the Upper Paraguay River Basin (Pantanal), 1966–1972*. Technical Report UNESCO/UNDP BRA66521, Paris.

Urban, H., Stribrny, B. & Lippholt, H. J. (1992) Iron and manganese deposits of the Urucum district, Mato Grosso do Sul, Brazil. *Economic Geology* **87**, 1375–1392.

Ussami, N., Shiraiwa, S. & Dominguez, J. M. L. (1999) Basement reactivation in a sub-Andean foreland flexural bulge: the Pantanal wetland, SW Brazil. *Tectonics* **18**, 25–39.

Vásquez, E. (1989) The Orinoco River: a review of hydrobiological research. *Regulated Rivers: Research and Management* **3**, 381–392.

Wantzen, K. M., da Silva, C. J., Figueiredo, D. M. & Migliácio, M. C. (1999) Recent impacts of navigation in the upper Paraguay River. *Revista Boliviana de Ecología* **6**, 173–182.

Welcomme, R. L. (1985) *River Fisheries*. FAO Fisheries Technical Paper 262, Rome.

Welcomme, R. L. (1990) Status of fisheries in South American rivers. *Interciencia* **15**(6), 337–345.

Wilcox, R. (1992) Cattle and environment in the Pantanal of Mato Grosso, Brazil, 1870–1970. *Agricultural History* **66**, 232–256.

Willink, P. W., Chernoff, B., Alonso, L. E., Montambault, J. R. & Lourival, R. (eds) (2000) *A Biological Assessment of the Aquatic Ecosystems of the Pantanal, Mato Grosso do Sul, Brasil*. RAP Bulletin of Biological Assessment 18, Conservation International, Washington, DC.

Worbes, M. (1985) Structural and other adaptations to long-term flooding by trees in Central Amazonia. *Amazoniana* **4**(3), 459–484.

CHAPTER 9

HYDROLOGY AND WATER QUALITY IN THE PIRACICABA RIVER BASIN, SÃO PAULO STATE, BRAZIL

Luiz A. Martinelli, Reynaldo L. Victoria, Epaminondas S. Ferraz, Plinio B. Camargo, Marcelo Z. Moreira, Alex V. Krusche, Jorge M. de Moraes, Maria V. Ballester, Luciene L. S. Lara, Marcelo C. Bernardes, Jean Pierre H. B. Ometto & Carlos Eduardo Cerri

The Piracicaba River basin is a mesoscale basin (12 400 km^2) located almost entirely in the State of São Paulo, in the southeast region of Brazil. This region is one of the most developed in the country, responsible for almost 10% of the Brazilian gross internal domestic product. It comprises 61 counties with a total of almost 3 million inhabitants. Its population growth was always larger than the average growth of São Paulo State due to the high number of industries (almost 7000) and to the presence of extensive sugarcane fields. There are three main rivers in this basin: the Atibaia, the Jaguari and the Piracicaba itself. These rivers receive 84% of the volume of raw urban sewage and 10% of the volume of untreated industrial effluents discharged in the basin. This heavy load of waste water has profoundly altered the rivers' biogeochemistry and biodiversity (Krusche *et al.*, 1997; Ometto *et al.*, 2000; Williams *et al.*, 2001). The large number of industries and periodic sugarcane burning has also changed the atmospheric chemistry, making atmospheric deposition of nitrogen similar to that in the northeastern United States (Lara *et al.*, 2001). In this chapter we review the socio-economics aspects of the Piracicaba basin, such as its demography and land use, as well as the natural aspects like climate, hydrology, and soil types. We then examine the effects of human actions on rivers and atmosphere.

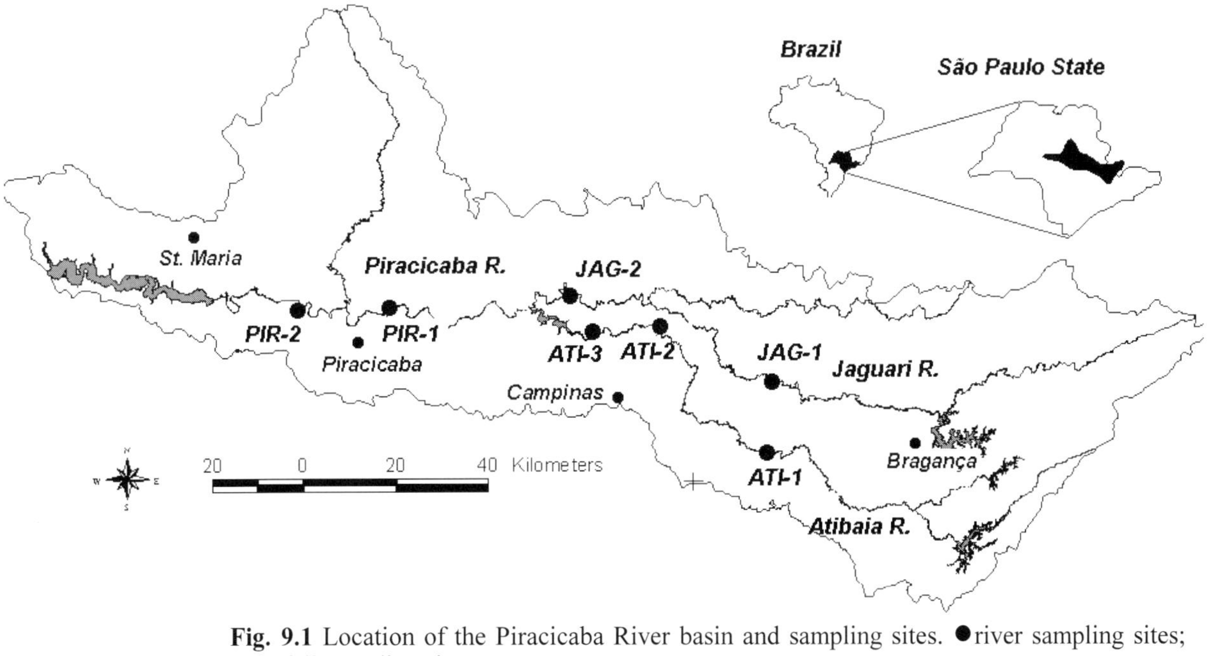

Fig. 9.1 Location of the Piracicaba River basin and sampling sites. ● river sampling sites; • rainfall sampling sites.

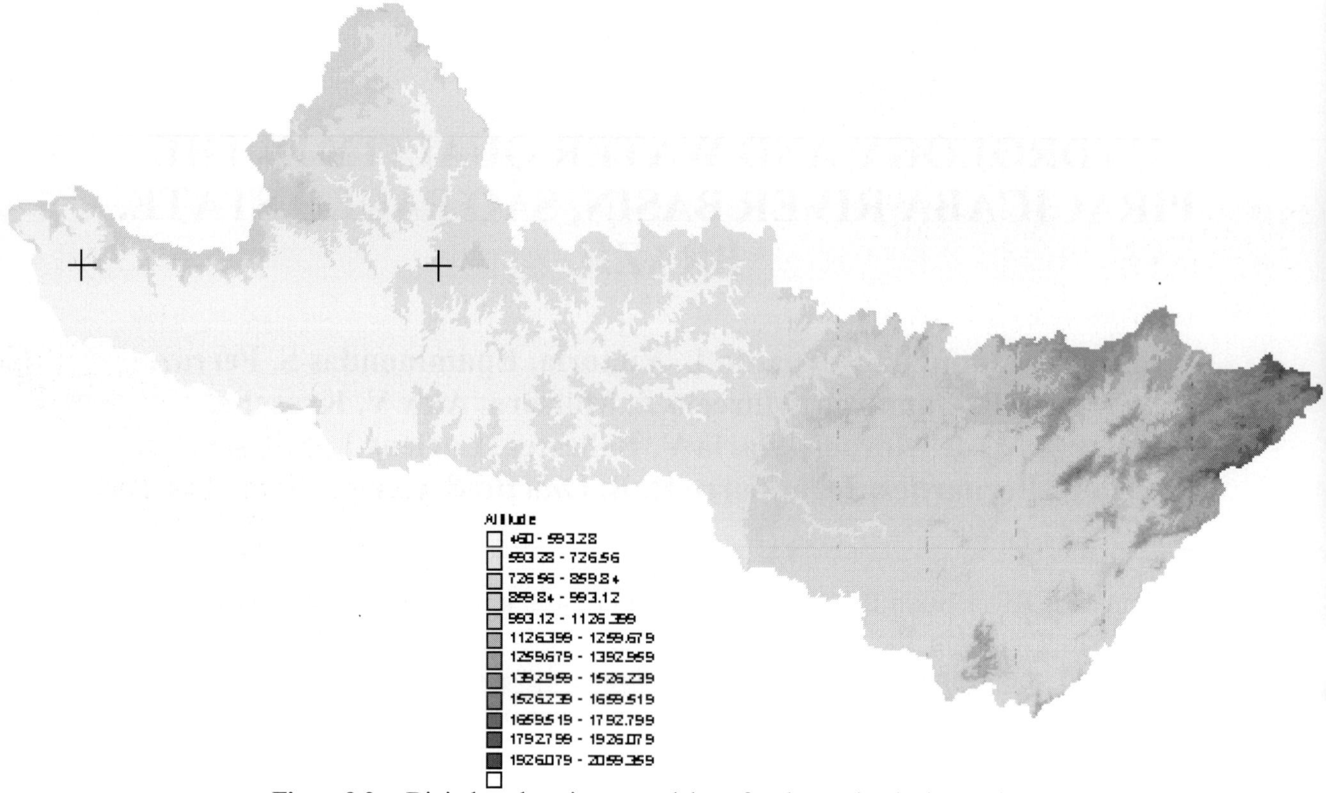

Fig. 9.2 Digital elevation model of the Piracicaba River basin (source: http://www.cena.usp.br/piracena).

THE PIRACICABA RIVER BASIN

Geographical location

The Piracicaba River basin (12 400 km^2) is located in the southeast region of Brazil, roughly between latitudes 22° and 23°S, and politically it is located almost entirely in São Paulo State (Fig. 9.1). The basin extends over three distinct regions (Fig. 9.2). The western and especially the eastern portions of the basin are hilly, with an average slope of 15% and the central region is mostly lowland. The main rivers of the basin flow from east to west in a stretch of approximately 250 km. The drainage network is particularly dense in the eastern side of the basin, where the highest elevations can be found. The density of the drainage network decreases in the central part of the basin, and increases again in the eastern portion of the basin (Fig. 9.3).

Soils

Most of soil types in the Piracicaba River basin are tropical soils, with low pH and low natural fertility (database available at http://www.cena.usp.br/piracena). Two types of soil predominate: *Latossolo Vermelho-Amarelo* (PV) and *Podzólico Vermelho-Amarelo* (LV) (Fig. 9.4). These two types of soils roughly correspond to oxisol and ultisol in the US soil taxonomy classification system. However, several different soil types also occur in the basin. The most important are: *Latossolo Vermelho-Escuro* (LE), *Areia Quartzosa* (AQ), *Latossolo Roxo* (LR), *Litossolo* (Li), *Glei Pouco Húmico/Húmico* (HPG/HG), *Terra Roxa Estruturada* (TE), *Podzolizados* soils (P) and *Regossolos* (R). There is a predominance of clay soils, with few sandy soils. Notable exceptions are the *Areia Quartzosa* and *Regossolo*, which by

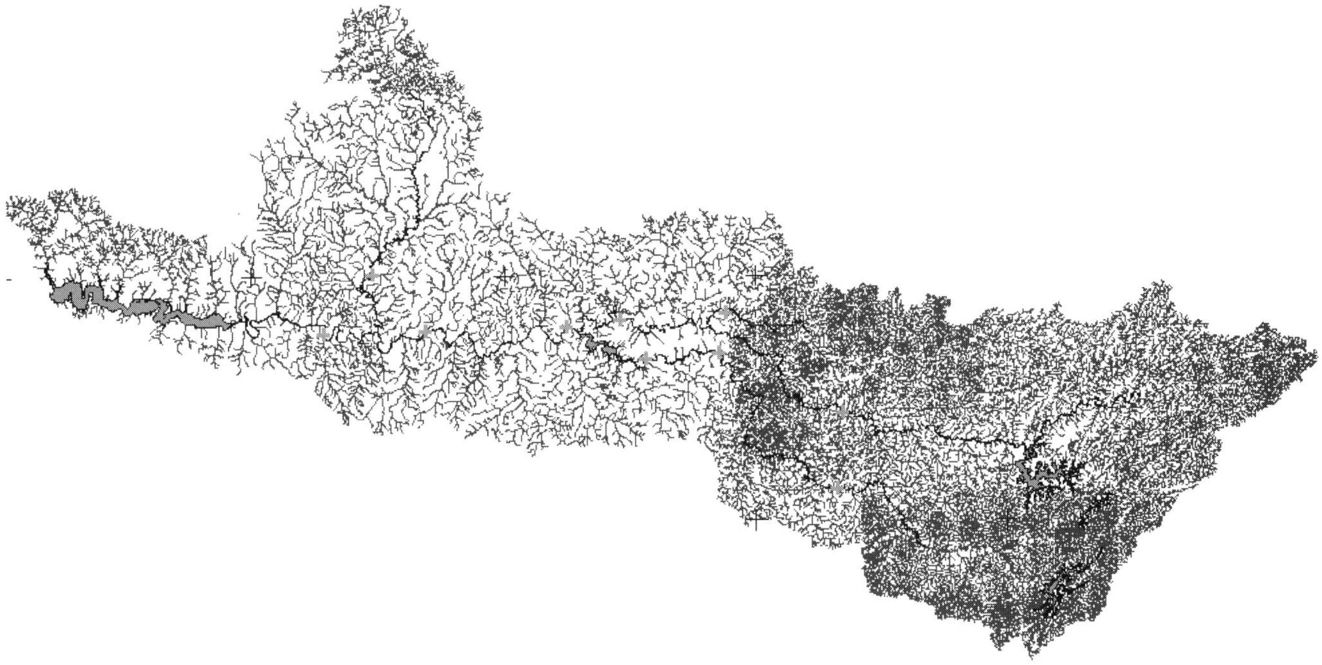

Fig. 9.3 Drainage network of the Piracicaba River basin (source: http://www.cena.usp.br/piracena).

definition are sandy soils (Fig. 9.5). Although variable, the sum of bases was generally low. The smallest values were observed in the AQ, HPG/HG, LE, LV and R soils, with values less than 10–12 $mmol_c\ kg^{-1}$. The highest values (~50 $mmol_c\ kg^{-1}$) were found only in TE soils. The soil types Li, LR, P and PV had intermediary values (10–30 mmolc kg^{-1})(Fig. 9.6).

Demographic information

There are almost 3 million people living in the Piracicaba River basin. In total there are 61 counties in the basin. Most of the population is concentrated in eight of these counties, which are

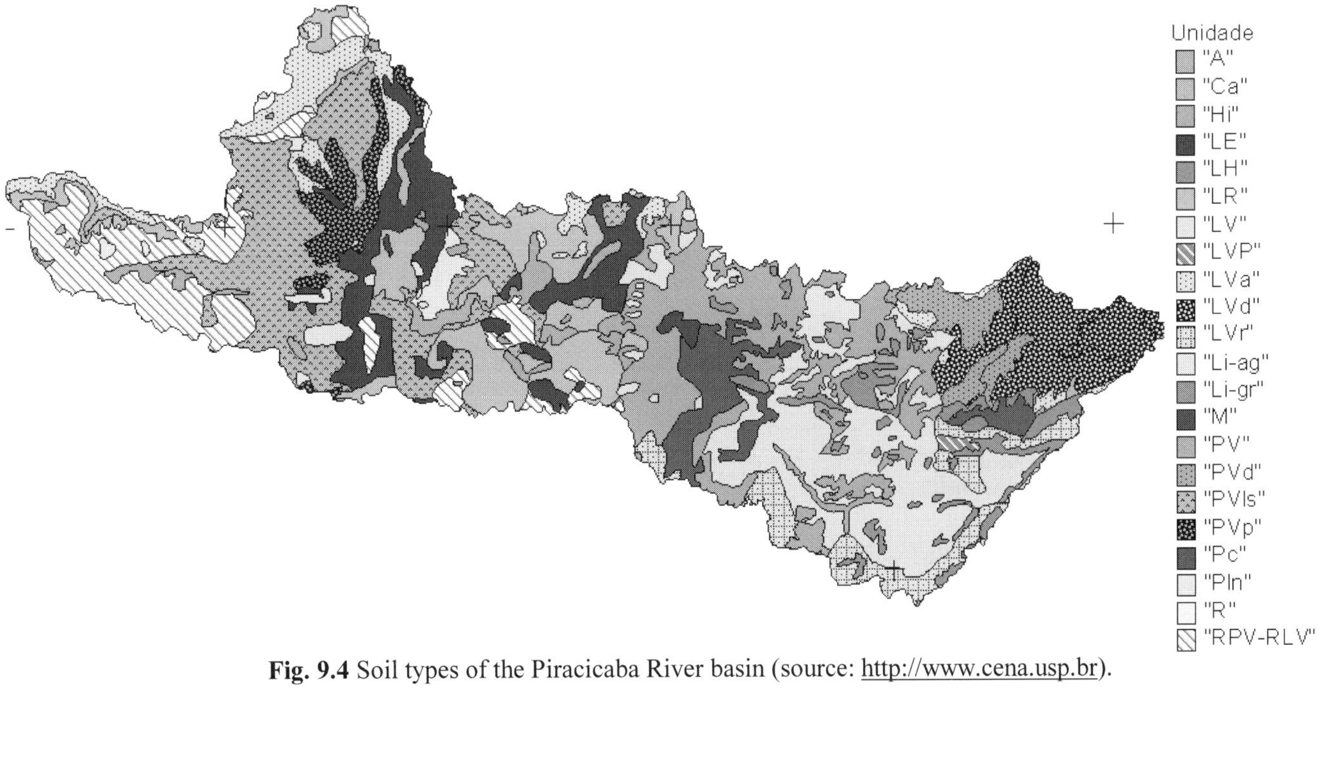

Fig. 9.4 Soil types of the Piracicaba River basin (source: http://www.cena.usp.br).

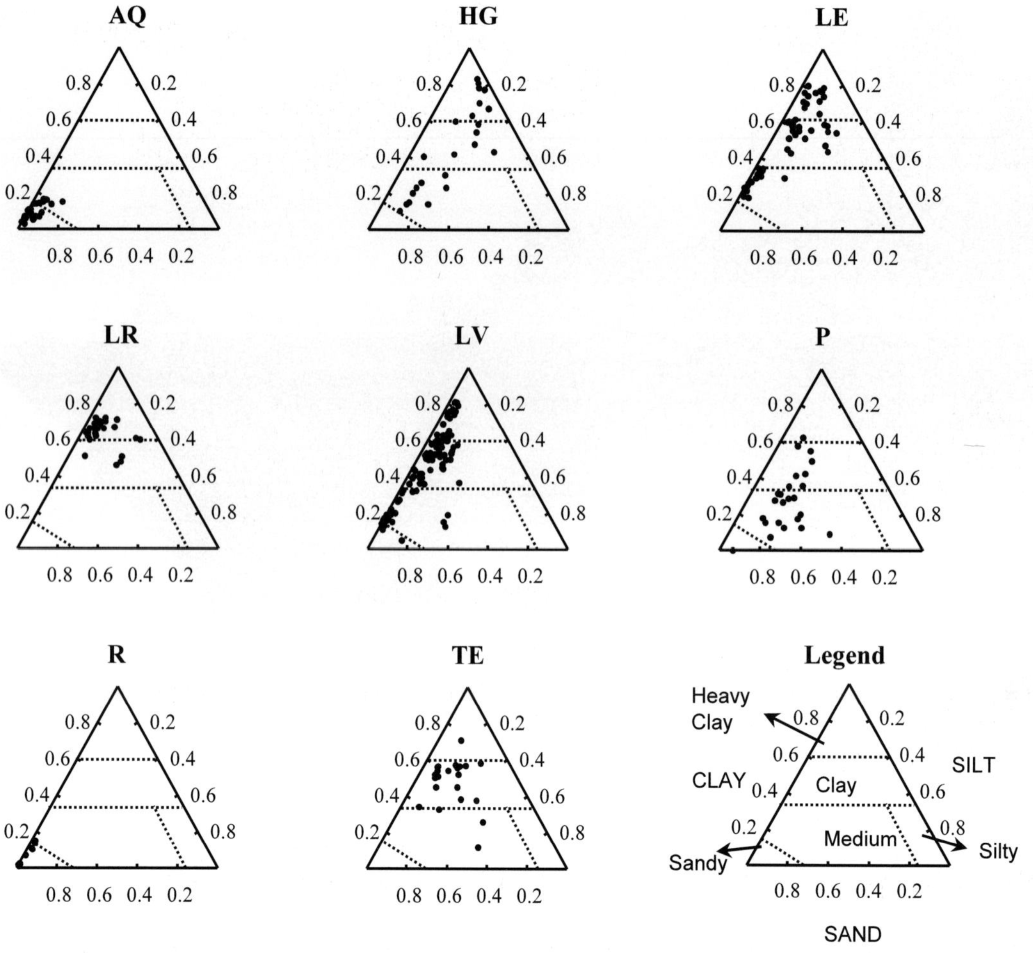

Fig. 9.5 Soil texture of 87 soil profiles sampled in the Piracicaba River basin.

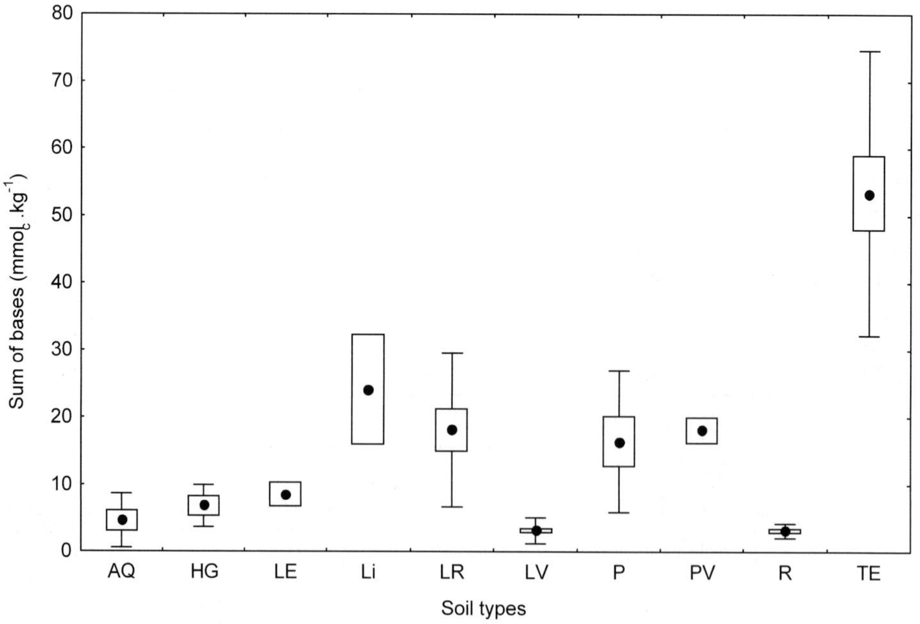

Fig. 9.6 Box and whisker plot of the sum of bases of the major soil types in the Piracicaba River basin.

located in the central and western side of the basin (Fig. 9.7). In this central region there are counties where the population density exceeds 1000 inhabitants km^{-2} (Table 9.1).

Historically, people were attracted to the Piracicaba region due to the large number of industries (almost 7000 in 1998), especially for the sugarcane agro-industry. Consequently,

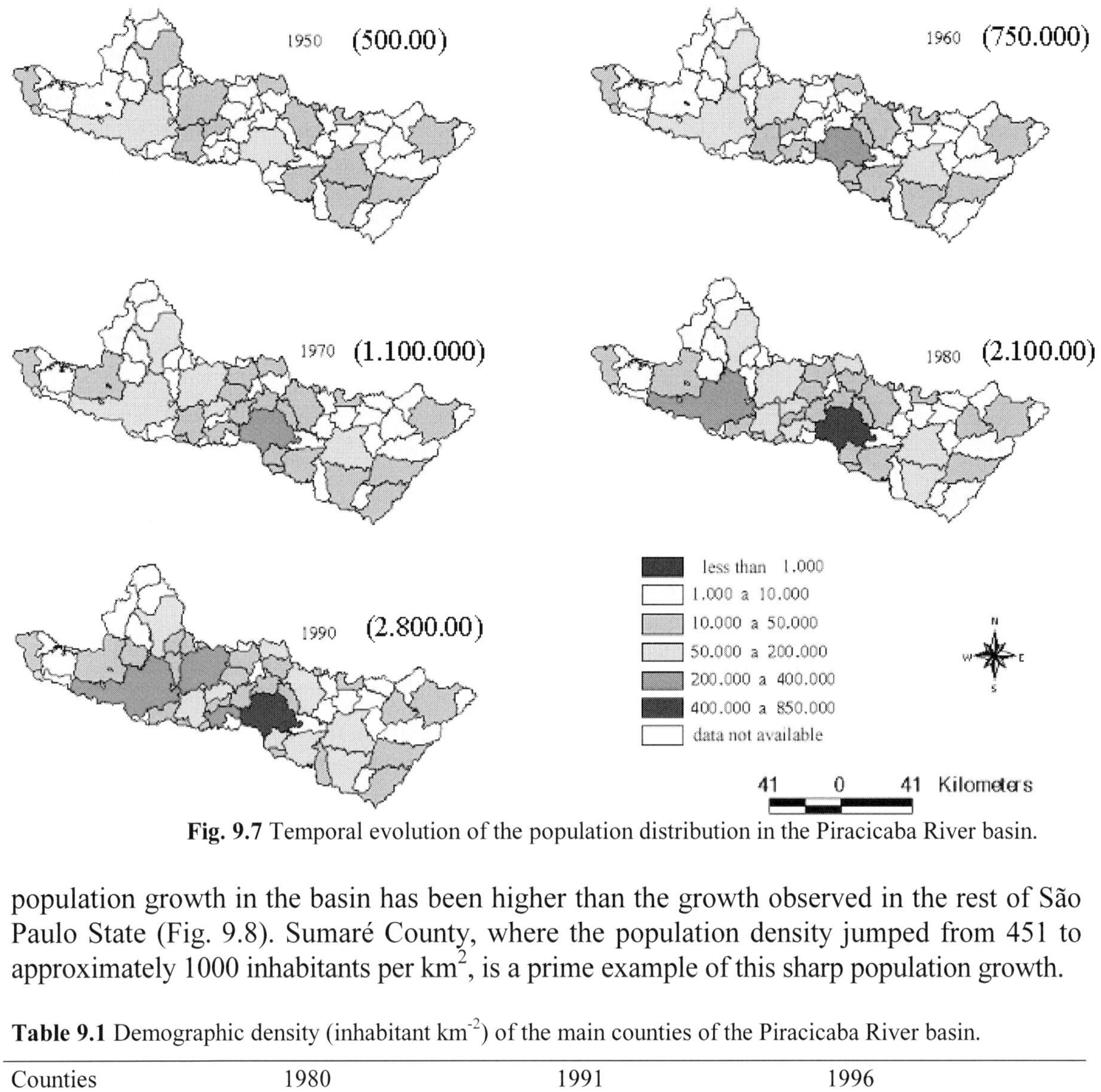

Fig. 9.7 Temporal evolution of the population distribution in the Piracicaba River basin.

population growth in the basin has been higher than the growth observed in the rest of São Paulo State (Fig. 9.8). Sumaré County, where the population density jumped from 451 to approximately 1000 inhabitants per km^2, is a prime example of this sharp population growth.

Table 9.1 Demographic density (inhabitant km^{-2}) of the main counties of the Piracicaba River basin.

Counties	1980	1991	1996
Hortolândia	-	-	1.844
Americana	847	1.068	1.164
Campinas	747	951	1.022
Sumaré	451	1.008	1.020
S. Bárbara d'Oeste	284	538	595
Limeira	260	358	397
Rio Claro	212	264	293
Piracicaba	148	196	223

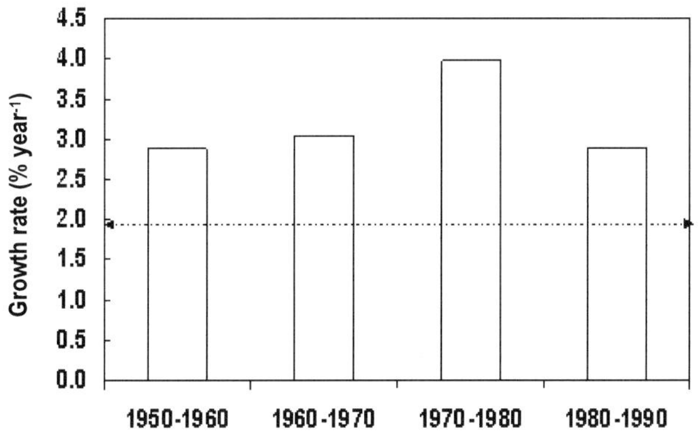

Fig. 9.8 Decadal growth population in the Piracicaba River basin. Dotted line indicates the average growth of the population of São Paulo State.

Historical changes in land use

We analysed historical changes in land use by looking at satellite images taken in 1978, 1985, 1993 and 1997. The main land use in the basin today is pasture, covering almost 45% of the region. The second most important use is sugarcane, covering approximately 30%. Trees cover almost 20% of the basin, 10% as planted forest and approximately 9% is natural forest.

The major change between 1978 and 1985 was an increase in the area occupied by sugarcane, especially in the western portion of the basin. However, an increase in the area cropped with sugarcane was also observed in the central part, where sugar replaced citrus orchards. Silviculture increased in the east and was consolidated in 1993 and 1997 (see http://www.cena.usp.br for maps). Another important pattern was the continual expansion of urban centres, especially cities such as Campinas, Americana and Santa Barbara d'Oeste.

Sources of pollution

The main type of pollution in the basin is organic-matter-rich urban sewage and industrial effluents dumped into the rivers (Martinelli *et al.*, 1999a). A total of approximately 120 t of biochemical oxygen demand (BOD) is produced daily by the cities in the basin. Of this total, only 16% is treated, therefore, almost 100 t of BOD per day of raw domestic sewage is dumped into the rivers. The volume of industrial effluents is much larger than the volume produced by the cities, reaching approximately 600 t of BOD per day. However, 92% of this volume is treated. Consequently, the volume of raw industrial effluent dumped into the rivers is approximately half of that dumped by the cities. Therefore, the total BOD dumped into the rivers is approximately 150 t day^{-1}.

HYDROLOGY OF THE PIRACICABA RIVER BASIN

Precipitation distribution: spatial and seasonal variation

The annual average precipitation ranges from 1300 to 1800 mm. The highest precipitation rates occur at higher elevations in the eastern part of the basin. There is a decrease in

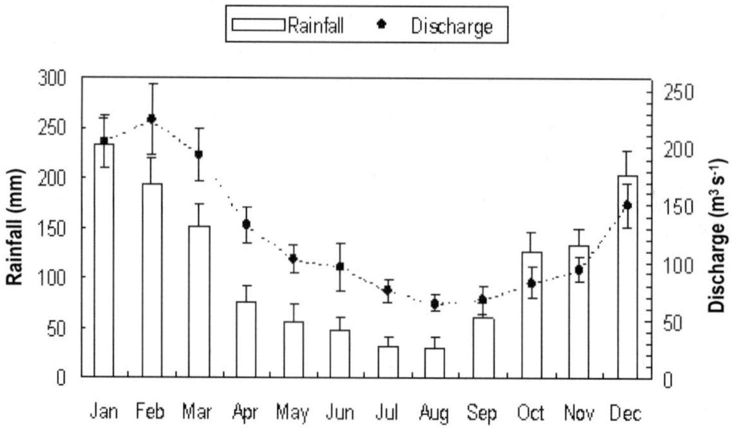

Fig. 9.9 Average monthly precipitation measured at the D4-061 gauging station (Departamento de Águas e Energia Elétrica do Estado de São Paulo). Source: http://www.cena.usp.br/piracena/html/postos/d4-061.htm. Average monthly discharge variability measured at the PIR-1 sampling site (full line). Source: Departamento de Águas e Energia Elétrica do Estado de São Paulo (http://www.cena.usp.br/piracena/html/daee.htm).

precipitation toward the central part of the basin, followed by an increase again to 1400–1500 mm in the northwest (Pellegrino *et al.*, 2001).

The whole basin is characterized by dry winters (April–August) and rainy summers (October–March). This seasonality is illustrated in Fig. 9.9, where the seasonal distribution of rainfall is illustrated in a gauge station located in the western part of the basin. An abrupt decrease in rainfall can be observed especially in July and August. Conversely, December, January and February are the months with highest rainfall (Pellegrino *et al.*, 2001).

Discharge distribution: spatial and temporal trends

The main rivers of the Piracicaba River basin are the Atibaia and Jaguari, forming together the Piracicaba River. Other important tributaries are the Camanducaia and Corumbataí. The final average discharge of the Piracicaba River is 143 $m^3 s^{-1}$ (average from 1947 to 1992). The final average discharge for the same period in the Jaguari River was 54 $m^3 s^{-1}$ and in the Atibaia River 36 $m^3 s^{-1}$. The seasonality of the river discharge follows that of rainfall (Fig. 9.9). The lowest river discharges are observed between July and August and the highest discharges are observed in December and January (Moraes *et al.*, 1998).

Water use in the basin: problems for the future

The main sources of water supply in the Piracicaba basin are rivers and creeks; only a small part is supplied by groundwater. Most of the population is served with treated water (80%). It is estimated that the population of the Piracicaba River basin has a daily *per capita* consumption of 200–250 litres of treated water, making a total consumption of 15 $m^3 s^{-1}$. From the total amount of water consumed per day, approximately 75–80% returns to rivers as waste waters. Practically the same amount of water that is used by the cities is also used by the industries of the basin. The same amount (75–80%) also returns to rivers as industrial effluents. Another important use of water in the basin is irrigation of agricultural fields, and approximately 6 $m^3 s^{-1}$ is used for this purpose. Due to losses by crop consumption and evaporation, only 25% of the water used for irrigation returns to rivers. Therefore, a total of approximately 37 $m^3 s^{-1}$ of water is used in the basin, and approximately 26 $m^3 s^{-1}$ return to rivers as waste waters. Besides the use of water in the basin itself, approximately 30 $m^3 s^{-1}$ of water is exported from the basin to the metropolitan area of São Paulo. The total water demand in the basin is equal to 67 $m^3 s^{-1}$. From this total, 23% is used by cities, 23% by industries, 8% by agriculture, and the major part (46%) is exported to São Paulo city. If the rate of population growth maintains its current pace, it is likely that in the near future, sectors of the basin will suffer water shortages, mainly during the winter time. It is also important to note that if the population increases at the same rate, the volume of waste water will also increase, jeopardizing moreover the integrity of rivers and creeks of the Piracicaba basin.

RIVER BIOGEOCHEMISTRY AND CHANGES CAUSED BY HUMAN IMPACTS

Here we examine the river biogeochemistry, emphasizing the effects of human impacts on river functioning. With regard to the biogeochemistry of the system we established several sampling points along the main rivers of the basin (Fig. 9.1). From June 1995 to September

1997 we performed 37 field excursions collecting water samples at each sampling site shown in Fig. 9.1.

Suspended solids (SS) distribution

Most of the SS in the Piracicaba River basin is transported as fine particles (<63 μm), only 15–20% is transported as coarse particles (Table 9.2). There was large seasonal variability in the total suspended solids (TSS) concentrations of all sampling sites (Fig. 9.10). At Atibaia-3, the average concentration during the low water period was 21.8 mg l^{-1}. This increased to 221.3 mg l^{-1} during the high water period, peaking at approximately 660 mg l^{-1} in February 1997. The TSS concentration increased downstream in the three sub-basins (Table 9.2). The mass balance indicates a net downstream gain of TSS in all river sectors. The highest increases were observed between the sampling sites Atibaia-1 and Atibaia-2, and in the Jaguari sub-basin (Table 9.2).

The fact that the sediment mass balance was always positive indicates that every river section sampled gained suspended materials. The source of this extra sediment could be re-suspension of previously deposited particles, surface runoff or ungauged smaller tributaries. It is difficult to separate natural effects from human impacts. However, considering that approximately 85% of the basin is agricultural fields, and that urban erosion is a serious problem in the basin, it is reasonable to hypothesize that at least part of the observed increase in the sediment flux is due to human impacts.

Major ions

For all ions and cations there was a clear downstream increase in concentration (Martinelli *et al.*, 1999a; Williams *et al.*, 2001). The most downstream sampling site had a much higher concentration than the two first sampling sites in the Atibaia and Jaguari rivers (Table 9.3). As for the suspended material, the mass balance indicates a net gain of dissolved ions in all rivers sections (Table 9.3). Therefore, it is evident that there is a source of ions that may be entering the rivers through small ungauged tributaries or point-source pipes discharging waste water. We found very good correlations between the area covered by urban centres in sub-basins delimited by sampling sites and average concentrations of major ions (Table 9.4). As already mentioned only 16% of the urban sewage is treated. Consequently, the area occupied by urban centres is a proxy for human impacts, especially the discharge of waste water. The good correlations we found suggest that waste waters are the source of major ions to rivers of the basin. In support of this hypothesis, seasonal variability between sampling sites less affected by waste waters is totally distinct from sites where pollution is

Table 9.2 Discharge (*Q*), coarse (CSS), fine (FSS) and total (TSS) suspended solids concentrations and flux of TSS in the sampling sites. The balance is the difference between the TSS flux of downstream sampling sites in relation to the next upstream sampling site. Positive values indicate a downstream increase in the TSS flux.

Sampling sites	Q ($m^3 s^{-1}$)	CSS (mg l^{-1})	FSS (mg l^{-1})	TSS (mg l^{-1})	TSS (t day^{-1})	Balance (%)
Atibaia-1	24.3	7.4	25.6	33.0	69	-
Atibaia-2	30.5	7.6	67.3	74.9	198	+65
Atibaia-3	36.3	16.8	82.7	99.4	312	+37
Jaguari-1	24.5	8.6	28.1	36.7	78	-
Jaguari-2	40.2	8.8	58.8	67.7	235	+67
Piracicaba-1	95.2	12.1	61.7	73.5	605	+10
Piracicaba-2	121.2	17.7	56.1	73.8	773	+22

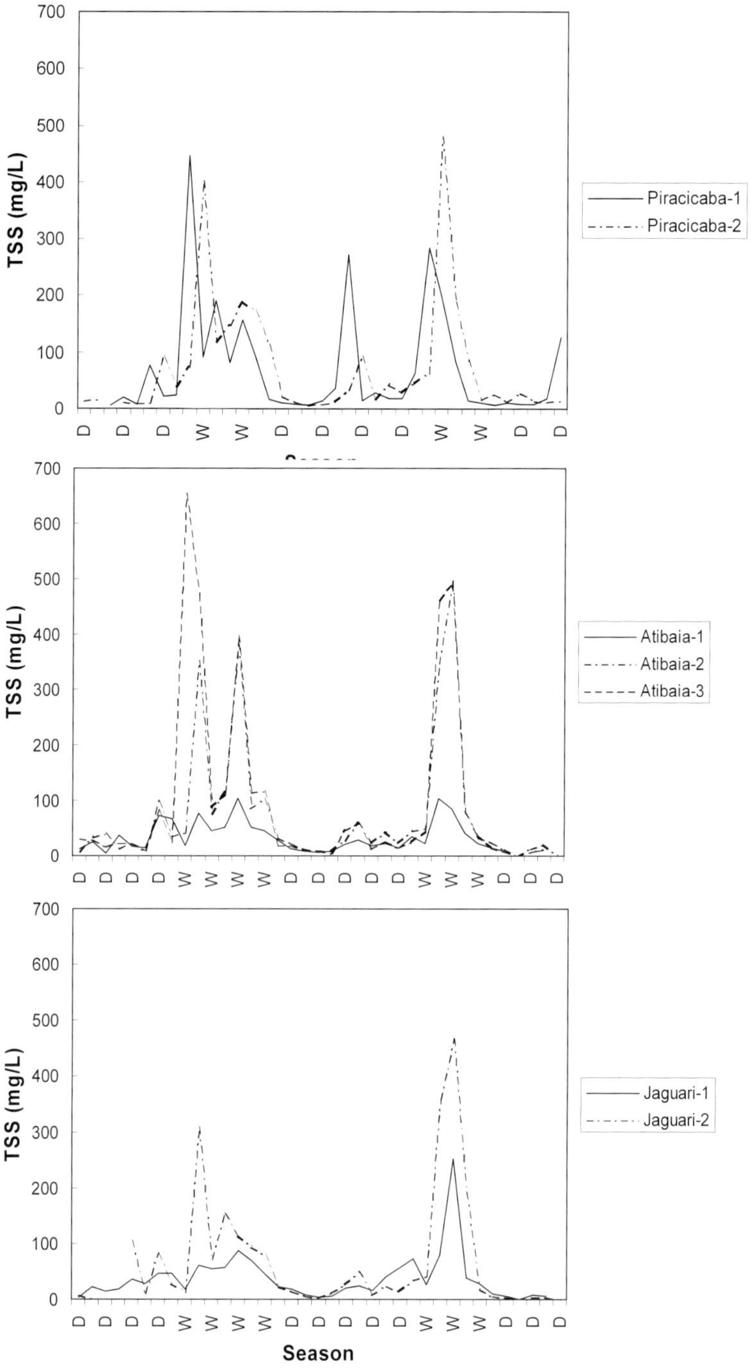

Fig. 9.10 Seasonal variability of total suspended solids (TSS) at several sampling sites along rivers in the Piracicaba River basin. W, wet season, D, dry season.

more pronounced (Martinelli *et al.*, 1999a). For instance, there was a significant seasonal variation in Cl concentration at the Piracicaba-1 sampling site. Conversely, at the Atibaia-1 sampling site, which is less affected by waste water, Cl concentrations were much lower and exhibited little seasonal variability (Fig. 9.11).

Carbon and nitrogen

Carbon and nitrogen are important in any riverine system. However, in the Piracicaba basin due to the constant input of organic-rich waste waters they are even more important (Ballester *et al.*, 1999; Martinelli *et al.*, 1999b). Although waste waters are mainly

Table 9.3 Average concentrations of major ions and sum of cations (TZ+) and anions (TZ–) at sampling sites in the Piracicaba River basin.

	SO_4	Cl	K	Na	Ca	Mg	TZ+	TZ–	Balance (%)
ATI-1	14.4	74.1	68.8	132.4	78.6	52.0	462.7	479.5	
ATI-2	36.8	129.4	84.7	290.6	116.3	67.2	741.9	714.1	+50
ATI-3	360.4	331.2	103.8	1088.4	173.9	81.0	1764.8	1768.8	+65
JAG-1	22.9	65.8	62.2	145.5	89.8	44.7	477.4	509.0	
JAG-2	38.5	111.5	72.7	272.1	114.3	68.6	710.7	688.6	+59
PIRA-1	245.7	392.7	101.0	969.9	201.7	86.8	1678.9	1630.7	+42
PIRA-2	229.8	360.7	99.6	875.4	215.7	116.9	1660.6	1597.3	+21

Table 9.4 Spearman correlation (*Rs*) coefficients yield from the correlations between urbanized area and major ions.

Variable	*Rs*
SO_4	0.82*
Cl	0.99**
K	0.89**
Na	0.86*
Ca	0.93**
Mg	0.93**
TZ+	0.86*
TZ–	0.86*

* ($P < 0.05$), ** ($P < 0.01$).

composed of organic carbon, almost half of the carbon in the rivers of the basin is transported as inorganic carbon (Table 9.5). The other half is equally divided between dissolved organic carbon (DOC) and particulate organic carbon (POC). There was no statistical difference between sampling sites. At most of the sampling sites, there was an increase in the relative contribution of dissolved inorganic carbon (DIC) and a decrease in the relative contribution of POC during low water (Fig. 9.13). Conversely, the relative contribution of POC increased and DIC decreased during low water, while the relative contribution of DOC remained practically constant during seasons (Fig. 9.13).

The dissolved organic carbon (DOC) concentration increased in the Atibaia River from sampling site ATI-1 to ATI-3, remained constant at the Jaguari sampling sites and reached the highest concentration at the Piracicaba River sampling sites (Fig. 9.13). There was a highly significant correlation between urbanized area and DOC (Table 9.6). This was expected, because DOC is the main form of carbon in waste waters of the Piracicaba basin. The DIC and dissolved inorganic nitrogen (DIN) concentrations behaved similar to DOC

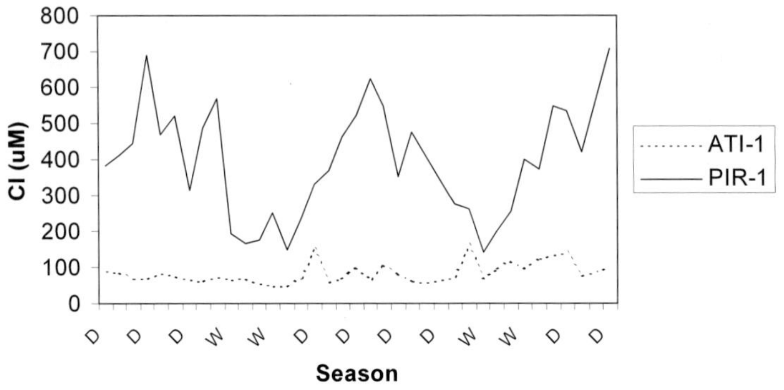

Fig. 9.11 Seasonal variability of Cl concentration at sampling sites ATI-1 and PIR-1. W, wet season; D, dry season.

Table 9.5 Relative contribution (%) of DOC, DIC, and POC in the total carbon distribution.

	DIC	DOC	POC
ATI-1	47	30	22
ATI-2	42	23	34
ATI-3	55	23	22
JAG-1	44	29	27
JAG-2	48	26	26
PIR-1	51	25	24
PIR-2	48	25	27
Average	48	26	26

(Fig. 9.14) and also had a strong correlation with the urbanized area (Filoso *et al.*, in press; Table 9.6). Finally, trends in POC concentration were not as clear as DOC and DIC. Sampling site ATI-2 had a high POC concentration, mainly due to surprisingly high total suspended solids concentrations. The correlation between POC and urbanized area was not strong ($Rs = 0.71$), and only significant at the 7% percentile. Conversely, particulate organic nitrogen (PON) correlated strongly ($Rs = 0.93$, $P < 0.01$) with urbanization.

River discharge also exerts a strong control on the distribution of particulate carbon and nitrogen (Fig. 9.14). As an example we analysed the seasonal variability of the weight percentage concentration of fine particulate organic carbon (FPOC wt % OC), C:N ratio and $\delta^{13}C$ values of this fraction (Fig. 9.15). There was an inverse relationship between river discharge and FPOC (wt % OC). The variability of C:N ratios was also high between seasons. There was a trend of high C:N ratios associated with high water discharge (Fig. 9.15), which corresponded to N depletion with increasing discharge. These elemental changes were accompanied by changes in $\delta^{13}C$ values. During the low water period both FPOM fraction contained ^{13}C-depleted organic matter, while during high water $\delta^{13}C$ values became heavier (Martinelli *et al.*, 1999b).

The same inverse relationship between FPOC and water discharge observed in many rivers (Meybeck, 1982), was also observed in the Piracicaba basin. For large rivers, this relationship is interpreted as resulting from autochthonous river material, rich in carbon and nitrogen, being diluted by land-derived allochthonous material, mainly soil particles (Meybeck, 1982). This could be the case for the Piracicaba rivers as well, since $\delta^{13}C$ values became lighter during the dry period and indicate that N-rich, isotopically depleted phytoplankton became an important source of organic matter (Martinelli *et al.*, 1999b). The

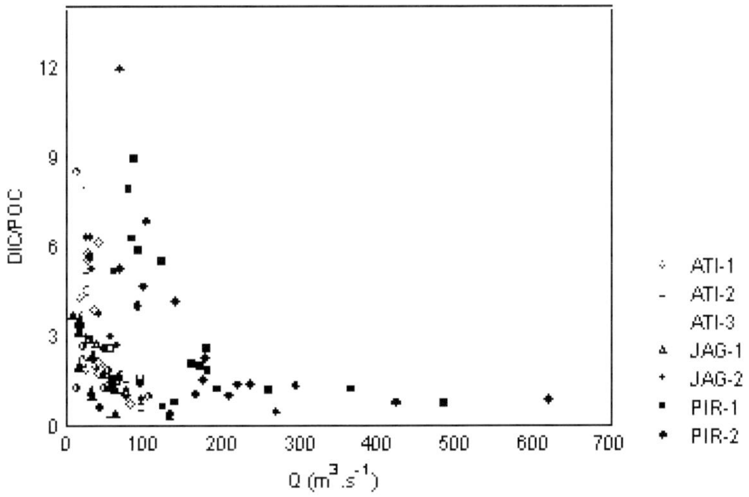

Fig. 9.12 Plot of river discharge (Q) *vs* the ratio dissolved inorganic carbon and particulate organic carbon (DIC/POC).

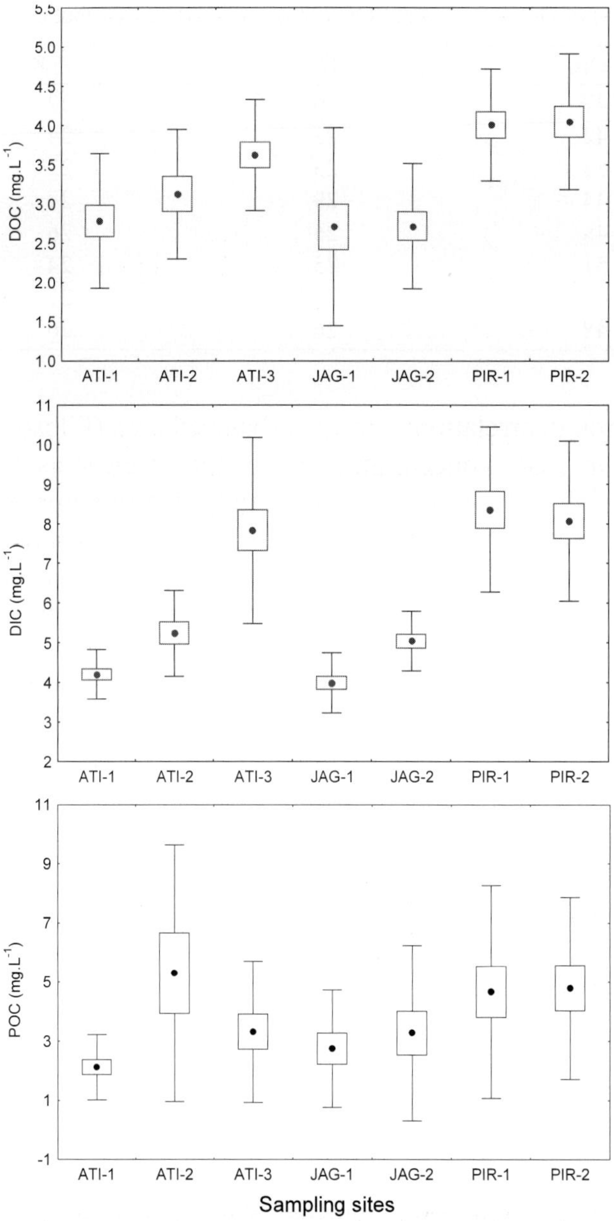

Fig. 9.13 Box and whisker plot of dissolved organic carbon (DOC), dissolved inorganic carbon (DIC), and particulate organic carbon (POC) at the sampling sites. Solid circles are averages, boxes are the standard error and bars represent the standard deviation.

high concentration of algae in rivers of the Piracicaba basin, especially during the dry season, support this hypothesis (Martinelli *et al.*, 1999b). Conversely, during high water periods, soil particles dominate with depleted C:N values and enriched $\delta^{13}C$ values due to the sugarcane contribution via soil erosion (Martinelli *et al.*, 1999b).

Table 9.6 Spearman correlation (*Rs*) coefficients yield from the correlations between urbanized area and major ions.

Variable	Rs
DOC	0.93**
DIC	0.99**
DIN	0.79*
POC	0.71
PN	0.93**

* ($P < 0.05$), ** ($P < 0.01$).

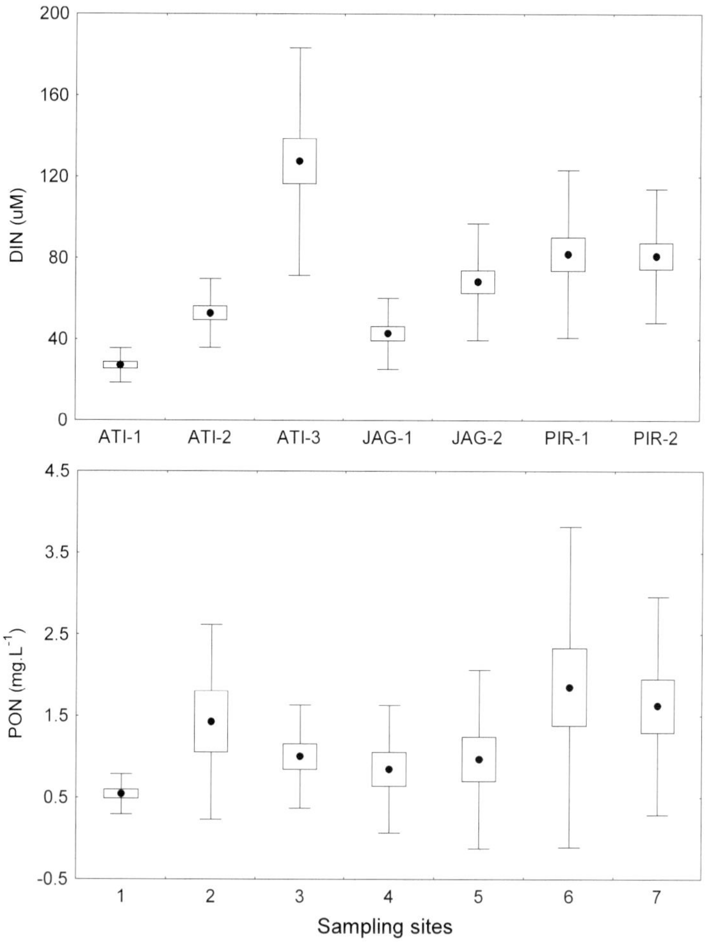

Fig. 9.14 Box and whisker plot of dissolved inorganic nitrogen (DIN), and particulate nitrogen (PN) at the sampling sites. Solid circles are averages, boxes are standard errors and bars represent the standard deviation.

Biogenic gases

Biogenic gases include dissolved oxygen (O_2) and free dissolved carbon dioxide (CO_2), which are involved in the photosynthesis–respiration process and define river metabolism (Ballester *et al.*, 1999). There were significant differences between the average dissolved oxygen concentrations of the sampling sites. Clearly, ATI-3, PIR-1 and PIR-2 had lower concentrations than other sites (Fig. 9.16). These sampling sites had, conversely, higher average concentrations of CO_2 and higher respiration rates (Fig. 9.16). Probably, in sampling sites where inputs of organic-rich waste waters are higher, respiration rates were higher, consuming more O_2 and producing more CO_2. Indirect evidence of this hypothesis is the significant inverse correlation between O_2 and DOC, which was used as a proxy of human impacts (Fig. 9.17). Although there was a lot of scatter, a significant direct correlation was also found between the respiration rate and DOC (Fig. 9.17). Another proxy of human impacts in the Piracicaba basin is the area occupied by urban centres. We found significant correlations between the urbanized area and the respiration rate ($Rs = 0.86$, $P < 0.01$), O_2 ($Rs = 0.71$, $P < 0.10$), and the DIC ($Rs = 0.96$, $P < 0.01$), supporting the hypothesis that untreated urban sewage and industrial effluents were responsible for changes observed in the distribution of biogenic gases (Ballester *et al.*, 1999).

The seasonal variability of biogenic gases also appears to be changed by human impacts. The O_2 seasonal variability is totally distinct between river sites more affected by waste water in comparison to those less affected (Fig. 9.18). There was small seasonal variability

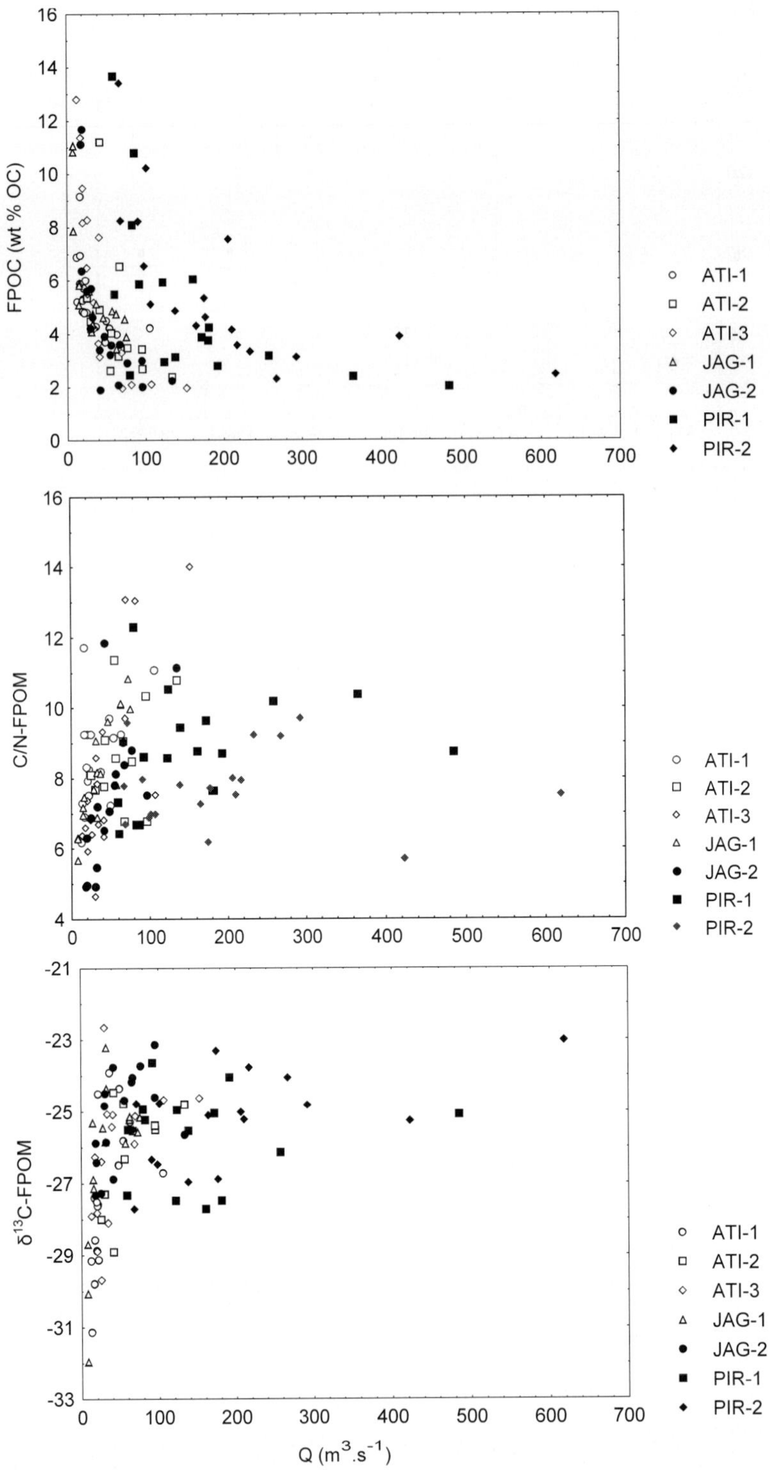

Fig. 9.15 Plot of river discharge (Q) *vs* fine particulate organic carbon (FPOC), C:N ratio and $\delta^{13}C$ values.

in the O_2 concentration at ATI-1, which is less affected by waste waters. Conversely, a strong seasonality in the O_2 concentration appeared at PIR-1, which is one of the most affected by waste waters (Fig. 9.18). During the dry season, when a smaller volume of water is available to dilute the load of waste waters, the O_2 concentration suffered a significant drop. During the wet season, river discharge increased and the sewage load was diluted, increasing the O_2 concentration (Fig. 9.18). The opposite trend occurs with CO_2 concentration, which is higher during the dry season and lower during the rainy season in sampling sites more affected by waste waters.

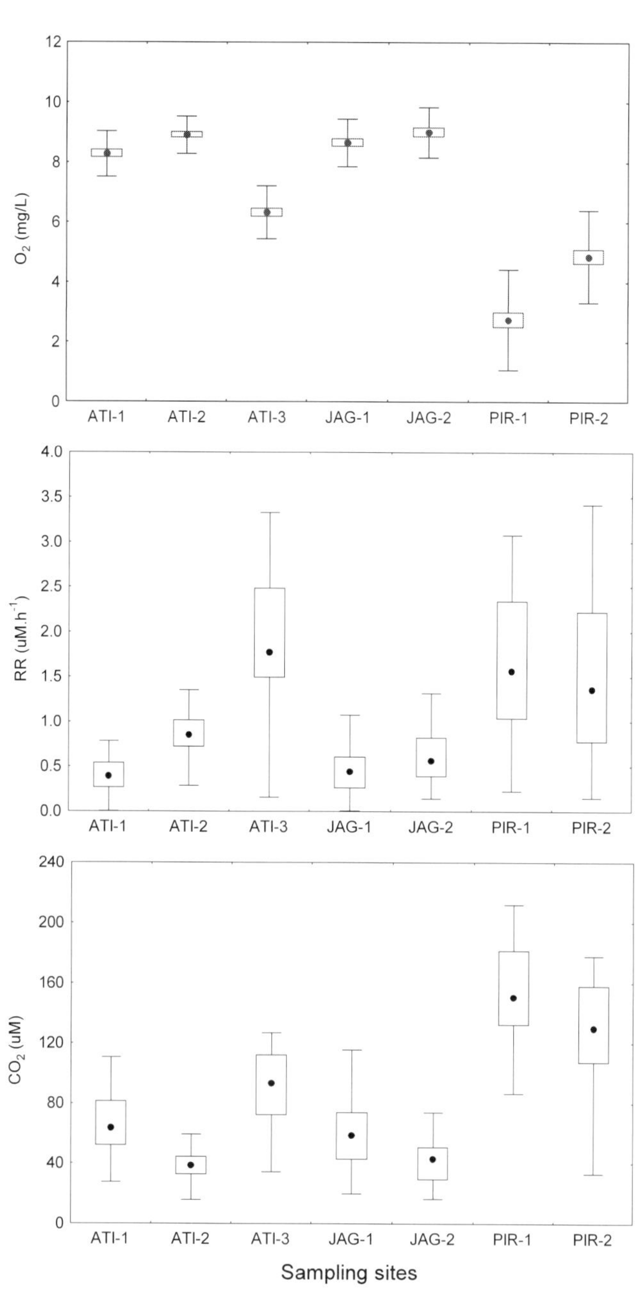

Fig. 9.16 Box and whisker plot of dissolved oxygen (O_2), respiration rate (RR), and free dissolved CO_2 in the sampling sites. Solid circles are averages, boxes are standard errors and bars represent the standard deviation.

RAINFALL COMPOSITION IN THE PIRACICABA RIVER BASIN

There are two factors that have the potential to alter the atmospheric composition in the Piracicaba basin (Lara *et al.*, 2001). One is agriculture, considering that approximately 80% of the basin is cropland. Approximately 31% of the total area is cultivated in sugarcane and burned every year, launching a large quantity of particles into the atmosphere. Second, there is strong industrial activity in the region, with more than 7000 industries. Using wet-only collectors we sampled rainfall for one year at four sites in the Piracicaba River basin (Fig. 9.1). The land use, urbanization and number of industries are distinct between sites (Table 9.7). The Campinas region had the highest number of industries and population density, followed by Piracicaba that also had a similar number of industries, but a

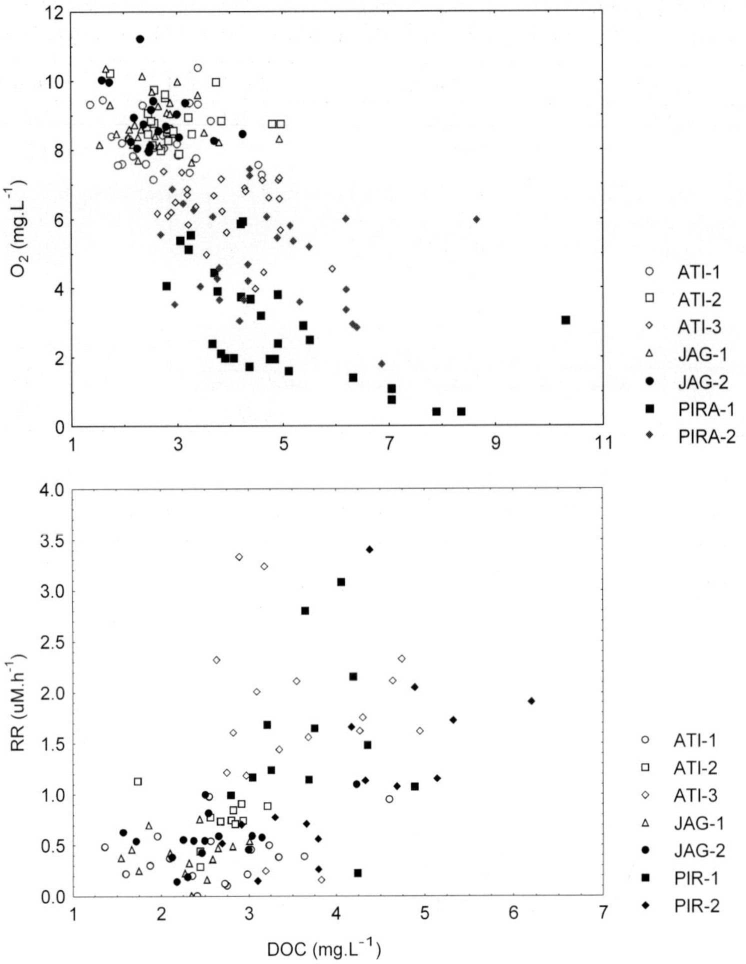

Fig. 9.17 Plot of dissolved organic carbon (DOC) *vs* respiration rate (RR).

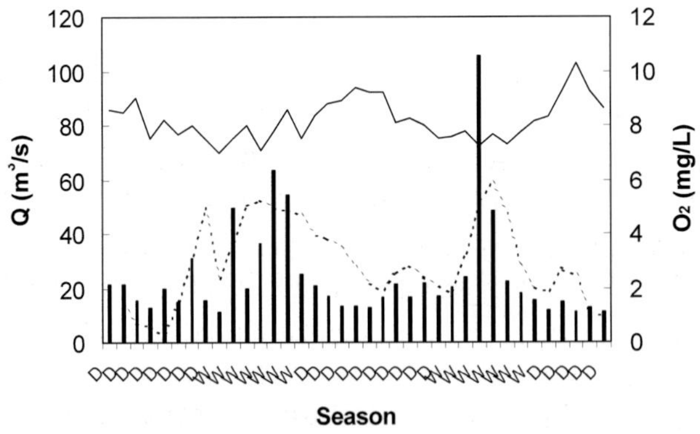

Fig. 9.18 Seasonal variability of dissolved oxygen concentration (O_2) at the sampling site ATI-1 (full line) and PIR-1 (dashed line). Bars represent the river discharge measured at the sampling site ATI-1.

population density equivalent to Bragança. Sugarcane dominates in Campinas, Piracicaba and Santa Maria, and pasture is the main land use in Bragança. This site also has the largest area covered by trees (forest and silviculture).

In all four sampling areas, the volume-weighted mean of pH was lower than 5.6, ranging from 4.4 to 4.6. These values are clearly characteristic of acid rain events. H^+ was the most abundant ion, followed in descending order by $NH_4^+ > SO_4^{2-} > NO_3^- > Cl^- > Ca^{2+} > K^+ > Na^+ > Mg^{2+}$. Annual deposition of total nitrogen was higher than the deposition measured in remote areas, similar to deposition values measured in the northeast region of

Table 9.7 Number of industries, population density (inhabitant km^{-2}) and land uses (%) in an area encompassed by a circumference centred on each sampling point and with a radius of 20 km (area of approximately 1200 km^2).

	Bragança	Campinas	Piracicaba	Santa Maria
Industries	623	1899	1088	79
Population density	212	1061	207	17
Urban	3.8	19.0	5.5	0.0
Forest	10.9	1.5	3.3	6.8
Sugarcane	0.0	61.1	79.4	63.0
Pasture	69.7	13.6	11.1	19.4
Silviculture	8.7	1.0	0.2	0.3
Annual crops	3.2	1.9	0.0	0.0
Water	2.5	1.6	0.5	10.6
Others	1.1	0.5	0.0	0.0

Table 9.8 Annual deposition rates of ions (meq m^{-2} $year^{-1}$) for each sampling site.

	Piracicaba basin (Lara *et al.*, 2001)	Northeastern USA (Munger & Eisenreich, 1983)	Remote area: northwest Europe (Galloway *et al.*, 1982)
H^+	39.6	70.0	12.0
NH_4^+	19.0	15.0	5.0
NO_3^-	18.5	30.0	5.0
SO_4^{2-}	20.0	65.0	15.0

the USA (Table 9.8). The annual deposition of H^+ was also higher than deposition measured in remote areas and similar to the deposition estimates found in Asia. Sulphate annual deposition was greater than in remote areas but less than other regions, like Asia and USA.

Principal component analysis indicated four qualitative sources to the precipitation in the Piracicaba basin:

(1) influence of soil dust associated with Ca^{2+} and Mg^{2+};
(2) biomass burning, more specifically sugarcane burning, which is associated with K^+ and Cl^-;
(3) industrial emissions, referred to as H^+ with significant correlations with SO_4^{2-} and NO_3^-;
(4) organic acidity factor, with DOC correlated with H^+ (Lara *et al.* 2001).

In the Bragança region, soil dust and organic acidity were the first and the second most important factors explaining the variability in rainfall composition. In Campinas, besides soil dust, industrial emissions were equally important, with sugarcane burning also present, but less important. Conversely, in Piracicaba and Santa Maria, sugarcane burning and soil dust were the most important factors. The fact that soil dust and sugarcane burning are important factors to explain the composition of rainfall in the Piracicaba basin is not surprising. Approximately 85% of the area in the basin is cropland. Consequently, every year, a relatively large amount of bare soil is exposed. In addition, urban areas are growing very fast in the basin. This activity exposes a good deal of bare soil, also facilitating the suspension of soil particles in the air. Sugarcane burning is estimated to input approximately 100 000 t of organic matter, or 50 000 t of carbon, to the atmosphere annually. Industries are another important and constant source of N and S to the atmosphere.

Although rain acidity was found in all four sampling sites, the origin of this acidity appears to be different in each site (Lara *et al.*, 2001). An association of DOC and H^+ is found at the Bragança station, in conjunction with the fact that most of the remaining original forests in the basin and extensive areas of silviculture are the dominant land use in this part of the basin. This suggests that biogenic emissions could be a significant source of acidity in this region. Conversely, at the Campinas station the main source of acidity appears to be SO_4^{2-} and NO_3^- provided by the intense industrial activity observed in this region. At the Piracicaba and Santa Maria stations, sugarcane burning is a significant source of NO_x, which, converted to NO_3^-, accounts for the acidity of rain.

CONCLUSIONS

The Piracicaba River basin has suffered profound disruptions to its natural cycles. Clearly, domestic sewage and industrial effluents inputs have overwhelmed the natural ability of the Piracicaba River to clean itself. As a consequence, significant changes in the riverine water composition in the more disturbed sites has occurred. Not only has the water composition changed, but also the distribution of elements during the year. These changes have also severely affected biodiversity in the basin. The Piracicaba River was known for its abundance of fish and mussel species. Today, few species of fish survive in the most disturbed areas of the Piracicaba. On the other hand, agricultural activities, which promote exposure of bare soils and annual burning of sugarcane fields, together with industrial emissions, are responsible for the alterations in the atmospheric composition. Acid rain is a common event all over the basin. Nitrogen deposition is already similar to the most developed regions in the world.

Environmental conditions in the Piracicaba River basin are far from acceptable. However, things are beginning to change through a new institutional structure for water resources management established in São Paulo State in 1991. The State was divided into 22 management units, which are roughly equivalent to natural basins. A committee manages each unit. The Piracicaba River basin committee was one of the first to be established (1995) and is deeply concerned about the environmental problems of the basin. One of the acts of this committee and others of the State was to establish a tax for the use of untreated water. This new tax is generating financial resources to be applied to environment mitigation. The water quality problems are relatively easy to solve, considering that point-source pollution is the main cause of the problems. Non-point source pollution, which is caused mainly by sugarcane and is more difficult to eliminate, is secondary in the Piracicaba River basin. Regarding the atmosphere, sugarcane burning is one of the main problems. The State government gave the sugar mill industries eight years to stop burning sugarcane. Overall, still more important it is the fact that society is demanding such actions. It is clear that society is mature enough and ready to start to pay the costly price of the cleanup process in the Piracicaba River basin.

REFERENCES

Ballester, M. V., Martinelli, L. A., Krusche, A. V., Victoria, R. L., Bernardes, M. & Camargo, P. B. (1999) Effects of increasing organic matter loading on the dissolved O_2, free dissolved CO_2 and respiration rates in the Piracicaba River basin, southeast Brazil. *Water Research* **33**, 2119–2129.

Filoso, S., Martinelli, L. A., Lara, L. B., Krusche, A., Ballester, M. V. & Krusche, A. (in press) The nitrogen budget of a developed meso-scale river basin. *Biogeochemistry* .

Galloway, J. N., Likens, G. E., Keene, W. C. & Miller, J. M. (1982) The composition of precipitation in remote areas of the world. *Journal of Geophysical Research* **87**, 8771–8786.

Krusche, A. W., de Carvalho, F. P., Moraes, J., de Camargo, P. B., Ballester, M. V., Martinelli, L. A. & Victoria, R. A. (1997) Spatial and temporal variability of selected parameters of water quality in the Piracicaba River basin. *Journal of the American Water Research Association* **33**, 1117–1123.

Lara, L. B. L. S., Artaxo, P., Martinelli, L. A., Victoria, R. L., Camargo, P. B., Krusche, A., Ayers, G. P., Ferraz, E. S. B. & Ballester, M. V. (2001) Chemical composition of rain water and anthropogenic influence in the Piracicaba River basin. *Atmospheric Environment* **35**, 4937–4945.

Martinelli, L. A., Krusche, A., Victoria, R. L., Camargo, P. B., Bernardes, M. C., Ferraz, E. S., Moraes, J. M. & Ballester, M. V. (1999a) Effects of sewage on the chemical composition of Piracicaba River, Brazil. *Water, Air and Soil Pollution* **110**, 67–79.

Martinelli, L. A., Ballester, M. V., Krusche, A., Victoria, R. L., Camargo, P. B., Bernardes, M. & Ometto, J. P. H. B. (1999b) Land-cover changes and $\delta^{13}C$ composition of riverine particulate organic matter in the Piracicaba River basin (southeast region of Brazil). *Limnology and Oceanography* **44**, 1826–1833.

Meybeck, M. (1982) Carbon, nitrogen and phosphorus transport by world rivers. *American Journal of Science* **282**, 401–405.

Moraes, J. M., Pellegrino, G. Q., Ballester, M. V., Martinelli, L. A., Victoria, R. L. & Krusche, A. V. (1998) Trends in hydrological parameters of a southern Brazilian watershed and its relation to human induced changes. *Water Resources Management* **12**, 295–311.

Munger, J. W. & Eisenreich, S. J. (1983) Continental-scale variations in precipitation chemistry. *Environmental Science and Technology* **17**, 32–42.

Ometto, J. P. H. B., Martinelli, L. A., Ballester, M. V., Gessner, A., Krusche, A., Victoria, R. L. & Williams, M. (2000) Effects of land use on water chemistry and macroinvertebrates in two streams of the Piracicaba River basin, southeast Brazil. *Freshwater Biology* **44**, 327–337.

Pellegrino, G. Q., Moraes, J. M., Guandique, M. E. G., Ballestre, M. V. R., Victoria, R. L. & Martinelli, L. A. (2001) Análise espaço-temporal de componentes hidroclimáticos na bacia do rio Piracicaba, SP. *Revista Brasileira de Agrometeorologia* **9**, 125–135.

Williams, M. R., Filoso, S., Martinelli, L. A., Lara, L. B.& Camargo, P. B. (2001) Precipitation and river water chemistry of the Piracicaba River basin, southeast Brazil. *Journal of Environmental Quality* **30**, 967–981.

CHAPTER 10

THE ECOHYDROLOGY OF THE PARAÍBA DO SUL RIVER, SOUTHEAST BRAZIL

Carlos Eduardo Veiga de Carvalho & João Paulo Machado Torres

HISTORICAL OCCUPATION OF THE PARAÍBA DO SUL DRAINAGE BASIN

The occupation of the Paraíba do Sul River basin is closely related to the history of Brazil. It began in the first part of the sixteenth century when the majority of the basin was occupied by several Indians tribes that were hunted to serve as slaves for the Portuguese colonists (MME *et al.*, 1995).

Small villages began to appear in the upper and in the lower basin near the river mouth at the beginning of the seventeenth century. Complete colonization was completed by the creation of Campos dos Goytacazes State and the foundation of the city of Campos in the middle seventeenth century. Sugarcane plantations began to develop in the region at this time, and more cities appeared in the upper basin in São Paulo State (MME *et al.*, 1995).

With the discovery of gold and precious gems in the interior of Minas Gerais State the population increased between the seventeenth and eighteenth centuries. During this gold rush period the Paraíba do Sul River basin remained almost unexplored, although it served as the main pathway between the mines and the shoreline where the gold was exported by ship to Portugal. To avoid gold and precious stones smuggling, the King of Portugal forbade the construction of new roads and also the navigation along the Paraíba do Sul River. The increase of gold mining activities in the eighteenth century, however, forced the opening better roads and established links between the mines and the cities of Rio de Janeiro and São Paulo.

Gold mining activities began to decline at the time Brazil gained its independence. Agriculture again assumed predominance in the region, with sugarcane and coffee being the favoured crops. Beginning at the end of the eighteenth century and intensifying through the nineteenth century, coffee plantations were responsible for the final occupation of the Paraíba do Sul basin and all land between the Serra do Mar and the Serra dos Órgãos (MME *et al.*, 1995). The abolition of slavery in 1888 and continual exhaustion of the region's soil fertility brought down the coffee culture and lead to a systematic substitution of agriculture by the extensive cattle farming. This change promoted a large decrease in the local population and decline of the local economy. Although the coffee plantations crisis impoverish the basin, a very developed transport infrastructure with railroads and roads remained to serve the basin (MME *et al.*, 1995).

In the beginning of the twentieth century industrial activities began in areas surrounding the basin. The hydrological potential of the basin also began to be exploited with the construction of small hydrological plants. With the opening of the National Siderurgic Plant (Companhia Siderúrgica Nacional) in 1946 at Volta Redonda city, and the expansion of the São Paulo industrial park toward the Paraíba do Sul River basin, the Rio de Janeiro/São Paulo industrial axis was established (MME *et al.*, 1995).

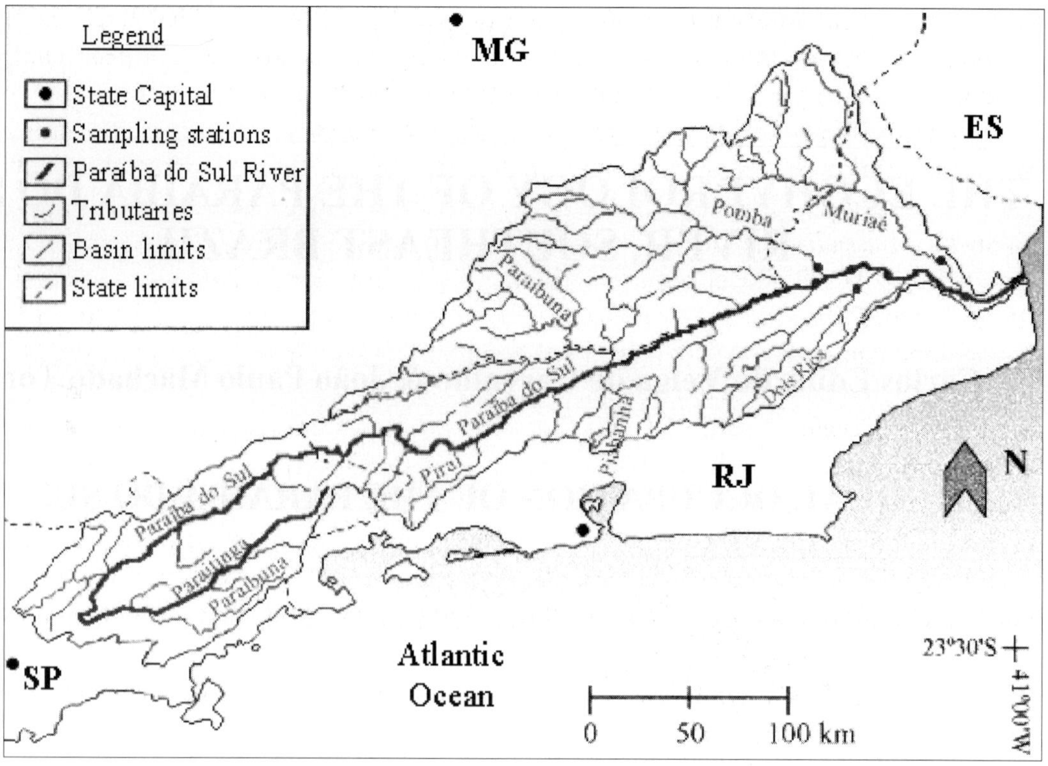

Fig. 10.1 The Paraíba do Sul River basin (modified from Figueiredo, 1999).

THE PARAÍBA DO SUL RIVER BASIN

The Paraíba do Sul River (PSR) is a medium sized river, approximately 1145 km long and with an area of 55 400 km^2. The river drains the most industrialized states of Brazil: Rio de Janeiro, São Paulo and Minas Gerais. The climate is subtropical with a well characterized precipitation regime where highest rainfall occurs between November and January. These heavy rains occasionally cause large floods of the Paraíba do Sul River. The period between June and August is the driest of the year. Maximum river flow is observed during summer (4380 m^3 s^{-1}) and the lowest discharges generally occur in the winter (180 m^3 s^{-1}) (DNAEE, 1983, Rosso *et al.*,1991). Water from the PSR is very important to the cities located in its basin and is used in agriculture, industry and for human consumption (over 20 million people live in the basin). On the other hand, the river receives effluent from all these activities, generally without efficient treatment.

The main tributaries of the lower PSR are the Paraítinga, Paraibuna (São Paulo State—SP), Piraí, Paraíbuna (Minas Gerais State—MG), Piabanha, Pomba, Dois Rios and the Muriaé rivers (Fig. 10.1). The geology of the PSR basin is dominated by ancient formations from the pre-Cambrian, mainly from the Arquean period. Modest occurrences of Algoquean formations are also present in the basin limits, in Minas Gerais State and in the upper Paraíbuna basin. Tertiary formations are also observed in a narrow area along the shoreline and in the medium portion near the city of Cachoeira Paulista in São Paulo State widening until the cities of Resende and Volta Redonda in Rio de Janeiro State (RJ) (Figueiredo, 1999). The lower basin is dominated by gneiss in the Itaocara region. Downstream of São Fidélis city the relief becomes flat and is composed of Quaternary (Holocene) fluvial deposits. Land use is basically urban development, agriculture, extensive cattle farming and small fragments of natural vegetation (Projeto RADAMBRASIL, 1983).

The original soils from the Paraíba do Sul River basin were considered to be very fertile, but they were easily exhausted (Figueiredo, 1999). Above these soils developed a very rich tropical forest known as "Floresta de Mata Atlântica", that today is practically extinct; only 5% of the original forest remains at steep scarps. Most of the land that was covered by this tropical forest in the past is totally devastated and now covered by pasture.

Along the upper basin the predominant soils are *Latossolos*, *Podzólico Vermelho-Amarelo* and *Cambissolos* (Brazilian soil classification system), with clay and sand/clay texture and variable thickness. The upper part of the medium portion is dominated by *Latossolos Vermelho-Amarelo*, the flood-plain areas are composed of clay/sand and clay/silt organic sediments (turf). The lower portion of the middle basin is dominated by *Latossolos Vermelho-Amarelo* and *Podzólico Vermelho-Amarelo*. Along the upper region of the lower PSR basin the predominant soil type is *Cambissolos*, although *Latossolos* were also observed. The Muriaé and Pomba River basins are dominated by *Latossolos Vermelho-Amarelo* and *Podzólico Vermelho-Amarelo*. At the mouth of the basin a delta complex composed of a group of sedimentary environments related to the last sea-level regression is observed. The Paraíba do Sul delta forms a vast plain with several abandoned and semi-abandoned channels, lakes, coastal lagoons, mangroves and swamps. The predominant soils are *Aluviais*, *Eutróficos*, *Álicos* and *Glei Pouco Húmico Álico*. The texture of all of these soil types is predominantly silt/clay. Along the shoreline the "restingas" are composed of *Regossol* soils, constituted of unconsolidated clastic materials, with grass and small bushes, other types of soils are also observed along the littoral-like *solos hidromórficos gleisados*, *Solonetz* and *Glei Húmico*.

In each of the three PSR basin regions (upper, medium and lower) there is a predominance of microclimates. According to DNAEE (1983), that used the Köppen classification, the upper basin is dominated by the climate types *Cwb* (tropical altitude with fresh and rainy summer) and *Cfb* (subtropical with fresh summer). Along the medium portion the *Cwa* (tropical altitude with hot and rainy summer) type dominates, although the predominant climate in mountain regions is *Cwb* and *Cfb*, and in lower regions *Cfa* (subtropical with hot summers) and *Aw* (hot and humid with rainy summer). In the lower basin *Aw* predominates, although in the upper basins of some of the PSR tributaries the *Cwa* type can be observed.

The temperature is higher in the medium and lower PSR basin, with maxima averaging between 32 and 34°C. Lower temperatures are observed in the upper basin with the minima averaging between 14 and 16°C. The average temperature for the entire basin is between 26 and 28°C (DNAEE, 1983).

Annual rainfall rates along the PSR basin occur over a wide range (1000–3000 mm). The lower basin and the bottom of the middle basin presented the lowest precipitation averages, with annual values ranging from 1000 to 1500 mm. The highest values were observed in the mountain region of Itatiaia and Serra dos Orgãos, both in the Rio de Janeiro State, and in the Serra do Mar in São Paulo State where the annual averages generally surpass 2000 mm, sometimes reaching 3000 mm (DNAEE, 1983).

PSR waters are widely used for human consumption, industrial use, irrigation, hydroelectric power plants and recreation proposes. The total water volume removed for domestic uses is estimated at 60 $m^3 s^{-1}$. Other uses like industry and agriculture did not have an official estimation of the volume utilized across the PSR basin.

HYDROLOGY AND PARTICULATE MATTER MASS BALANCE

Mass transfer within the lower Paraíba do Sul River drainage basin (Fig. 10.2) and the contribution of its tributaries has been the focus of several recent studies. The highest metal loads were related with the highest water flows and consequently with the highest suspended particulate matter loads (Table 10.1). The month of highest total transported load

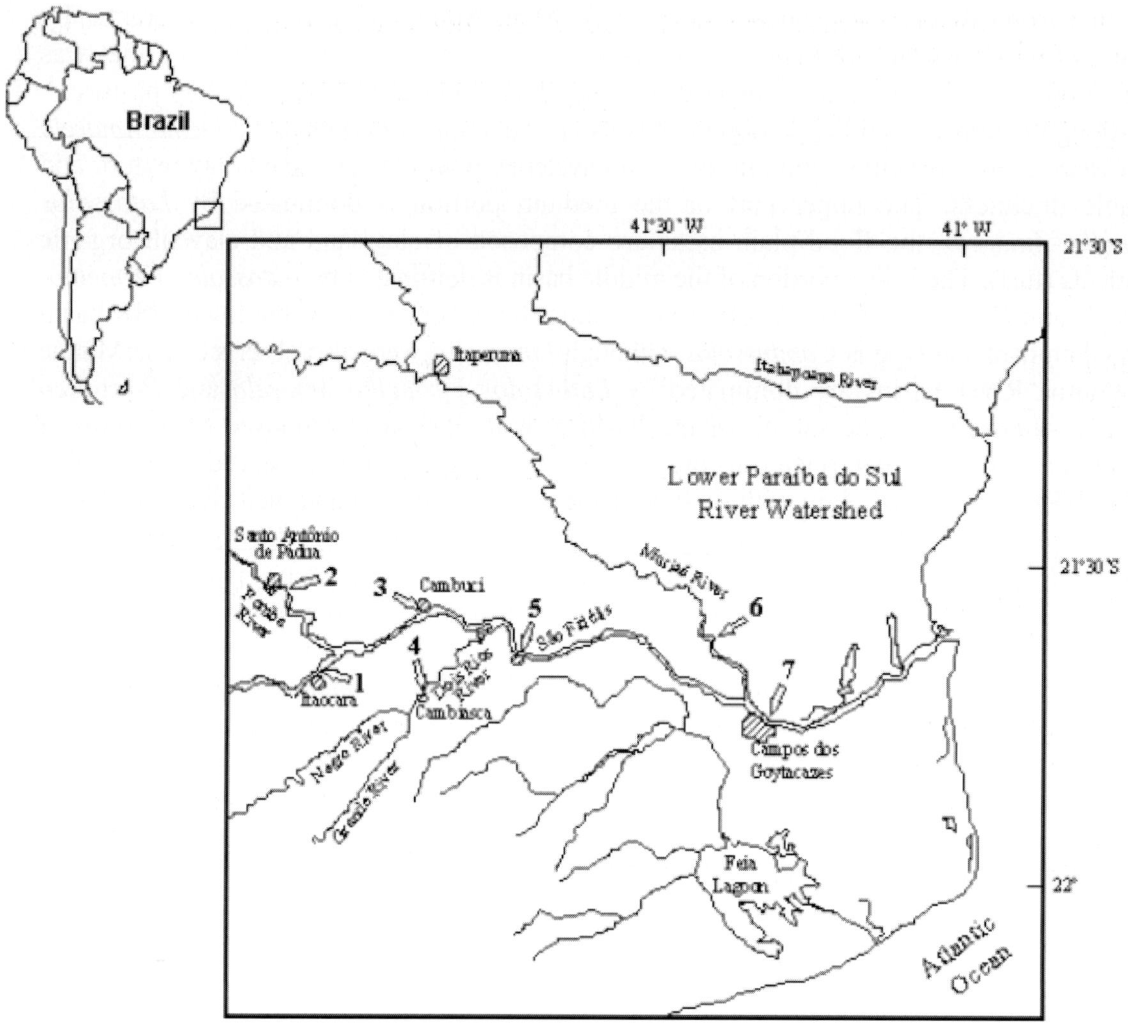

Fig. 10.2 Sampling sites at the lower Paraíba do Sul River basin.

represented 34–66% of the total metal transported per year. The total load transported during the entire rainy season ranged from 82 to 93% of the total transported load. The importance of the tributary input for the total transported load of the Paraíba do Sul River was generally lower than 10%. Although in some months, where the Paraíba do Sul River presented low water flux and strong local rain, this importance reached 40% (Carvalho *et al.*, 1999; Salomão *et al.*, 2000).

The total flow of heavy metals transported by the Paraíba do Sul River drainage basin was based on the flow calculations at Station 7 (Campos). Comparing the two periods it is possible to observe that in the second period (April 1995–March 1996) loads were more than 100% higher than in the first period (April 1994–March 1995). This fact was observed not only for suspended particulate matter but also for all the studied metals. The explanation is probably linked to annual rainfall. The first period (April 1994–March 1995) was characterized by very little rainfall (733 mm) and could be considered a dry year. Although the second period (April 1995–March 1996) presented higher rainfall (779 mm), it was also below the average rainfall for the region (890 mm $year^{-1}$). This observation can be confirmed from Fig. 10.3 where the maximum, average and minimum discharge curves calculated by DNAEE, based on information obtained in measurements during the period of 1934–1978, are compared with the values measured by this study. Based on these two observations it is possible to conclude that the sampling period of this study is representative of years of low rainfall and river discharge (Carvalho *et al.*, 1999; Salomão *et al.*, 2000).

Table 10.1 Percentile distribution of heavy metals and suspended particulate matter loads transported by the Paraíba do Sul River during 1995–1996.

Year 1995/96	Cu (%)	Cr (%)	Zn (%)	Mn (%)	Fe (%)	SPM (%)
April[a]	5.5	21.5	26.9	18.7	14.1	13.8
May[b]	1.6	3.8	4.8	5.3	4.7	4.2
June[b]	2.1	4.0	3.4	5.4	4.5	4.3
July[b]	1.5	2.8	1.4	2.7	1.9	2.1
August[b]	0.9	1.5	1.2	1.2	0.4	1.0
September[b]	0.3	0.6	1.1	2.0	0.6	0.9
October[b]	0.6	1.1	0.8	1.0	0.5	0.8
November[a]	5.3	3.8	4.0	4.9	3.9	4.5
December[a]	7.7	8.9	12.3	5.7	5.3	6.5
January[a]	6.7	4.9	7.8	6.1	5.6	5.3
February[a]	66.0	42.8	34.0	42.8	55.5	52.2
March[a]	1.7	4.0	2.3	4.1	3.0	4.4
Rainy season	92.9	85.9	87.3	82.3	87.4	86.7
Dry season	7.1	14.1	12.7	17.7	12.6	13.3

[a] Rainy season months.
[b] Dry season months.

The variation in monthly loads was strongly correlated with rainfall and consequently with surface runoff. In some months the increase in load between two consecutive Paraíba do Sul River sampling stations (e.g. stations 3 and 5) could not be explained by tributary contributions (e.g. Dois Rios). Thus bottom sediment re-suspension must also be an important process (Salomão *et al.*, 2000).

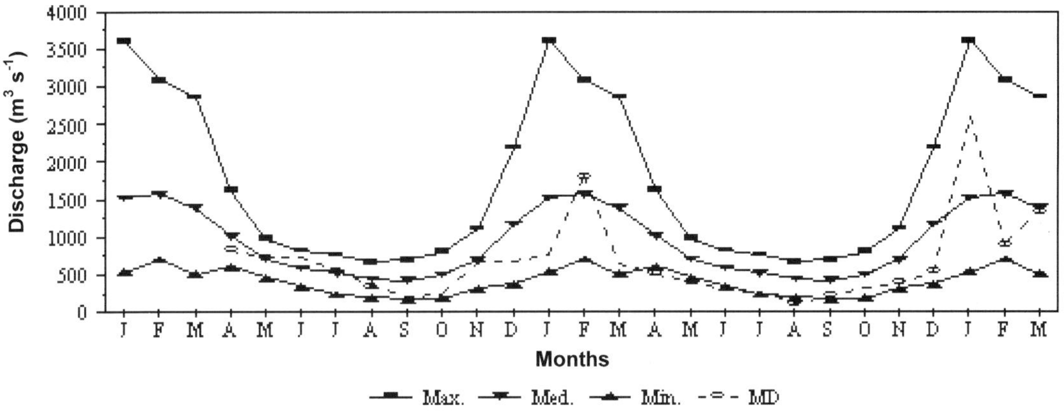

Fig. 10.3 Maximum, average and minimum discharges curves calculated by Departamento Nacional de Águas e Energia Elétrica (DNAEE, 1983) for station 7 (1934–1978) compared with the measured discharges during this study (MD).

NUTRIENT AND PHYSICO-CHEMICAL PARAMETER DYNAMICS

Dissolved organic carbon (DOC) distributions along the lower portion of the Paraíba do Sul River were initially studied by Calasans *et al.* (1995). This work evaluated the dynamics and transport of bulk organic carbon by the Paraíba do Sul River. Temporal variations did not show a clear seasonal trend. Nevertheless a higher homogeneity was observed during the periods of high-water discharge (summer) in contrast to the periods of lower discharge that showed a higher heterogeneity (winter). During this preliminary study losses and gains of DOC between two sampling sites along the river were observed. This pattern was associated with riverine processes and with the diffuse organic carbon sources (anthropogenic and natural). The Paraíba do Sul River presented DOC concentrations between 3.02 and 6.18 mg C l^{-1} with an average of 4.29 mg C l^{-1}. The observed values were inside the wide range described for tropical rivers (2–15 mg C l^{-1}) according to Martins *et al.* (1991). This study also calculated DOC fluxes over a six-month period. The highest flux was observed during the summer (February) when fluxes were up to 7898 g s^{-1} and the lowest flux was observed in the winter time (July) when fluxes dropped to 782 g s^{-1}, only 10% of the maximum observed flux.

Inorganic nutrient dynamics in the lower Paraíba do Sul basin and its tributaries were initially studied by Carvalho *et al.* (1995). This work described the variation of inorganic nutrients (NO_2, NO_3, NH_4, SiO_2, PO_4) and physico-chemical parameters (DO, pH, Eh and conductivity) in three different water discharge periods: low, medium and high (180, 700 and 1800 m^3 s^{-1}, respectively).

Ammonia and phosphate presented similar distributions with the lowest concentrations at lowest discharge. The highest concentrations were observed in the medium water discharge period (10.85 and 1.90 μM, respectively) and in the high-water discharge a small decrease was observed (9.94 and 1.36 μM, respectively). This behaviour was possibly due to the input of these elements from the drainage basin soils by surface runoff. At the beginning of the rainy season the surface runoff input is amplified by the lower water volume of the river. During the higher flows, although the surface runoff is still transporting these elements to the river, the water volume is much higher causing a dilution of the surface runoff input signal (Carvalho *et al.*, 1995).

Nitrite presented an opposite trend with the highest concentrations in the low discharge period (2.52 μM) and the lowest values in the high and medium discharge periods (0.25 μM). This behaviour was attributed to the urban sewage source of this nutrient (Carvalho *et al.*, 1995). Considering that during the low discharge regime the water volume is small, the effect of this source is amplified; this trend was also observed in another drainage basin in the south of Rio de Janeiro State by Figueiredo (1995).

Nitrate did not present any clear trends and was nearly constant during the three studied periods. Concentration peaks were observed only at some stations; the highest concentration was measured in the medium discharge period (41.9 μM) and the lowest in the lower discharge period (22.45 μM). The low observed values were probably due to the retention of this element by riverine sediments, a process described by Svendensen & Kronvang (1993). The observed nitrate peaks were attributed to the runoff from fertilized soils and/or to the flood of adjacent areas (Carvalho *et al.*, 1995).

Silica presented a special distribution pattern with lowest concentrations occurring during the low- and medium-water discharge periods (91 μM) and the highest concentrations during the high-water discharge period (223 μM). The authors associated this behaviour with the groundwater cycle. During the dry season groundwater has more time to interact with the soil/bedrock system, consequently SiO_2 concentrations increase. During periods of high rainfall this accumulated SiO_2 is carried to the river system (Carvalho *et al.*, 1995).

Physico-chemical parameters presented the following temporal variations. Dissolved oxygen presented the lowest (4.2 mg l^{-1}) and highest (7.5 mg l^{-1}) concentrations in the up-stream station (Itaocara) in the low and high period of discharge respectively. The pH varied from 7.09 to 9.16; the highest values were observed during the low discharge period. The Eh was always positive indicating an oxidizing system. Higher values were measured during the high discharge period. This behaviour probably explains the low nitrite concentrations observed, considering the high rate of the nitrification processes. Conductivity, on the other hand, presented similar values for the low and medium discharge periods with smaller values during the high discharge period; the observed range varied from 48 to 101 μS cm^{-1}. This behaviour was attributed to dilution effects caused by the rainfall increase (Carvalho *et al.*, 1995).

Recently, Figueiredo (1999) studied in more detail carbon and nitrogen transport in the lower portion of the Paraíba do Sul River basin. According to this work the main sources are: (a) forests, (b) urban sewage, (c) sugarcane plantations, (d) sugarcane industry, (e) aquatic macrophytes, (f) flood plains, and (g) flood river islands. Seasonal differences were observed in the C and N behaviour and transport in the lower Paraíba do Sul River basin as a function of discharge magnitude. These differences were identified as determinant factors in the C and N transport, fluxes and associated biogeochemical processes.

During the low discharge period phytoplankton biomass and its contribution to the suspended particulate matter (SPM) grow, driving up dissolved inorganic carbon (DIC) and nitrite concentrations and driving down reduced inorganic nitrogen concentrations as well as the dissolved organic carbon (DOC) and nitrogen (DON). The fine sediment contribution to the total SPM bulk is augmented in these period. This behaviour suggests that during the lower discharges: (a) the fluvial hydrochemistry is influenced of the groundwater flux enriched in dissolved inorganic forms of C and total N, (b) significantly higher organic matter decomposition rates occur, decreasing the described organic forms and increasing alkalinity, (c) sediment adsorption processes, coarse SPM sedimentation and phytoplankton biomass increase, which probably explains the particulate carbon and nitrogen increase, and (d) the sugarcane industry and agricultural activities input (e.g. the use of fertilizers, sugarcane industry sewage, etc.) is intensified and cannot be neglected when C and N input is considered, with special attention to nitrate (Figueiredo, 1999).

During the high discharge period less phytoplankton biomass was observed. At the same time, coarse particulate matter became more important to the total SPM transported. In this period C and N concentrations increased in the dissolved organic phases as well as the reduced inorganic N forms. On the other hand, the inorganic dissolved C and N forms decreased in concentration. There was also a drop in the total particulate carbon and nitrogen concentrations, although the SPM fluxes increased significantly. This trend suggested a higher flux during the medium- and high-water discharge period—up to 90% of the annual flux is transported during these periods. This scenario probably indicates that: (a) the surface runoff carries more organic matter and younger water (with shorter residence time in the soils) to the river, (b) the erosion of agricultural soils cannot be neglected, as it increases the SPM transported, (c) the river water interacts strongly with the flood plains and islands, indicated by the inorganic reduced nitrogen forms, (d) C and N contributions to the estuarine region are higher as shown by the larger fluxes, and (e) the sugarcane input seems to be less important, probably due to dilution effect caused by the high water volume and/or to the reduction of the sugarcane industry and agriculture activities (Figueiredo, 1999). The medium discharge period was characterized by an intermediate behaviour between the high and low discharge periods.

These results also indicate that the influence of sugarcane and alcohol production increase the C and N fluxes in the lower basin, mainly during the lower discharge periods. Although the influence of urban domestic sewage input was detected in the vicinities of urban centres, this influence could not be estimated. The terrestrial inputs from the surrounding soils were larger than those from insular soils, which seem to act mainly as biogeochemical barriers for C and N fluxes. The role of macrophytes in the input and metabolization of C and N was unclear. Phytoplankton was identified as an important component of the biogeochemical processes mainly during the low discharge period. There were some indications that nitrification processes related to agricultural management may be mitigating the high nitrification rates in the flood plains and agricultural soils which are subjected to fire and irrigation (Figueiredo, 1999).

HEAVY METAL DYNAMICS

Previous work in the middle portion of the Paraíba do Sul River showed critical concentrations of some elements (Pb, Cu, Cr and Zn) in suspended particulate matter (Malm, 1986). The lower Paraíba do Sul River drainage basin, however, has little or no metal contamination in the particulate matter and sediments (Carvalho *et al.*, 1999, Molisani *et al.*, 1999). The main process that seems to regulate the seasonal variation is dilution caused by the change in the source of the suspended particulate matter (runoff input *vs* industrial input).

Temporal and spatial variability of particulate metal concentrations (Cu, Cr, Zn, Mn and Fe) were investigated in the lower drainage basin of the Paraíba do Sul River. The results showed that spatial variability was not significant for any of the studied metals. Temporal variations, however, seem to be considerable. In general, two distinct types of behaviour were observed for particulate heavy metals: (a) metal concentration increases together with water discharge (Fe and Cu), and (b) concentrations decrease with increasing water discharge (Zn, Cr and Mn) (Carvalho *et al.*, 1999). Fe and Cu behaviour is probably due to the strong association of these metals with surface runoff, although their sources seem to be distinct. Iron probably originates from the regional soils rich in iron oxides, and Cu is possibly associated with the large-scale use of copper fungicides in the sugarcane plantations. The opposite trend observed for Zn, Cr and Mn probably reflects the importance of industrial and urban effluents as a secondary source of these elements for the system. Their behaviour is probably associated with the dilution caused by the input of a metal-poor suspended matter coming from surface runoff during the rainy season (Carvalho *et al.*, 1999).

Recently Salomão (1999) studied temporal and spatial variation of dissolved and particulate heavy metals in the lower basin of the Paraíba do Sul River. Higher concentrations of Cd and Zn were found in the particulate phase at the upstream station of Itaocara than at the downstream Campos station. This fact was probably related to the influence of the pollution sources from the middle Paraíba do Sul basin.

HEAVY METAL CONTAMINATION IN SEDIMENTS

Torres (1992) studied Cd in the Paraibuna River, the first important tributary of the left margin of the Paraíba do Sul River. According to this work, the Paraibuna has received effluents contaminated with Cd since 1980, when a non-ferrous smelting plant (Paraibuna de Metais) was built in Juiz de Fora (MG). More than 15 ppm of Cd were measured in the small creek near the plant that receives most of the liquid effluents (Torres, 1992). Downstream, contamination is progressively reduced. This contamination, when expressed using the geoaccumulation index, proposed by German Muller in 1979, shows a typical point source pattern (class 5–6). However, when the Paraibuna reaches the Paraíba do Sul River, the Cd concentrations in the sediments are only slightly above the natural background (0.3–0.6 ppm, classes 0–1 not polluted to moderately polluted).

Carvalho (1997) and Molisani *et al.* (1999) studied the sediment heavy metal geochemistry in the lower PSR. The total concentration of those elements showed a decrease in bottom sediment concentrations when compared to the middle basin sediments (Azcue, 1987). Molisani (1999) used sequential extraction to identify the chemical forms of heavy metals accumulation in riverine, estuarine and flood-plain sediments along the lower PSR. The total concentration of the heavy metals confirmed the decrease of values of metals compared to the middle basin, except for Cd that had high values at the São Fidélis station. Compared to other work in this area total Fe concentrations were low in bottom sediments. The sequential extraction of the <63 μm size fraction samples showed that: Cd and Mn were mainly associated with exchangeable (60%) and reducible phases (18%), Pb with oxidable (44%) and residual phases (38%), and Zn was distributed in all phases but the mobile phases prevail (53%). Fe, Ni, Cu and Cr occurred in the sediments mainly in the residual phase (>75%). The high occurrence of mobile phases for Cd, Zn and Pb in bottom sediments can be associated with pollution sources in the middle basin but we must also consider local contributions. The distribution of Mn reflects its geochemical mobility.

The weakly bound fraction of all analysed metals was also smaller than in the middle PSR. According to Azcue (1987), the middle PSR (PSR–Guandú river system) receives high anthropogenic inputs, presenting the following weakly bound percentages: Cu 70%, Cr 45%, Zn 70%, Mn 70%, Fe 20%, and our results showed only: Cu 17%, Cr 7%, Zn 3%, Mn 11%, Fe 1.5%. Although the methodologies used in the studies differ, this downstream decrease strengthens the observation that sediments of the lower basin are mostly uncontaminated and of low risk for the environment and human health. However, they suggest that a significant amount of heavy metals is being trapped in the river.

The results showed that, with exception of Cr, all metals presented the same distribution pattern, with higher concentrations in the fluvial portion, decreasing toward the estuary of the river. Comparing the obtained values with the upper basin of the PSR, under higher anthropogenic influence, a decrease of some metal concentrations was observed, mainly in the weakly bound fraction. This observation and the absence of large metal pollution sources results in a decrease of heavy metal concentrations in the lower basin and estuary, and also explains the lower heavy metal concentrations found by Carvalho *et al.* (1993) in the adjacent continental shelf sediments.

ORGANIC MICROPOLLUTANTS IN SEDIMENTS OF THE MIDDLE PARAIBA DO SUL RIVER

Since the middle 1980s, the use of organochlorine pesticides has been restricted in Brazil. Some studies during the last decades showed that in Brazil, as in other parts of the world, organic micropollutant contamination derives not only from agricultural fields but also urban areas (Tommasi, 1985; Luchini *et al.*, 1981). Polychlorinated biphenyls (PCBs) are used in closed systems like electric transformers as a dielectric fluid. Although their release to the environment is not frequent, accidents involving leakage and direct contamination of surface water and groundwater, especially near electrical facilities and big smelters, do occur. In 1988, trying to protect the transformers from a fire in the Thyssen Foundry, more than 200 kg of ASKAREL (a mineral oil containing PCBs) were dumped on the Paraíba do Sul River by the factory workers (Coelho, 1990). All of the water treatment plants situated below this place had to stop pumping until the toxic wave passed.

Polycyclic aromatic hydrocarbons (PAH) are a class of compounds in which two or more benzene rings are attached to each other and are formed by the partial combustion of organic matter. Some of them are highly mutagenic (Harvey, 1982). The carcinogenic potential of these molecules caused them to be listed by the US Environmental Protection Agency as priority pollutants. All of these substances share some common characteristics. Their octanol/water partition coefficients are always very high, meaning that they are insoluble in water (IEOPOPS, 1990). On the other hand, as they are very lipophilic, they accumulate in animal tissues (Larsson & Sodergren, 1987; Neff, 1984; Mackay & Paterson, 1982). Once present in the environment, PAHs thus biomagnify along food chains.

Sediments were sampled and analysed between 1994 and 1998. PCBs were detected sparsely on the surface of the collected sediments. No residues were found at Itatiaia or Porto Real. After Volta Redonda, low-level PCB contamination was detected (Σ of PCBs = 19 ppb). The relative contribution of the lower chlorinated congeners was evident, especially at Pinheiral (Σ of PCBs = 67 ppb). DDT and its metabolites, op'-DDT and pp'-DDE, were detected between Volta Redonda (densely populated) and Barra do Pirai. The sum of DDTs near Volta Redonda was around 225 ppb. Traces of HCB near the first town were below 5 ppb, while near Barra do Piraí the level was 5 times greater (25 ppb). At Volta Redonda low levels of the cyclodienes heptachlor-epoxide and aldrin were detected (5–20 ppb)(Torres, 1998). The PAH level before and near Barra Mansa was low (Σ of 12 PAHs < 1 ppm). At Volta Redonda, however, PAH values could be more than 40 times greater (Σ of 12 PAHs > 40 ppm), indicating the presence of an important source in this city. Volta Redonda owns the largest steelwork plant in Latin America, the CSN (Companhia Siderúrgica Nacional). The ratio of benzopyrene to phenanthrene is less than one. Only at Santo Agostinho, a station very close to the refuse site of the steelworks, and at the Santana Reservoir, an important sedimentation area, does the ratio reach 1.2. Large amounts of coal have been used since the start of the operation of the CSN plant in 1942. Its well known that coal combustion generates a complex mixture of PAHs, particularly rich in heavier PAHs (Edwards, 1983).

BIODIVERSITY

The study of the fauna along the Paraíba do Sul River began in the nineteenth century when several expeditions were organized to collect and study species distributions and diversity (Bizerril, 1998). The most important was the Nathaniel Thayer Expedition that began in 1865. This expedition was coordinated by Louis Agassiz and described by Eigenmann (1917). The presence of biologists, ornithologists, geologists and artists made this expedition one of the most complete ever made in the region. The samples obtained, together with past samplings, produced a comprehensive early database of the biodiversity of the Paraíba do Sul River basin.

The most studied taxonomic group is fish, with several studies that described very well the biodiversity of the Paraíba do Sul River basin (Ribeiro, 1902; Britski, 1972; Nunann *et al.*, 1983; Araújo 1983, 1985, 1996; Langeani, 1990; Caramaschi & Caramaschi, 1991; Costa, 1992; Mazzoni *et al.*, 1993; Bizzeril, 1995; Bockmann *et al.*, 1996; Espírito-Santo *et al.*, 1996; Oliveira, 1997; Pereira, 1997). Bizzeril (1998) summarized the knowledge about fish biodiversity along the Paraíba do Sul River basin. This work described the occurrence of 169 fish species, although this number could probably be enlarged if more studies were performed. The species were divided in three major groups: (a) freshwater species, (b) saltwater species, and (c) introduced species. A complete list of species can be found in Bizerril (1998).

Although there have been several recent studies concerning the present fish biodiversity along the basin, there are very few data and studies from before the deforestation and occupation of the river basin began. This lack of data makes it impossible to evaluate the magnitude of the anthropogenic impact on local fish populations.

CONCLUSIONS

Considering the wealth of research in different fields done within the Paraíba do Sul basin, it is possible to paint a relatively complete picture of the river's environmental condition.

Pristine conditions persist in the uppermost part of the river where the river is formed, but water quality decreases along the middle portions of the river. Conditions improved again near the river mouth. Two processes seem to account for decreasing contaminant concentrations in sediments downstream: (a) sedimentation behind dams, and (b) dilution caused by the input of uncontaminated particulate suspended matter and sediments from the lower basin.

Although we may consider the Paraíba do Sul River as a well studied compared to other rivers in undeveloped countries of the tropics, much remains to be done. There are very few data on contamination by pathogenic microorganisms and the role of cities contaminating river water. We also do not have very consistent time series measurements to integrate all of the environmental variability. The other problem is that although there are several recent studies addressing the present river situation, there are very few data or studies of the area before deforestation and the occupation of the river basin began. This lack of data makes it very difficult to evaluate the magnitude of the anthropogenic impact on the river basin.

Another troubling aspect is the absence of information exchange between the different groups working along the river basin. This lack of collaboration only exacerbates the difficulties in understanding and managing the system. There should be an attempt, perhaps from the government, to joint efforts and knowledge about the basin and to develop a plan to regulate the sustainable use of the water resources of the Paraíba do Sul River. Otherwise this precious resource will be very soon lost forever.

REFERENCES

Araújo, G. F. (1985) *Levantamento Preliminar dos Organismos Aquáticos do Rio Paraíba do Sul—RJ. Relatório Final.* Convênio FINEP/Posto de Aqüicultura/UFRRJ, Rio de Janeiro.

Araújo, G. F. (1996) Composição e estrutura da comunidade de peixes do médio e baixo rio Paraíba do Sul, RJ. *Revista Brasileira de Biologia* **56**(1), 111–126.

Araújo, J. R. S. (1983) *Projeto de Biodetecção de Tóxicos em Sistemas Fluviais de Utilização em Captação de Água para Sistemas Públicos de Abastecimento.* FEEMA, Rio de Janeiro.

Azcue, J. M. P. (1987) Possíveis vias de acesso de metais pesados provenientes do sistema hídrico Rio Paraíba do Sul —Rio Guandú, RJ, até o homem. MSc Thesis, Instituto de Biofísica Carlos Chagas Filho, Universidade Federal do Rio de Janeiro, Brazil.

Bizzeril, C. R. S. F. (1995) Descrição de uma espécie de Neoplecostomus (Loricariidae, Neoplecostominae), com uma sinopse da composição taxonômica dos Loricariidae no leste do brasileiro. *Arquivos de Biologia e Tecnologia* **38**(3), 693–704.

Bizerril, C. R. S. F. (1998) A Ictiofauna: diversidade biológica e padrões biogeográficos. In: *Contribuição ao Conhecimento da Bacia do Rio Paraíba do Sul: Coletânea de Estudos* (ed. by C. R. S. F. Bizerril, L. M. N. Araujo & P. C. Tosin). ANEEL: CPRM, Rio de Janeiro.

Bockmann, F. A., Castro, R. M. C. & Bizerril, C. R. S. F. (1996) Sistemática e biogeografia dos bagres do gênero Rhamdiopsis Haseman, 1911 (Siluriformes, Pimelodidae). In: *XXI Congresso Brasileiro de Zoologia, Resumos.* Universidade Federal do Rio Grande do Sul, Porto Alegre, Brazil.

Britski, H. E. (1972) Peixes de água doce do estado de São Paulo. In: *Poluição e Pisicultura* (ed. by Comissão Interestadual da Bacia do Paraná-Uruguai), 79–108. USP/Instituto de Pesca da CPRN, São Paulo.

Calasans, C. F., Carvalho, C. E. V., Ovalle, A. R. C., Correia, V. G. & Rezende, C. E. (1995) Distribuição de carbono orgânico dissolvido na porção inferior do Rio Paraíba do Sul, Rio de Janeiro. In: *Anais do V Congresso Brasileiro de Geoquímica e III Congresso de Geoquímica* dos países de língua portuguesa. Niterói, Rio de Janeiro.

Caramaschi, E. P. & Caramaschi, U. (1991) Taxonomic status of the trichomycterid catfish Trichomycterus itatiayae. *Copeia* (1), 222–224.

Carvalho, C. E. V. (1997) Distribuição espacial, temporal e fluxo de metais pesados na porção inferior da Bacia de Drenagem do Rio Paraíba do Sul, RJ. PhD Thesis, Universidade Federal Fluminense, Brazil

Carvalho, C. E. V.; Lacerda, L. D.; Rezende, C. E., Ovalle, A. R. C. & Abrão, J. J. (1993) The fate of heavy metals in sediments of the northeastern and southeastern Brazilian continental shelf. In: *Proceedings of the 9th Conference on Heavy Metals in the Environment* (Toronto, 1993), 153–156. CEP Consultants Ltd, Edinburgh, UK.

Carvalho, C. E. V., Ovalle, A. R. C., Rezende, C. E., Aragon, G. T., Castro, S. S. G., Figueiredo, R. O., Suzuki, M. S., Salomão, M. S. M. B., Pereira, E. A., Silva, C. F., Souza, W. P., Gama, D. M. & Kandler, I. (1995) Dinâmica de amônia, nitrito, nitrato, fosfato e sílica em três situações de vazão (alta, média e baixa) do Rio Paraíba do Sul, RJ, em sua porção inferior em 1994/95. In: *Anais do V Congresso Brasileiro de Geoquímica e III Congresso de Geoquímica* dos países de língua portuguesa. Niterói, Rio de Janeiro.

Carvalho, C. E. V., Ovalle, A. R. C., Rezende, C. E., Salomão, M. S. M. B., Molisani, M. M. & Lacerda, L. D. (1999) Seasonal variation of particulate heavy metals in the lower Paraíba do Sul River drainage basin, RJ, Brazil. *Environmental Geology* **37**(4), 297–302.

Coelho, V. (1990) *O Acidente com Oleo Ascarel no Rio Paraiba do Sul*. FEEMA Report.

Costa, W. J. E. M. (1992) Description de huit nouvelle espèces du genre Trichomycterus (Siluriformes: Trichomycteridae) du Brésil oriental. *Revue Française d'Aquariologie* **18**, 101–110.

DNAEE—Departamento Nacional de Águas e Energia Elétrica (1983) *Boletim Fluviométrico F-5.02, Bacia do Paraíba do Sul.*

Edwards, N. T. (1983) Polycyclic aromatic hydrocarbons (PAHs) in the terrestrial environment—a review. *Journal of Environmental Quality* **12**(4), 427–441.

Eigenmann, C. H. (1917) The American Characidae 1. *Memoirs of the Museum of Comparative Zoology* **43**(1), 1–102.

Espírito-Santo, C. E.; Oyakawa, O. T.; Manna de Deus, J. R. & Camargo, F. A. (1996) Manual de peixes da Serra da Mantiqueira, vale do Paraíba, SP, com ocorrências na Fazenda São Sebastião do Ribeirão Grande, Pindamonhangaba. In: *XII Encontro Brasileiro de Ictiologia. Resumos*. Universidade de São Paulo, São Paulo.

Figueiredo, R. O. (1995) Fluxos hidrogeoquímicos e sua relação com fatores ambientais nas bacias de drenagem do sistema fluvio-lagunar Maricá-Guarapina, RJ. MSc Thesis, Universidade Federal Fluminense, Brazil.

Figueiredo, R. O. (1999) Transporte de carbono e nitrogênio no baixo Paraíba do Sul: fluxos e processos. PhD Thesis, Universidade Estadual do Norte Fluminense, Horto, Campos dos Goytacazes, RJ, Brazil.

Harvey, R. G. (1982) Polycyclic hydrocarbons and cancer. *American Scientist* **70**, 386–393.

IEOPOPS (International experts on POPs) (1990) Meeting background report *International Experts Meeting on Persistent Organic Pollutants: Towards Global Action.* Ministry of Environment, Ottawa.

Langeani, F. (1990) Revisão do gênero Neoplecostomus Eigenmann & Eigenmann 1888, com descrição de quatro novas espécies do sudeste brasileiro (Ostariophysi, Siluriformes, Loricariidae). *Comunicações do Museu de Ciências e Tecnologia da Pontifícia Universidade Católica do Rio Grande do Sul, série Zoologia* **3**(1), 3–31.

Larsson, P. & Sodergren, A. (1987) Transport of polychlorinated biphenyls (PCBs) in freshwater mesocosms from sediment to water and air. *Water, Air and Soil Pollution* **36**, 33–46.

Luchini, L. C., Lord, K. A. & Ruegg, E. F. (1981) Sorption and desorption of pesticides on Brazilian soils. *Ciência e Cultura* **33**(1), 97–101.

Mackay, D. & Paterson, S. (1982) Fugacity revisited. *Environmental Science and Technology* **16**(12), 654a–660a.

Malm, O. (1986) Estudo da poluição ambiental por metais pesados no sistema Rio Paraíba do Sul-Rio Guandú (RPS-RG) através da metodologia de abordagem pelos parâmetros críticos. MSc Thesis, Instituto de Biofísica Carlos Chagas Filho, Universidade Federal do Rio de Janeiro.

Martins, O. et al. (1991) In: *Biogeochemistry of Major World Rivers*, 127–155. John Wiley, Chichester, UK.

Mazzoni, R., Caramaschi, U. & Weber, C. (1993) Taxonomical revision of the species of Hypostomus Lacépè (1803) (Siluriformes, Loricariidae) from the lower Paraíba do Sul, State of Rio de Janeiro, Brazil. *Revue Suisse de Zoologie* **101**(1), 3–18.

MME—Ministério das Minas e Energia; Departamento Nacional de Águas e Energia Elétrica (1995) *Projeto Paraíba do Sul, Fase B, Implantação da Agência Técnica e Diagnóstico da Bacia.*

Molisani, M. M. (1999) Geoquímica de metais pesados associados ao sedimento de fundo da porção inferior e estuário interno do Rio Paraíba do Sul, RJ. MSc Thesis, Universidade Estadual do Norte Fluminense, Horto, Campos dos Goytacazes, RJ, Brazil.

Molisani, M. M., Carvalho, C. E. V., Ovalle, A. R. C., Rezende, C. E., Salomão, M. S. M. B. & Lacerda, L. D. (1999) Heavy metals distribution in sediments of the lower Paraíba do Sul River and Estuary—RJ, Brazil. *Bulletin of Environmental Contamination and Toxicology* **63**(5), 682–690.

Neff, J. M. (1984) Bioaccumulation of organic micropollutants from sediments and suspended particles by aquatic animals. *Frezenius Z. Analytical Chemistry* **319**, 132–136.

Nunann, G.W., Cardoso, L. W. & Bandeira, W. D. (1983) Levantamento da ictiofauna do rio Paraíba do Sul. Trecho Represa do Funil-Cidade de Barra do Piraí, Estado do Rio de Janeiro. In: *X Congresso Brasileiro de Zoologia, Resumos*. Universidade Federal de Minas Gerais, Belo Horizonte.

Oliveira, J. C. (1997) Redescrição de Hemipsilichthys garbei, Ihering, 1911, com designação do lectótipo e comentários sobre sua distribuição e a de Hemipsilichthys gobio (Lütken, 1874) (Teleostei, Siluriformes, Loricariidae). *Papeis Avulsos de Zoologia* **40**(6), 113–126.

Pereira, E. H. L. (1997) Revisão taxonômica das espécies de Hemipsilichthys Eigenmann & Eigenmann, 1889 (Teleostei, Siluriformes, Loricariidae). Dissertação de Mestrado, Pontifícia Universidade Católica do Rio Grande do Sul Porto Alegre, Brazil.

Projeto RADAMBRASIL (1983) *Folhas SF.23/24 Rio de Janeiro/Vitória; Geologia, Geomorfologia, Pedologia, Vegetação e Uso Potencial da Terra.* RADAMBRASIL, Rio de Janeiro.

Ribeiro, A. M. (1902) Oito espécies do rio Pomba. *Boletim da Sociedade Nacional de Agricultura* **8**.

Rosso, T. C. A., Neves, C. F. & Rosman, P. C. C. (1991) O estuário do Rio Paraíba do Sul: perspectivas em um cenário de variação do nível do mar. In: *Simpósio Brasileiro de Recursos Hídricos 9 & Simpósio Luso-Brasileiro de Hidráulica e Recursos Hídricos*, vol. 5, 578–586. Editora da ABRH 3, Rio de Janeiro.

Salomão, M. S. M. B. (1999) Dinâmica de metais pesados nas frações particulada e dissolvida na porção inferior do Rio Paraíba do Sul, RJ. MSc Thesis, Universidade Estadual do Norte Fluminense, Horto, Campos dos Goytacazes, RJ, Brazil.

Salomão, M. S. M. B., Molisani, M. M., Ovalle, A. R. C., Rezende, C. E., Lacerda, L. D. & Carvalho, C. E. V. (2000) Particulate heavy metal transport in the lower Paraíba do Sul River basin, southeastern, Brazil. *Hydrological Processes* **15**(4), 587–593.

Svendsen, L. M. & Kronvang, B. (1993) Retention of nitrogen and phosphorus in a Danish lowland river system. Implications for the export from the watershed. *Hydrobiologia* **251**, 123–135.

Tommasi, L. R. (1985) Resíduos de praguicidas em águas e sedimentos de fundo do sistema estuarino de Santos (SP). *Ciência e Cultura* **37**(6).

Torres, J. P. M. (1992) Ocorrência e distribuição de metais pesados no Rio Paraibuna, Juiz de Fora, MG. MSc Thesis, Instituto de Biofísica Carlos Chagas Filho, Universidade Federal do Rio de Janeiro.

Torres , J. P. M. (1998) Ocorrência de micropoluentes orgânicos (organoclorados e hidrocarbonetos policíclicos aromáticos) em sedimentos fluviais e solos tropicais. PhD Thesis, Instituto de Biofísica Carlos Chagas Filho, Universidade Federal do Rio de Janeiro.

CHAPTER 11

THE APPLICATION OF ECOHYDROLOGICAL PRINCIPLES FOR BETTER WATER RESOURCES MANAGEMENT IN SOUTH AMERICA

Michael E. McClain

Clean surface water is fundamental to the health and welfare of South Americans and to the preservation of the continent's stunning natural ecosystems. The majority of water for domestic and industrial use in South America is taken from surface sources, while groundwater is important in arid regions and in some more industrialized agricultural areas to the south. Overall the continent is estimated to have 13 400 km^3 in renewable water supplies, with 21% of this flowing annually in rivers. Total annual withdrawals were estimated at 166 km^3 in 1995, with 60% going to agriculture, 20% going to domestic use, and the remainder going to industrial and other uses. South Americans also rely on surface waters for food. Recent estimates place the annual freshwater fish harvest at nearly 400 000 metric tonnes (WRI, 1998).

Currently, people in South America have unequal access to what the United Nations defines as improved drinking water and sanitation services (WHO/UNICEF, 2002). Urban inhabitants' access to improved drinking water ranges from 85% in Venezuela to 99% in Chile. However, for rural inhabitants these values fall to 70% and 58%, respectively. Suriname is reported to have the lowest percentage of rural inhabitants with access to improved drinking water (50%), while Uruguay has the highest (93%). Closely related to the issue of clean drinking water is the availability of improved sanitation services. The lowest availability of sanitation services for urban inhabitants is reported for Venezuela (71%), while the highest is for Chile (97%). As with improved drinking water, the availability of sanitation services tends to drop for rural inhabitants, with the minimum reported for Brazil (43%) and the maximum for Chile (97%).

The condition and future of South America's surface water quality are strongly influenced by the interaction of the region's social and economic systems with its natural resource base. This chapter will devote considerable attention to the natural resource and ecohydrological aspects of the problem, but it is important to highlight certain socio-economic factors that will undoubtedly play a role in determining management options. The countries of South America may be classified as developing nations and as such they have relatively few financial resources to address water quality management problems (Table 11.1). Only Argentina and Uruguay have *per capita* gross national products (GNP) above the world average of US$5020 and all are well below the average (US$26 440) GNP of the high-income OECD member states (World Bank, 2001). Pressing problems reflected in population growth, illiteracy, unacceptable infant mortality rates, and high external debt will continue to be top priorities for expenditure in these nations. Water resource management themes that relate directly to economic development, such as dams and irrigation works, are also likely to be priorities. However, management of *in situ* water quality in surface waters, used to support natural ecosystems, protect artisan fisheries, and reduce the occurrence of waterborne illness among rural peoples, is not likely to garner much financial support in the coming years, nor is it likely to make its way onto any national priority action lists.

Table 11.1 Demographic and economic data for South American countries for 1999. Data from World Bank (2001)*.

	Total population (millions)	Urban population (% of total)	Infant mortality rate†	Overall budget deficit‡	GNP *per capita*, Atlas method (1999 US$)
Argentina	36.6	89.6	18.4	–1.49	7550
Bolivia	8.1	61.9	58.8	–2.27	990
Brazil	168	80.72	32.2	–7.34	4350
Chile	15	85.44	9.98	–1.47	4630
Colombia	41.5	73.48	22.76	–7	2170
Ecuador	12.4	64.3	28.4	-	1360
Guyana	0.85	37.64	56.8	-	760
Paraguay	5.4	55.28	23.8	-	1560
Peru	25.2	72.42	39.2	–2.09	2130
Uruguay	3.3	91.04	14.5	–3.72	6220
Venezuela, RB	23.7	86.62	20.2	–2.27	3680

*Relevant data for French Guiana and Suriname were not available.

†Per 1000 live births.

‡Including grants (% of GDP). Budget deficit data from Argentina are from 1998 and for Brazil are from 1997.

In this chapter I present an ecohydrological approach to surface water quality management that I believe is viable given both the needs and limitations in South America. The premise for this approach is that rivers and wetlands, especially natural rivers and wetlands, possess a strong capacity for self purification that can be explicitly incorporated into water management programmes. Processes to eliminate organic wastes, strip excess nutrients, filter sediments, neutralize acidity, and reduce metal concentrations are intrinsic to free-flowing rivers with intact biological communities. Many of these processes have been noted in the preceding chapters of this book, and I will draw on those examples. Ironically, the very socio-economic problems that hamper water quality management today also precluded widespread development and intensive resource use in many South American basins historically. Consequently, South American rivers and wetlands remain in a more natural state than many of their northern counterparts and exhibit stronger self-purification capacities.

SOUTH AMERICAN DRAINAGE SYSTEMS AND WATER RESOURCES

By world standards South America has abundant available water resources, although as indicated in the introduction the safety of these resources varies between countries and social groups. South America's *per capita* water availability is 38 000 m^3 $year^{-1}$ (Gleick, 2000). That is twice the availability of water in North and Central America and nearly an order of magnitude higher than the availability of water in Europe and Asia. South America's 10 largest river basins account for 70% of the total continental area and an even greater percentage of the continent's runoff (Fig. 11.1, Table 11.2). The Amazon is the world's largest basin and alone accounts for 35% of the continent's area and nearly 70% of its total runoff.

South America contains diverse landscapes resulting from the wide range of latitude (12°N–55°S) and elevation (0–6959 m a.s.l.) found on the continent. The continent's largest basins, the Amazon, Paraná, and Orinoco, each include varied terrain but distinctions are

Fig. 11.1 South America's major river basins and wetlands.

evident. The Amazon and Orinoco rivers drain large tracts of rainforest and tropical savannas, and both are fed by mountain rivers from the eastern flank of the Andes cordillera. The Paraná River, by contrast, drains the vast plains and grasslands of the Gran Chaco and the

Table 11.2 Details on the characteristics and condition of South America's major river basins.

River basin	Area (km^2)	Average discharge (m^3 s^{-1})*	Populat'n density (ind. km^{-2})	No. of fish species	Percent wetland	Percent arid	Percent forest	Percent crop-land	Percent grass-land†	Deforest. rate/loss‡	Eroded area (%)
Amazon	6 144 727	175 000 (205 000)	4	3000	8	4	73	15	10	5/13	1
Chubut	182 631	-	2	-	0	61	25	1	67	-/28	0
Magdalena	263 858	7 500 (6 840)	79	149	0	7	37	39	13	26/88	10
Orinoco	953 598	30 000 (35 000)	13	318	15	9	50	19	27	12/23	2
Paraná	2 582 672	18 000 (19 200)	23	355	11	10	12	45	39	18/71	4
Parnaíba	322 823	825	10	90	19	42	5	52	40	14/27	1
Río Colorado	402 956	-	10	-	2	71	1	10	71	-/100	0
São Francisco	617 812	2 900 (2 800)	18	-	10	32	1	61	33	9/64	1
Tocantins	764 183	11 000	6	-	19	0	9	63	27	10/50	2
Uruguay	297 199	5 000	25	160	4	0	7	44	45	12/92	20

Primary source: Revenga *et al.* 1998. A dash (-) is shown where data are unavailable.
*Discharge data from Oki *et al.* (1995), (and Meybeck (1988)).
†Percent grassland includes both grasslands and shrublands.
‡Deforestation values are reported as annual percent deforestation rate on the left and total percent of forest lost on the right.

coastal hills of central Brazil, including the metropolis of São Paulo. The next largest basins, the Tocantins and São Francisco, are both predominantly north-flowing rivers that drain the eastern extension of Brazil. The Chubut, Colorado, and Parnaíba rivers drain more arid parts of the continent; though their basins are large, their discharges are relatively small. The Uruguay River drains the humid chaco between the Paraná and the Atlantic coast, and the Magdalena River drains the moist central valleys of the northern Andes.

The data in Table 11.2 reveal something of the degree of development and intensity of human impacts in these major river basins. Of the major basins the Magdalena is most densely populated, with multiple large cities including Bogotá, Cali, and Medellin. The Magdalena also exhibits the highest rate of deforestation, with 88% of the basin already deforested. The Uruguay basin is the second most densely populated and nearly completely deforested. The basins of the São Francisco, Colorado, and Parnaíba also exhibit significant levels of alteration. The Paraná basin has lost 71% of its original forest, 45% of its area is under cultivation, and along with the Uruguay basin its population density is second highest on the continent. The least impacted rivers on the continent are the Amazon, Orinoco, and Chubut. The great extent and relative inaccessibility of the Amazon and Orinoco no doubt contribute to their less-impacted condition, while similar remoteness and an inhospitable climate likely explain the low intensity of human activity in the Chubut.

Expansive lowlands are characteristic of South America's largest river basins. Approximately 50% of the Amazon basin's 6 million km^2 are below 200 m elevation. In fact, one can travel more than 5000 km up the Amazon River and remain below 200 m a.s.l.! One can also travel thousands of kilometres up the Paraná and Orinoco rivers before rising above the 200 m contour line. The combination of broad lowlands and significant rainfall creates ideal conditions for the formation of large wetlands. Aselmann & Crutzen (1989) estimated that 1.52 million km^2 of freshwater wetlands occur in South America, but this number is highly conservative. Junk (1993) estimated that nearly 2 million km^2 of wetlands occur in the tropical portion of South America alone, including everything from the massive flood plains along major rivers to narrow fringing wetlands in low-order stream segments. The most significant of these are the seasonally-flooded wetlands of the mainstem Amazon River in Brazil (~300 000 km^2), the seasonal savanna wetlands of eastern Bolivia (~150 000 km^2), and the Pantanal wetland (~140 000 km^2), which lies mostly in Brazil. The Pantanal in particular is regarded as one of the largest and most pristine contiguous wetlands in the world. Considerable international attention is focused on the Pantanal today as conservationists and national agencies work to prevent damage by encroaching development (Swarts, 2000).

Another distinctive and globally important freshwater environment of South America is the endorheic basin of Lake Titicaca, which covers an area of approximately 56 000 km^2 (Montes de Oca, 1989). At 3800 m a.s.l. this large lake and its associated wetlands is the only significant body of freshwater on the arid Bolivian altiplano. As such it hosts a number of endemic fish species and provides important habitat for Andean migratory birds. The lake is now feeling the effects of uncontrolled sewage discharges from the Peruvian city of Puno and other industrial and mining waste discharges.

With some exceptions water may be considered to be abundant in South America, and consequently the main factor limiting water availability in many areas is water quality rather than quantity. In the next section I present select data to illustrate the types of contamination impacting South American surface waters. It will become apparent that the most widespread water quality problems are those that ecohydrological processes can effectively address provided that the self-purification capacity of rivers and wetlands is not compromised by human alterations.

THE CURRENT QUALITY OF SOUTH AMERICA'S SURFACE WATERS

As might be expected, water quality in South American rivers and wetlands ranges from absolutely pristine to dangerously poor. While data are limited, it is clear that the most polluted waters occur in and downstream of major population centres, and pollution levels tend to be more intense in smaller rivers where dilution is less effective. The Bogota and La Paz rivers are prime examples, as they are virtually lifeless immediately downstream of the cities that share their names. By contrast, the mainstem sections of the continent's largest rivers, the Amazon, Orinoco, and Paraná, have very high water quality.

Currently it is impossible to present a truly comprehensive assessment of water quality in South America. Existing data are relatively few and thinly scattered in the scientific literature, obscure reports of national agencies, and student theses from South American universities. Of the 715 monitoring stations contained in the United Nations' Global Environment Monitoring System Freshwater Quality Programme (GEMS/WATER), only 50 (or 7%) are from South America. By contrast, South America accounts for more than 20% of recharge and runoff from the continents (WRI, 1998). UNESCO's International Hydrology Programme has an effort underway in South America to compile data on water quality, but this effort has thus far produced no publicly available results. Consequently, in the following sections I will simply highlight the predominant forms of water contamination in South America.

Organic wastes

The most widespread contamination problem is organic waste entering water bodies in the form of untreated sewage and industrial discharges. Presently less than 20% of industrial and domestic waste water is treated prior to discharge in Latin America (CEPIS, 2001). Martinelli *et al.* (2002) describe a severe example of organic contamination in the Piracicaba River, which drains more than 10 000 km^2 of the Brazilian State of São Paulo (Fig. 9.1 in Martinelli *et al.*, 2002) and ultimately empties into the Paraná River. This basin is among the most economically advanced areas of South America, with 7000 industries that account for nearly 10% of the entire Brazilian gross national product. Despite the relative affluence of this basin and its cities, the Piracicaba River receives 84% of the raw sewage and 10% of the untreated industrial effluents generated in the basin. This amounts to approximately 150 t of biochemical oxygen demand (BOD) per day, as well as unknown quantities of metals, salts, and suspended solids. Consequently, this river which was once renowned for its fish and mussel diversity is now biologically barren, with some sections exhibiting pollution so extreme as to eliminate nearly all aquatic life.

In Colombia, the Institute for Hydrology, Meteorology and Environmental Studies (IDEAM) maintains a network of 26 water quality monitoring stations in the vicinity of the cities of Bogotá and Pasto. Near Bogotá, 10 stations lie in the META river basin, which flows into the Orinoco, and 12 stations lie in the headwaters of the Magdalena River basin. The four stations located near Pasto lie in the headwaters of the Patia River basin, which flows westward into the Pacific ocean. By far the dominant form of contamination in these rivers is excess organic matter from domestic discharges. During October 2000, dissolved oxygen deficits reached 100% at two stations passing through heavily urbanized areas (IDEAM, 2000). Deficits of greater than 25% were reported in an additional four stations. Several stations within this monitoring network also reported medium to high values of turbidity, chemical oxygen demand, NH_3, and NO_3. Nationwide in Colombia, an estimated

690 kt of BOD and 1500 kt of suspended solids were discharged into the country's surface waters during 1996 (Rodrigo *et al.*, 1998). Of these national totals, 76% of BOD discharges and 93% total suspended solids discharges are thought to derive from domestic sources, with the remainder coming from industrial and agricultural sources. It should be noted that inputs of BOD from domesticated animals is not considered, but may amount to an additional 2000 kt or more of BOD per year.

Pathogens

Inputs of organic wastes, which give rise to BOD may also introduce pathogens. Illnesses related to poor water quality are the number one health problem in rural South America. During a series of health surveys conducted between 1990 and 1995, 31% of urban children less than 5 years old in Bolivia were found to be suffering from diarrhea. In rural areas the number was 28% (WRI, 1998). High frequencies of diarrhea among rural and urban children were also reported in Brazil (17%, 13%), Colombia (16%, 17%), and Peru (16%, 22%). In the La Paz River of Bolivia, the high incidence of diarrhea has been directly linked to high bacterial concentrations, which persist for more than 40 km downstream of the city (Ohno *et al.*, 1997). Many more alarming forms of viral contamination may also be observed, such as the Oropouche virus detected near the city of Iquitos in the lowland Amazon of Peru. This virus was found to be transmitted to the population of the city and surrounding areas (Watts *et al.*, 1997). In a survey of households in the rural Pachitea basin of central Peru, Aparicio Alcázar (1999) found that greater than 50% of households reported some form of health problem linked to water. Hospitals in the two towns within the basin recorded a total of 8125 cases of illness related to water between January and October 1998 (Aparicio Alcázar, 1999). Acute diarrhea was the most commonly reported illness (4172 cases) along with other helminchic infections (intestinal worms—3578 cases), but 62 cases of viral hepatitis and 18 cases of cholera were also reported. Pathogens in water are especially troubling because dilution does not reduce their potency. A single pathogenic organism can cause illness.

Nutrients

Eutrophication of surface waters has been reported in many parts of South America. Most problems are reported in lake and reservoir systems receiving significant inputs of raw sewage from nearby urban areas (Ceballos *et al.*, 1998). Estuarine waters along urbanized coasts are also sometimes heavily impacted (Braga *et al.*, 2000; Huszar *et al.*, 1998). Problems of eutrophication in rivers have been reported mainly in highly developed areas such as the Piracicaba basin (Martinelli *et al.*, 2002). Concerns about excess nutrient concentrations have also been expressed in field studies from the upper São Francisco basin of southeastern Brazil (Galdean *et al.*, 2000) and the lower portion of the Chubut (Sastre *et al.*, 1998). Modelling studies from the agricultural region of Chile's Central Valley also suggest a real potential for eutrophication (Donoso *et al.*, 1999). The threat of stream and river eutrophication exists throughout the heavily agricultural regions of Chile, Argentina, and Brazil, where fertilizer use continues to rise. By contrast, nutrient levels in the Amazon, Orinoco, and Paraná river systems appear unaffected by anthropogenic processes operating in their basins. Hamilton (2002) notes that nutrient levels in the Paraná River are even lower than those in the Amazon and Orinoco and suggests that greater river–flood-plain exchange in the Paraná may explain the difference.

Sediments

Natural suspended sediment concentrations vary enormously in South American river systems, mainly as a function of basin relief and climate. Virtually no data are available, however, to document anthropogenically driven changes in suspended sediment concentrations. Some of the highest natural sediment loads have been found in rivers draining the Andean cordillera. Guyot *et al.* (1995) reported average suspended sediment concentrations in excess of 10 000 mg l^{-1} for multiple high gradient rivers in the Bolivian Andes, while downstream concentrations in the Madera basin of the Brazilian Amazon were more commonly between 10 and 500 mg l^{-1}. In the Magdalena River Restrepo & Kjerfve (2000) reported mean sediment concentrations of 670 mg l^{-1} during rising water and 420 mg l^{-1} during falling water.

The lack of data to indicate increased sediment loads in South American rivers is somewhat surprising given the considerable attention given to erosion on the continent. Table 11.2 indicates significant areas of eroded lands in the Uruguay and Magdalena river basins. Small basin studies in the Brazilian Amazon (Williams *et al.*, 1997), the Peruvian Amazon (Alegre & Rao, 1996), French Guyana (Fritsch & Sarrailh, 1986), and the basaltic plateaux of southern Brazil (Dos Reis Castro *et al.*, 1999) have all documented increased erosion resulting from deforestation and agriculture. In the bauxite mining area of the Río Trombetas, initial research has also documented the reductions in bottom habitat and reductions in water column photosynthesis due to influxes of mining tailings (Callisto *et al.*, 1998; Roland *et al.*, 1997). A good source of indirect data on changing sediment loads is sediment accumulation rates determined using ^{210}Pb. At least two studies have documented increased sedimentation rates in basins undergoing large-scale development. Forsberg *et al.* (1989) reported that, following the construction of highway BR-364 in 1961 along the Jamari River in Rondonia, Brazil, sedimentation rates increased exponentially in Lago Paca near the river's mouth. By the end of 1985 sedimentation rates had increased 10-fold. Godoy *et al.* (1998) also made use of ^{210}Pb to document increased sediment fluxes in the Taquari River, which flows into the Pantanal.

Metals

In mineral-rich zones of the Andean cordillera where mining is common, surface waters receive sometimes alarming levels of metals and acidity. Near critical situations have also been reported in high Andean lakes (Roca, 1995). Metal contamination from mining operations generally affects smaller rivers at higher altitudes. Resulting problems include harmful levels of heavy metals, increased sediment loads, and acidification. Roca (1995) also reports that agricultural soils have been contaminated with heavy metals by irrigation waters drawn from contaminated rivers. More than 21 000 ha of farmland are said to have been contaminated using irrigation waters from the Mantaro River. This medium-sized river draining the high Peruvian Amazon is one of the country's most polluted water bodies.

There have been few reports of significant levels of metal contamination in larger rivers. A case in point is the Paraíba do Sul River (55 400 km^2 basin) in Brazil, which flows from São Paulo and across the States of Rio de Janeiro and Minas Gerais (Fig. 10.1 in Carvalho & Torres, 2002). Like the Piracicaba River mentioned above, the Paraíba do Sul receives large quantities of untreated (or incompletely treated) waste waters from domestic, industrial, and agricultural sources. However, unlike the Piracicaba, contamination of the Paraíba do Sul River has not risen to such ecologically destructive levels (Carvalho & Torres, 2002). Isolated occurrences of organic contaminants such as PCBs and DDT have been detected downstream of heavily populated areas. Likewise, unacceptably high levels of heavy metals like cadmium (>15 ppm) have been detected in bed sediments adjacent to industrial sites (Torres, 1992), but metal concentrations tend to decrease rapidly downstream to values of less than 1 ppm. Similarly, metal concentrations in sediments across the entire lower portion of the basin contain only low levels of heavy metals.

Trace metal concentrations in the largest South American rivers are generally near background levels despite sometimes significant influxes of industrial effluents. There are certainly no indications of contamination in the mainstem sections of the Amazon and Orinoco rivers, and even in the more highly impacted Paraná system mainstem metal concentrations are low (Facetti *et al.*, 1998). In the much more industrialized tributaries near the mouth of the Paraná and along the southern margin of the Río de la Plata estuary, concentrations are higher but still not alarmingly high (Villar *et al.*, 1998). A significant individual source of contamination to the Río de la Plata estuary is the Reconquista River, which drains 1670 km^2 of the Buenos Aires province. This river suffers from strong hypoxia near its mouth and carries the effluent of many industrial sites (Castañé *et al.*, 1998). The Magdalena River contains the highest metal concentrations of the large South American rivers, which is not surprising given the intensity of human activity within its basin. Perdomo *et al.* (1998) reported concentrations of Cd, Cu, Ni, and Zn in sediments of the lower Magdalena that exceeded international standards; other areas of concern were identified nearly a decade earlier by HIMAT-INGEOMINAS (1991).

Mercury

Without a doubt the most publicized water quality issue in the Amazon during recent years is mercury contamination. Martinelli *et al.* (1988) were among the first to raise this issue in the scientific literature and to tie it to gold mining operations in the Amazon. More recently investigators have shown that naturally high concentrations of mercury occur in Amazon soils as a result of long-term, low-level inputs from volcanism and other atmospheric sources (Roulet & Lucotte, 1995). Consequently, even in remote portions of the river basin erosion of adjacent soils may lead to elevated mercury concentrations in river waters (Maurice-Bourgoin *et al.*, 2000). In the Madera sub-basin of the Bolivian Amazon, levels of mercury in river water are elevated through both natural and anthropogenic processes. Recent investigations there have found that mercury is accumulating in fish tissue and has been reported at concentrations more than 4-times higher than the World Health Organization's (WHO) suggested limit of 0.5 mg g^{-1} (Maurice-Bourgoin & Quiroga, 2002). The highest levels were reported for piscivorous species (2.304 mg g^{-1} for *Pseudoplatystoma tigrinum*). As a result of the biomagnification of this metal, even higher levels of mercury were found in the hair of humans living along the river who ate abundant fish. It is interesting to note that mercury levels in the hair of gold miners (who burned mercury but seldom ate fish) were a factor of 4–20 below those of fish-eating riparian inhabitants (Maurice-Bourgoin & Quiroga, 2002). Even higher concentrations of mercury have been reported in the hair of fishermen and their families from the Tapajós basin in the Brazilian Amazon (Malm *et al.*, 1995) and around the Tucuruí reservoir in the Tocantins basin of the Brazilian State of Pará (Leino & Lodenius, 1995).

The question of what factors control the location of Hg contamination in Amazon fish and the people who eat them, is made more intriguing by the work of Silva-Forsberg *et al.* (1999), who reported that Hg concentrations in fish and people correlated more with the pH and dissolved organic carbon (DOC) concentration of river waters, than its proximity to gold-mining operations. These findings highlight the positive influence of DOC in mobilizing mercury through complexation and the negative influence of pH in inhibiting methylation and bioaccumulation of mercury.

Mercury contamination is certainly not a problem unique to the Amazon. Problems of elevated mercury in fish have also been reported from the flood plain of the Paraná River (Moraes *et al.*, 1997), and Hylander *et al.* (1994) reported mercury levels above WHO guidelines in the Alto Pantanal region. These authors also noted the correlation of mercury levels with DOC and pH. Higher trophic level fish from the Magdalena River system were also found to contain mercury concentrations in excess of WHO guidelines (Olivero *et al.*, 1998), but concentrations decreased in the same species collected from downstream marsh systems (Olivero *et al.*, 1997).

NATURAL ATTENUATION IN RIVER AND WETLAND SYSTEMS

The waste removal capacities of rivers and wetlands have been recognized from the very beginning of human civilization, and consequently these water bodies have forever been the receptacles of our wastes. The most immediate advantage of a river was its flow, which transported wastes from the dumping point, but dilution also quickly reduced liquid waste concentrations, ideally to levels that were harmless to aquatic organisms and downstream human communities. The use of dilution as a means of reducing contamination has remained fixed in the minds of waste managers and even the general public, as evidenced in the popular rhyme "dilution is the solution to pollution". Dilution is fundamental to quickly reducing contaminant levels in rivers, but there are also a number of ecohydrological processes that work to reduce or eliminate contaminants. I will highlight several processes useful in water quality management by presenting three examples of forms of contamination that affect South American river basins and wetlands: (a) organic wastes and associated nutrient releases, (b) excess sediment loads, and (c) metal wastes.

Processes reducing organic matter and nutrient levels

Many South American rivers are net heterotrophic and therefore rely on inputs of organic matter (OM) from upstream to fuel downstream productivity. In these rivers bacteria and fungi form the base of the food chain, and higher trophic levels are adapted to feeding on these organisms. Mayorga & Aufdenkampe (2002) cite rates of community respiration in the rivers of the Amazon basin that range from 0.2 to 2 mmol C l^{-1} h^{-1}. Amon & Benner (1996) go further and, based on glucose-addition experiments, show that respiration rates may be increased by more than three-fold. From this perspective, modest increases in non-pathogenic OM concentrations should not negatively impact riverine food webs, as the additional OM will be rapidly processed into heterotrophic organisms.

In addition to uptake and oxidation by organisms, or more likely prior to it, OM may be stripped out of river water by physical processes such as adsorption onto suspended or bed sediment surfaces, especially the surfaces of Al and Fe oxides. In the presence of certain substances, organic matter may also coagulate and settle out of the water column. Mounier *et al.* (2002) report a loss of 20×10^6 t $year^{-1}$ of organic matter along the mainstem of the Amazon River between the mouth of the Río Negro and the city of Obidos. These losses were observed only at low water stages, however, and removal processes were far less efficient during high water. Particulate forms of organic matter may also be trapped by filtration through bank and bed sediments (just as with fine sediment grains). These various processes serve to immobilize organic matter, which may be subsequently decomposed by benthic fungi and bacteria.

Decomposition of excess organic matter is accompanied by the mineralization of potentially limiting nutrients like nitrogen and phosphorus. In small excess amounts, these nutrients are likely to be taken up by microorganisms or by plants living along the banks of the river. Phosphate is very effectively sorbed onto mineral surfaces and is generally maintained at low levels in tropical rivers. Elevated levels of nitrate may be reduced through denitrification reactions in anoxic bed and bank sediments, or within the anoxic zones that may form in accumulations of organic litter behind obstacles such as fallen trees. Engle & Melack (1993) reported on the efficiency of nutrient uptake from river water on the Amazon flood plain. Water entering the lake at its upstream connection to the river had NO_3^- and PO_4^{3-} concentrations of approximately 7 and 2 mM, respectively. By contrast water being returned to the river 3 km downstream had NO_3^- and PO_4^{3-} concentrations well below 1 mM.

In oligotrophic (or nutrient poor) water bodies, the extra nutrients released from decomposing organic matter may fuel increased rates of productivity. In British Columbia, Canada, oligotrophic lakes have been fertilized with nitrogen and phosphorus for more than two decades to enhance lake productivity and to therefore increase the survival rate of juvenile sockeye salmon (Stockner & Macisaac, 1996). As a result juvenile survival rates increased two-fold and average smolt weight increased by greater than 60%.

Intact riparian forests have been shown to be especially effective buffers in protecting against excess nutrient levels in rivers (Haycock *et al.*, 1997). These forests are most effective at stripping sediments and nutrients from waters flowing from upland sources, but Triska *et al.* (1989) demonstrated that nitrate could be stripped from rivers waters flowing in a hyporheic zone that extends beneath adjacent riparian forest. McClain *et al.* (1994) and Williams *et al.* (1997) reported efficient removal of inorganic nitrogen from runoff waters beneath riparian forest in the central Amazon basin, but otherwise very little work on riparian zones as buffers has been carried out in South America. Nothing is know as yet about their capacity to eliminate higher nitrogen fluxes.

Again the processes operating in rivers to eliminate organic matter and nutrients are identical to those employed in many water treatment systems, including filtration, coagulation, sedimentation, and sorption. The decomposition of organic matter by microorganisms in the river may also be viewed as a sort of bioreactor.

Processes reducing sediment loads in rivers

Sediments entering rivers will initially be partitioned according to dominant size classes and the level of turbulence of the river flow. Large sediment grains will be deposited near the point of erosion, while finer grain sizes will be carried further downstream. Suspended sediments may initially reduce water quality and riverine productivity by limiting light penetration. However, free-flowing rivers that are properly graded contain varying levels of turbulence that serve to strip out excess sediments. In smaller mountainous river systems, riffle and pool structures develop, with pools serving as sinks for excess sediments derived from upstream. Similarly, larger meandering rivers develop low-turbulence zones on the inside curve of meanders, which also serve as sediment sinks. Sedimentation is a commonly applied technique in water treatment facilities, and the process of sedimentation in rivers is fundamentally the same. Furthermore, finer sediments will be filtered from water passing through bank or bed sediments, again emulating processes operating in sand filtration systems or water treatment plants.

Wetlands are very efficient sediment traps with strong ecohydrological ramifications because very often aquatic vegetation in wetlands acts to slow water flow and promote sedimentation. Hamilton (2002) highlights the decisive role of Pantanal wetlands in reducing sediment loads carried by upstream rivers and delivering nearly sediment-free waters to the Paraguay River downstream. Engle & Melack (1993) also reported a dramatic decrease in sediment loads as Amazon River water moved onto the flood plain. Water flowing onto the flood plain contained nearly 150 mg l^{-1} of total suspended solids (TSS), while flood-plain water flowing back into the river 3 km downstream contained less than 10 mg l^{-1} TSS. While both depositional and erosional processes operate on the flood plains of South America's major rivers, deposition likely dominates. Dunne *et al.* (1998) calculated the exchanges of sediment between the mainstem Amazon and its flood plain along a 2010-km reach of the river. They noted an accumulation of sediment along the reach at a rate of 209 (±167) Mt $year^{-1}$, and calculated a net transfer of sediment to the flood plain amounting to 500 Mt $year^{-1}$.

Processes neutralizing acids and stripping metals from river systems

The reduced forms of elements (e.g. Fe and S) that typically weather from mining slag heaps are rapidly oxidized upon interaction with meteoric waters. These oxidation processes generally liberate H^+ ions and therefore drive pH down to more acidic levels. Levels of pH below 2 have been recorded in mine runoff, with correspondingly high concentrations of metals (Förstner & Wittmann, 1983). The sensitivity of rivers to acidification depends on acid buffering capacity, or alkalinity. Acid buffering capacity increases in rivers as a function of carbonate ion concentrations (CO_3^{2-} and HCO_3^-). Thus, rivers draining basins containing carbonate rocks generally have higher acid buffering capacity and are better able to neutralize acid drainage waters.

The effectiveness of ecohydrological processes in reducing metal contamination depends on both the metal considered and even the form (oxidation state) of individual metals. For example, the oxidized form of chromium, Cr(VI), is highly mobile and toxic in aquatic systems, while the more reduced from, Cr(III), is less mobile and less toxic (NAS, 2000). In fact, Cr(III) precipitates from solution at pHs above 5 and is thus rendered harmless. Many microorganisms reduce Cr(VI) to Cr(III), but they all rely on the presence of environmental conditions that favour this reduction. In other words, this reaction is likely to take place only in anoxic zones within the river bed, wetland, or flood-plain sediments, within riparian soils, or within accumulations of organic debris.

Microorganisms will also concentrate metals in their tissues and thereby strip them from solution. These metals may then be passed to higher trophic levels, they may be remineralized when the organisms decompose, or under certain conditions they may be stored in deposited sediments. If deposited in anoxic environments, these metals may be reduced to less mobile forms that can be more or less permanently removed from the system. Konhauser & Fyfe (1993) found 17 trace metals that had been concentrated in the tissue of benthic diatoms from some of the large tributaries of the Amazon River system. For Ti and Zr concentration factors exceeded 1 million, while Hg was concentrated by a factor of roughly 1000. Given such efficient concentration from aqueous phases, it is easy to understand why microorganisms are increasingly used in mining operations to concentrate metals leached from ores.

Thus water purification processes such as dilution, filtration, sedimentation, acid-base reactions and pH buffering, redox reactions (oxidation and reduction), precipitation and dissolution, complexation/coagulation, chemical sorption, hydrolysis, and plant uptake all occur naturally in river and wetland systems, and when properly quantified may be used as effective tools in the management of water quality.

INTEGRATING ECOHYDROLOGICAL PRINCIPLES INTO MANAGEMENT PLANS

The management of water resources is a complex subject that responds to and emerges from a number of international agreements and national and state laws. Specific management actions are generally divided between multiple governmental agencies, with varying levels of interaction with nongovernmental organizations and citizen groups. Management plans may be also highly engineered and strongly regulated in the case of urban areas or they may be essentially non-existent in the very remote rural areas still common on the South American continent. The preceding sections have reviewed the water quality problems in South American surface waters and shown how ecohydrological processes function to counteract them. How, then, can these natural processes be integrated into the heterogeneous mix of management plans and actions that operate on the continent? The answer, I suggest, is to first formally recognize natural processes (e.g. dilution, filtration, sedimentation, oxidation, etc.) as explicit tools in water management plans and then to prescribe management actions that seek to preserve and even enhance these processes. Management actions are likely to be primarily conservation and restoration oriented, where upon development or destruction of crucial components of natural systems (e.g. riparian zones, natural channel forms, fringing wetlands, etc.) are prohibited and restoration of already degraded systems is ordered. They may also be procedural as with maintaining a degree of natural flow variability in highly regulated river regimes (Poff *et al.*, 1997). Focus must be placed on conserving and restoring the integrity of aquatic systems within an overall plan that is consistent with the development goals of the given river basin.

The effort to gain formal recognition of natural processes as explicit tools in water management will require a rather dramatic change in perspective and may be best introduced to South American countries through international agreements. The governments of South America have been very receptive to the visions of water management proclaimed in major international agreements. The 1992 International Conference on Water and the Environment, held in Dublin, Ireland, called for multiple changes in the way water resources were managed. Most notably it advocated the use of river basins as the geographic unit of management and called for a decentralization of decision-making through the establishment of committees composed of government and nongovernment representatives within each river basin. These concepts were amplified and further articulated in subsequent conferences such as the 1992 United Nations Conference on Environment and Development (the Earth Summit) in Rio de Janeiro, the 1994 Summit of the Americas in Miami, the 1995 Pan-American Conference on Health and Environment in Sustainable Development in Washington DC, and the 1996 Summit Conference on Sustainable Development in Santa Cruz, Bolivia. Subsequently, several nations (perhaps most notably Brazil; ANA, 2001) in South America revised their national plans to incorporate these concepts.

There is already considerable attention being given to the maintenance of ecological integrity in water management planning at the international level. The commission report of the recently prepared World Water Vision (WWC, 2001) makes the point eloquently. "There are those who see water only by use: water for municipalities, for industry, for irrigation, for the environment, as though the last were a competing use, not an inherent part of maintaining the entire ecological system on which all water services depend." This and other international statements make the point clearly, but the required follow-up actions remain vague and water resource managers in different countries struggle with the question of how to reconcile conservation of ecological integrity with concrete management actions. It is here where I believe it is imperative to revise the language to refer to specific components of intact ecosystems as tools of water management. In this way riparian zones, wetlands, and channel forms may be explicitly incorporated into management plans and managed with the same concrete purpose as waste-water treatment plants, holding ponds, and other more traditional "tools" of water resources management. This task is made easier in South America because water, by law, remains the property of the state, and in many countries there is a fiscal buffer along the edges of rivers that must remain undeveloped.

Another valuable outcome of the formalization of ecohydrological processes in water quality management plans is that it generates management tools in remote rural settings that have little or no engineering infrastructure and are unlikely to see any in the coming years. Even in these remote settings organized water quality management is essential and best achieved through the formation of basin committees. By recognizing riparian zones, wetlands, and natural channel forms as management tools, these rural committees will be able to develop concrete plans that take into consideration the quantity and configuration of these intrinsic water purification systems. Rural communities should recognize that they have at their disposal an assemblage of highly effective water purification services that are virtually free and already perfectly positioned in the hydrological system. The specific management actions required to implement the developed plans will also illuminate the individual responsibilities of institutions and households in the basins.

At the level of implementation, the conservation of ecohydrological processes clearly should not be viewed as an independent and stand-alone solution to water quality management problems. Natural ecohydrological processes should be viewed as a fundamental component of more multifaceted solutions, including additional systems to treat water for domestic supply and to treat wastes prior to disposal. Wherever possible, the varied components should be interconnected to form a continuum of water quality management solutions. In urban and industrial areas, ecohydrological processes should be maintained to fortify nearby aquatic systems and increase the likelihood that other, engineering-based controls will achieve water quality goals. Amplifying the chances that more technically complicated management activities will succeed is a wise strategy and a guiding principle within the UNESCO Ecohydrology Programme (Zalewski *et al.*, 1997). In rural areas, natural ecohydrological processes may be recognized as primary tools for controlling surface water quality in conjunction with simple drinking water treatment and waste treatment systems (e.g. solar latrines). Small rural communities may even construct simple engineering systems to interact with natural components of river basins that are hot spots of ecohydrological processes. For example, sewage from small communities may be piped into wetlands designated for the purpose of cleaning the waste water prior to discharge to an adjoining river. Of course, in this example the value of the wetland as a waste-water purification system must be weighed against other services (e.g. fish nursery habitat) which may be lost as a result of sewage inputs.

It is clear that the effective use of ecohydrological processes in water quality management plans (especially in rural areas) will require the complete and voluntary participation of local communities and landholders. As recommended by the World Water Council and other international initiatives, local people should play a leading role in developing water quality management so that they share a sense of ownership of the plan. This will not only increase the likelihood of compliance among the population. It will also foster a mechanism for community-driven enforcement when individuals do not comply. Long-term community participation will perhaps be best accomplished by ensuring that respected local community leaders, elders, and normal individuals serve on basin committees.

Finally, I do not wish to suggest that South American water managers, in viewing natural ecosystem components as tools, lose sight of the broader ecosystem services provided by riparian zones, wetlands, and channel forms. Many services related to fisheries, transportation, tourism, and other intangible benefits stem from intact river and wetland systems. These too must be weighed in management decisions.

CONCLUDING REMARKS

Although research to date has shown that many ecohydrological processes act to purify natural waters in South America, virtually no studies have specifically addressed the utility of these processes in actually mitigating pollution inputs. Water managers require solid information on the specific capabilities of the tools at there disposal. What processes are effective against which kinds of pollution? Where exactly in the system (i.e. in which landscape features) are these processes most efficient? What is the throughput of water in these features (e.g. riparian zones or wetlands)? What are the concentration and flow thresholds where the efficiency of these processes drop off or where other services are compromised? How do efficiencies and thresholds vary with season and time? There are also considerations that must be addressed regarding the economic trade-offs of conserving landscape features for water quality management instead of developing them for other purposes. A great deal of research remains to be done.

Chapters in this book have illuminated a variety of good research that touches upon these needs, but these studies were originally mounted with other goals in mind such as simple system characterizations or academically-driven research questions on natural ecological processes. Real progress on the employment of ecohydrological understanding for improved water quality management will require coordinated and focused research into the questions posed above. To be successful, this research must combine the expertise of ecologists, hydrologists, and water quality and sanitary engineers. With its expansive surface water systems and relatively low degree of disturbance, South America is an ideal region in which to develop and apply ecohydrological approaches to water quality management.

REFERENCES

Alegre, J. C. & Rao, M. R. (1996) Soil and water conservation by contour hedging in the humid tropics of Peru. *Agriculture, Ecosystem and Environment* **57**, 17–25.

Amon, R. M. W. & Benner, R. (1996) Photochemical and microbial consumption of dissolved organic carbon and dissolved oxygen in the Amazon river system. *Geochimica, Cosmochimica Acta* **60**, 1783–1792.

ANA (Agência Nacional de Águas, Brazil) (2001) The recently created Brazilian National Water Agency charged with the implementing that nation's new water law. Information available at http://www.ana.gov.br.

Aparicio Alcázar, L. M. (1999) Diagnóstico del uso actual de los recursos hídricos de la cuenca del Río Pachitea. Ing. Thesis, Facultad de Ciencia Forestales, Universidad Nacional Agraria La Molina, Lima.

Aselmann, I. & Crutzen, P. J. (1989) Global distribution of natural freshwater wetlands and rice paddies, and their net primary productivity, seasonality and possible methane emissions. *Journal of Atmospheric Chemistry* **8**, 307–358.

Braga, E. S., Bonetti, C. V. D. H., Burone, L. & Bonetti Filho, J. (2000) Eutrophication and bacterial pollution caused by industrial and domestic wastes at the Baixada Santista estuarine system—Brazil. *Marine Pollution Bulletin* **40**, 165–173.

Callisto, M., Gonçales, J. F. & Fonseca, J. J. L. (1998) Benthic macroinvertebrates of four Amazonian streams influenced by bauxite mining (Brazil). *Verh. Internat. Verein. Limnol.* **26**, 983–985.

Carvalho, C. E. V. & Torres, J. P. M. (2002) The ecohydrology of the Paraíba do Sul River, southeast Brazil. Chapter 10 in: *The Ecohydrology of South American Rivers and Wetlands* (ed. by M. McClain). IAHS Special Publication no. 6 (this volume).

Castañé, P. M., Topalián, M. L., Rovedatti, M. G. & Salibián, A. (1998) Impact of human activities on the water quality of the Recoquista River, Buenos Aires, Argentina. *Verh. Internat. Verein. Limnol.* **26**, 1206–1208.

Ceballos, B. S. O., König, A. & Oliveira, J. F. (1998) Dam reservoir eutrophication: a simplified technique for a fast diagnosis of environmental degradation. *Water Research* **32**, 3477–3483.

CEPIS (Centro Panamericano de Ingeniería Sanitaria y Ciencias del Ambiente) (2001) Proyecto Regional Sistemas Integrados de Tratamiento y Uso de Aguas Residuales en América Latina: Realidad y Potencial. Data available at http://www.cepis.ops-oms.org/bvsaar/e/proyecto/proyecto.html.

Donoso, G., Cancino, J. & Magri, A. (1999) Effects of agricultural activities on water pollution with nitrates and pesticides in the Central Valley of Chile. *Water Science and Technology* **39**, 49–60.

Dos Reis Castro, N. M., Auzet, A. V., Chevallier, P. & Leprun, J. C. (1999) Land use change effects on runoff and erosion from plot to catchment scale on the basaltic plateau of southern Brazil. *Hydrological Process* **13**, 1621–1628.

Dunne, T., Mertes, L. A. K., Meade, R. H., Richey, J. E. & Forsberg, B. R. (1998) Exchanges of sediment between the flood plain and channel of the Amazon river in Brazil. *Geological Society of America Bulletin* **110**, 450–467.

Engle, D. L. & Melack, J. M. (1993) Consequences of riverine flooding for seston and the periphyton of floating meadows in an Amazon floodplain lake. *Limnology and Oceanography* **38**, 1500–1520.

Facetti, J., Dekov, V. M. & Van Grieken, R. (1998) Heavy metals from the Paraguay River: a preliminary study. *The Science of the Total Environment* **209**, 79–86.

Forsberg, B. R., Godoy, J. M., Victoria, R. L. & Martinelli, L. A. (1989) Development and erosion in the Brazilian Amazon: a geochronological case study. *GeoJournal* **19**, 402–405.

Förstner, U. & Wittmann, G. T. W. (1983) *Metal Pollution in the Aquatic Environment*. Springer-Verlag, Berlin, Germany.

Fritsch, J. M. & Sarrailh, J. M. (1986) Les transports solids dans l'écosystème forestier tropical humide guyanais: effets du défrichement et de l'aménagement de pâturages. *Cahiers de ORSTOM, Sér. Pédologie* **22**, 209–222.

Galdean, N., Callisto, M. & Barbosa, F. A. R. (2000) Lotic ecosystems of Serra do Cipó, southeast Brazil: water quality and tentative classification based on benthic macroinvertebrate community. *Aquatic Ecosystem Health and Management* **3**, 545–552.

Gleick, P. (2000) *The World's Water 2000–2001*. Island Press, Washington DC, USA.

Godoy, J. M., Padovani, C. R., Pereira, J. C. A. & Vieira, L. M. (1998) Aplicabilidade da geocronologia da deposição de sedimento com ^{210}Pb como ferramenta na avaliação do assoreamento do Río Taquari, Pantanal, MS. *Geochimica Brasiliensis* **12**(1/2), 113–121.

Guyot, J. L., Quintanilla, J., Cortes, J. & Filizola, N. (1995) Les flux de matières dissoutes et particulaires des Andes de Bolivie vers le Río Madeira en Amazonie Brésilienne. *Bulletin de l'Institut Français d'Etude Andines* **24**, 415–423.

Hamilton, S. K. (2002) Hydrological controls of ecological structure and function in the Pantanal wetland (Brazil). Chapter 8 in: *The Ecohydrology of South American Rivers and Wetlands* (ed. by M. McClain). IAHS Special Publication no. 6 (this volume).

Haycock, N. E., Burt, T. P., Goulding, K. W. T. & Pinay, G. (1997) *Buffer Zones: Their Processes and Potential in Water Protection*. Quest Environmental, Harpenden, UK.

HIMAT-INGEOMINAS (1991) *Estudio de la contaminación del Río Magdalena por metales traza, su relación con parametros hidrológicos, fisico-quimicos y su incindencia en la salud humana*. HIMAT-INGEOMINAS, Bogotá.

Huszar, V. L. M., Silva, L. H. S., Domingos, P., Marinho, M. & Melo, S. (1998) Phytoplankton species composition is more sensitive than OECD criteria to the trophic status of three Brazilian tropical lakes. *Hydrobiologia* **369/370**, 59–71.

Hylander, L. D., Silva, E. C., Oliveira, L. J., Silva, S. A., Kuntze, E. K. & Silva, D. X. (1994) Mercury levels in Alto Pantanal: a screening study. *Ambio* **23**, 478–484.

IDEAM (Instituto de Hidrología, Meteorología y Estudios Ambientales) (2000) Condiciones ambientales en Colombia. Data available at http://www.ideam.gov.co.

Junk, W. J. & Furch, K. (1993) A general review of South American floodplains. *Wetlands Ecology and Management* **2**, 231–238.

Konhauser, K. O. & Fyfe, W. S. (1993) Biogeochemical cycling of metals in freshwater algae from Manaus and Carajás, Brazil. *Energy Sources* **15**, 595–608.

Leino, T. & Lodenius, M. (1995) Human hair mercury levels in Tucuruí area, State of Pará, Brazil. *The Science of the Total Environment* **175**, 119–125.

Malm, O., Branches, F. J. P., Akagi, H., Castro, M. B., Pfeiffer, W. C., Harada, M., Bastos, W. R. & Kato, H. (1995) Mercury and methylmercury in fish and human hair from the Tapajós River basin, Brazil. *The Science of the Total Environment* **175**, 141–150.

Martinelli, L. A., Ferreira, J. R., Forsberg, B. R. & Victoria, R. L. (1988) Mercury contamination in the Amazon: a gold rush consequence. *Ambio* **17**, 252–254.

Martinelli, L. A., Victoria, R. L., Ferraz, E. S., Camargo, P. B., Moreira, M. Z., Krusche, A. V., de Moraes, J. M., Ballester, M. V., Silveira, L. L., Bernardes, M. C., Ometto, J. P. H. B. & Cerri, C. E. (2002) Hydrology and water quality in the Piracicaba River basin, São Paulo State, Brazil. Chapter 9 in: *The Ecohydrology of South American Rivers and Wetlands* (ed. by M. McClain). IAHS Special Publication no. 6 (this volume).

Maurice-Bourgoin, L., Fraizy, P., Alanoca, L., Seyler, P. & Guyot, J. L. (2000) Hydrological control on the temporal variability of mercury in surface waters of the upper Madeira basin, Bolivia. In: *25th International Conference on Heavy Metals in the Environment* (ed. by J. Nriagu), contribution #1282. University of Michigan, School of Public Health, Ann Arbor, Michigan (CD-ROM).

Maurice-Bourgoin, L. & Quiroga, I. (2002) Total mercury distribution and importance of the biomagnification process in rivers of the Bolivian Amazon. Chapter 4 in: *The Ecohydrology of South American Rivers and Wetlands* (ed. by M. McClain). IAHS Special Publication no. 6 (this volume).

Mayorga, E. & Aufdenkampe, A. (2002) Processing of bioactive elements in the Amazon River system. Chapter 1 in: *The Ecohydrology of South American Rivers and Wetlands* (ed. by M. McClain). IAHS Special Publication no. 6 (this volume).

McClain, M. E., Richey J. E. & Pimentel T. P. (1994) Groundwater nitrogen dynamics at the terrestrial–lotic interface of a small catchment in the Central Amazon Basin. *Biogeochemistry* **27**, 113–127.

Meybeck, M. (1988) How to establish and use world budgets of riverine materials. In: *Physical and Chemical Weathering in Geochemical Cycles* (ed. by A. Lerman & M. Meybeck). Kluwer Academic Publishers, Dordrecht, The Netherlands.

Montes de Oca, I. (1989) *Geografía y Recursos Naturales de Bolivia*. Academia Nacional de Ciencias de Bolivia, La Paz.

Moraes, L. A. F., Lenzi, E. & Luchese, E. B. (1997) Mercury in two fish species from the Paraná River floodplain, Paraná, Brazil. *Environmental Pollution* **98**, 123–127.

Mounier, S., Benedetti, M., Benaïm, J. Y. & Boulègue, J. (2002) Organic matter size dynamics in the Amazon River. Chapter 2 in: *The Ecohydrology of South American Rivers and Wetlands* (ed. by M. McClain). IAHS Special Publication no. 6 (this volume).

NAS (National Academy of Sciences) Committee on Intrinsic Remediation (2000) *Natural Attenuation for Groundwater Remediation*. National Academy Press, Washington DC, USA.

Ohno, A., Marui, A., Castro, E. S., Benitez, A. A., Elio-Calvo, D., Kasitani, H., Ishii, Y. & Yamaguchi, K. (1997) Enteropathogenic bacteria in the La Paz River of Bolivia. *American Journal of Tropical Medicine and Hygiene* **57**, 438.

Oki, T., Musiake, K., Matsuyama, H. & Masuda, K. (1995) Global atmospheric water balance and runoff from large river basins. *Hydrological Processes* **9**, 655–678.

Olivero, J., Navas, V., Perez, A., Solano, B., Acosta, I., Arguello, E. & Salas, R. (1997) Mercury levels in muscle of some fish species from the Dique channel, Colombia. *Bulletin of Environmental Contamination and Toxicology* **58**, 865–870.

Olivero, J., Solano, B. & Acosta, I. (1998) Total mercury in muscle from fish in two marshes in goldfields, Colombia. *Bulletin of Environmental Contamination and Toxicology* **61**, 182–187.

Perdomo, L., Ensminger, I., Espinosa, L. F., Elster, C., Wallner-Kersanach, M. & Schnetter, M. L. (1998) The mangrove ecosystem of the Cienaga Grande de Santa Marta (Colombia): observations on regeneration and trace metals in sediment. *Marine Pollution Bulletin* **37**, 393–403.

Poff, N. L., Allan, J. D., Bain, M. B., Karr, J. R., Prestegaard, K. L., Richter, B. D., Sparks, R. E. & Stromberg, J. C. (1997) The natural flow regime: a paradigm for river conservation and restoration. *Bioscience* **47**, 769–784.

Restrepo, J. D. & Kjerfve, B. (2000) Magdalena River: interannual variability (1975–1995) and revised water discharge and sediment load estimates. *Journal of Hydrology* **235**, 137–149.

Revenga, C., Murray, S., Abramovitz, J. & Hammond, A. (1998) *Watersheds of the World*. The World Resources Institute and Worldwatch Institute, Washington DC, USA.

Roca, R. (1995) Uso de agua y medio ambiente. In: *Estudio de Reconocimiento del Uso del Recurso Hidrico por los Differentes Sectores Productivos en el Peru*, 297–305. Instituto Nacional de Recursos Naturales, Lima.

Rodrigo, L., Montaña, C., Cuervo, M. P., Gómez, J. & Toro, M. A. (1998) Emisiones al ambiente en Colombia. In: *El Medio Ambiente en Colombia* (ed. by P. Leyva), 477–490. Instituto de Hidrología, Meteorología y Estudios Ambientales, Santa Fe de Bogotá.

Roland, F., Esteves, F. A. & Barbosa, F. A. R. (1997) The influence of bauxite tailings on the light regime and its consequence on phytoplankton primary production in an Amazonian floodplain lake. *Verh. Internat. Verein. Limnol.* **26**, 765–767.

Roulet, M. & Lucotte, M. (1995) Geochemistry of mercury in pristine and flooded ferrallitic soils of a dense tropical forest in French Guiana. *Water, Air and Soil Pollution* **80**, 1079–1088.

Sastre, A. V., Santinelli, N. H., Otaño, S. H. & Ivanissevich, M. E. (1998) Water quality in the lower section of the Chubut River, Patagonia, Argentina. *Verh. Internat. Verein. Limnol.* **26**, 951–955.

Silva-Forsberg, M. C., Forsberg, B. R. & Zeidemann, V. K. (1999) Mercury contamination in humans linked to river chemistry in the Amazon basin. *Ambio* **28**, 519–521.

Stockner, J. G. & Macisaac, E. A. (1996) British Columbia lake enrichment programme: two decades of habitat enhancement for Sockeye salmon. *Regulated Rivers: Research and Management* **12**, 547–561.

Swarts, F. A. (ed) (2000) *The Pantanal of Brazil, Bolivia and Paraguay*. Hudson MacArthur Publishers, Gouldsboro, Pennsylvania, USA.

Torres, J. P. M. (1992) Ocorrência e distribuição de metais pesados no Rio Paraibuna, Juiz de Fora, M.G. MSc Thesis, Instituto de Biofísica Carlos Chagas Filho, Universidade Federal do Rio de Janeiro.

Triska, F. J., Kennedy, V. C., Avanzino, R. J., Zellweger, G. W. & Bencala, K. (1989) Retention and transport of nutrients in a third order stream: channel processes. *Ecology* **70**, 1877–1892.

Villar, C., Tudino, M., Bonetto, C., de Cabo, L., Stripeikis, J., d'Huicque, L. & Troccoli, O. (1998) Heavy metal concentrations in the lower Paraná River and right margin of the Río de la Plata estuary. *Verh. Internat. Verein. Limnol.* **26**, 963–966.

Watts, D. M., Phillips, I., Callahan, J. D., Griebenow, W., Hyams, K. C. & Hayes, C. G. (1997) Oropouche virus transmission in the Amazon river basin of Peru. *American Journal of Tropical Medicine and Hygiene* **56**, 148.

WHO/UNICEF Joint Monitoring Programme for Water Supply and Sanitation Coverage Estimates 1980–2000. Available at http://www.childinfo.org/eddb/water/.

Williams, M. R., Fisher, T. R. & Melack, J. M. (1997) Solute dynamics in soil water and groundwater in a central Amazon catchment undergoing deforestation. *Biogeochemistry* **38**, 303–335.

World Bank (2001) *World Development Indicators Database*. Available at http://www.worldbank.org/data

World Water Council (WWC) (2001) *World Water Vision*. Available at http://www.worldwatercouncil.org.

WRI (World Resources Institute) (1998) *World Resources 1998–1999: A Guide to the Global Environment*. Oxford University Press, New York, USA.

Zalewski, M., Janauer, G. A. & Jolankaj, G. (1997) *Ecohydrology: a New Paradigm for the Sustainable Use of Aquatic Resources*. Technical Documents in Hydrology no. 7, UNESCO, Paris.